Primary Succession and Ecosystem Rehabilitation

Natural disturbances such as lava flows, landslides and glacial moraines, and human-damaged sites such as pavements, road edges and mine wastes, often leave little or no soil or biological legacy. This book provides the first comprehensive summary of how plant, animal and microbial communities develop under the harsh conditions following such dramatic disturbances. The authors examine the basic principles that determine ecosystem development and apply the general rules to the urgent practical need for promoting the reclamation of damaged lands. Written for those concerned with disturbance, landscape dynamics, restoration, life histories, invasions, modeling, soil formation and community or population dynamics, this book will also serve as an authoritative text for graduate students and a valuable reference for professionals involved in land management.

Lawrence R. Walker is Professor of Biological Sciences at the University of Nevada, Las Vegas. His research focuses on the mechanisms that drive primary succession and the applications of succession to restoration.

Roger del Moral is Professor of Botany at the University of Washington. His research relates observed patterns of vegetation recovery to ecological theory.

Primary Succession and Ecosystem Rehabilitation

LAWRENCE R. WALKER
University of Nevada, Las Vegas

and

ROGER DEL MORAL
University of Washington, Seattle

CAMBRIDGE
UNIVERSITY PRESS

PUBLISHED BY THE PRESS SYNDICATE OF THE UNIVERSITY OF CAMBRIDGE
The Pitt Building, Trumpington Street, Cambridge, United Kingdom

CAMBRIDGE UNIVERSITY PRESS
The Edinburgh Building, Cambridge CB2 2RU, UK
40 West 20th Street, New York, NY 10011-4211, USA
477 Williamstown Road, Port Melbourne, VIC 3207, Australia
Ruiz de Alarcón 13, 28014 Madrid, Spain
Dock House, The Waterfront, Cape Town 8001, South Africa

http://www.cambridge.org

First published 2003

Printed in the United Kingdom at the University Press, Cambridge

Typeface Bembo 11/13 pt   *System* LATEX 2_ε [TB]

A catalogue record for this book is available from the British Library

Library of Congress Cataloging in Publication data
Walker, Lawrence R.
Primary succession and ecosystem rehabilitation / Lawrence R. Walker and Roger del Moral.
    p.   cm.
Includes bibliographical references (p.).
ISBN 0 521 80076 5 – ISBN 0 521 52954 9 (pb.)
1. Ecological succession.   2. Ecosystem management.   I. Moral, Roger del, 1943-   II. Title.
QH541 .W25   2003
577'.18–dc21   2002073586

ISBN 0 521 80076 5 hardback
ISBN 0 521 52954 9 paperback

Contents

Preface and acknowledgements

We wrote this book for many reasons. First, we wanted to share our excitement about some of the wild places on this planet. The wilds we consider are places where new land is being formed, whether by dramatic natural forces (volcanoes, glaciers, landslides) or by the steady, unobtrusive forces of wind and water (dunes, beaches, soil erosion). We are equally intrigued by successional processes following abandonment of human artifacts (pavement, mines, waste dumps). Our careers started with our entrancement about natural disasters and how natural processes or regeneration follow. The next logical step was to extend our studies to disturbances of human origin, applying the same curiosity and scientific methodology. Now we summarize what we have learned and what we believe still needs study.

Our second reason for writing this book was because we feel the urgency of understanding the natural and human-assisted processes involved in ecosystem rehabilitation. With the spiraling challenges of over-population and resource depletion, including a startling loss in arable land, rehabilitation of severely damaged terrestrial and aquatic systems is just as essential as recycling of waste products into useful resources. We maintain that the best approach to rehabilitation is the merger of science and management. This book aims to forge links between successional theory and potential applications of that knowledge. Communication between scientists and land managers, theorists and practitioners of rehabilitation must improve. Theory can be helpful, but hands-on practical experience, particularly when combined with appropriate field experiments, is essential to addressing local problems. It is the theorist's challenge to develop a general framework from the accumulated local experience.

We also wrote this book in order to provide the first summary of a global and growing literature on primary succession. We hope our readership will include professional scientists and students of ecology looking for synthesis of ideas from research on succession, competition, facilitation, ecosystem assembly, conservation and rehabilitation. This book will easily serve as a textbook and summary of a complex yet interrelated collection of ideas. We also hope that land managers from government, corporate and non-profit organizations dedicated to the repair of degraded habitats will find this book of practical significance.

We are both terrestrial plant ecologists biased toward easily visible, sedentary green things that do not flee when we study them. Yet we have included research on aquatic systems, soil microbes and other fauna whenever it elucidates principles of primary succession. Fascinating examples include succession of tubeworms around thermal vents on the ocean floor, algae growing inside rocks in Antarctica, or the sequence of decomposers on rotting carcasses. There is also a rich body of research on land–water interfaces such as dunes and shorelines that we evaluate in terms of primary succession. However, the vast majority of studies on primary succession deal with terrestrial plants, although more emphasis is beginning to be placed on the sum of the interactions of plants, animals and microbes.

We gratefully acknowledge the support of our wives, Elizabeth Powell and Beth Brosseau, our home institutions and the editorial staff of Cambridge University Press. We both thank our mentors for inspiration and our many colleagues and students for stimulating discussions. Lawrence Walker particularly recognizes the positive influences of Terry Chapin and Peter Vitousek. Roger del Moral became an ecologist because C. H. Muller showed him how compelling were the problems of biotic interactions, and was subsequently influenced by Robert H. Whittaker, Joe Connell and Eddy van der Maarel. Several books have also served as models for us to emulate, especially Pickett & White (1985), which clarified the importance of disturbance, and Matthews (1992), which provided an impressively thorough compendium of one type of primary succession. We are also indebted to McIntosh (1985) for his insightful summaries of the history of ecological thought, to Luken (1990) for providing the first successful bridge between theory and practice and to Glenn-Lewin *et al.* (1992) for their cogent summaries of successional ideas.

Lawrence Walker was supported in part by grants BSR-8811902 and DEB-9411973 from the National Science Foundation to the Institute for Tropical Ecosystem Studies, University of Puerto Rico and the International Institute for Tropical Forestry, as part of the Long Term Ecological Research Program in the Luquillo Experimental Forest. The Forest Service (U.S. Department of Agriculture), the University of Puerto Rico, the University of Nevada, and Landcare Research of New Zealand provided substantial additional support.

Roger del Moral was supported by Grants DEB 94-06987 and DEB 00-87040 from the National Science Foundation for long-term studies of Mount St. Helens. Support while writing this book was provided by the University of Washington and by the Instituto di Biologia ed Ecologia Vegetale, Università di Catania. Prof. Emilia Poli Marchese contributed greatly to facilitating Roger's study of Mount Etna and was an excellent guide to Sicilian ecology.

We appreciate the assistance of Paula Jacoby-Garrett with the figures and M. Kay Suiter with the references and figures. We are indebted to several critical readers whose comments helped us improve all or parts of the book. In addition to one anonymous reviewer, these readers included Peter Bellingham, John Bishop, Nicholas Brokaw, Ray Callaway, Charles Cogbill, Charlie Crisafulli, Chris Fastie, Tara Fletcher, Charles Halpern, Richard Hobbs, Chad Jones, Craig Palmer, Duane Peltzer, Mohan Wali, Alan Walker, Margery Walker and David Wardle.

The publisher has used its best endeavors to ensure that URLs for external websites referred to in this book are correct at the time of going to press. However, the publisher has no responsibility for the websites and can make no guarantee that a site will remain active or that the content is or will remain appropriate.

1 · *Introduction*

1.1 Why learn about primary succession?

This book is for anyone interested in the consequences of disturbance. What happens after the lava cools, or when the muddy floodwaters recede or an old road is abandoned? Primary succession is the process of ecosystem development on barren surfaces where severe disturbances have removed most vestiges of biological activity. It includes the development of complex systems from simple biotic and abiotic (non-biological) components. Primary succession starts when plants, animals and microbes colonize new surfaces. The process is influenced by local conditions, context and site history. All new surfaces are initially devoid of life, so primary succession has been crucial throughout Earth's history. Today, all communities of plants, animals and soils are the result of primary succession. It is this process of recovery of ecosystems after disturbance that provides the clean air and water and fertile soils that humans and all organisms need to survive.

Ecosystem development on initially barren surfaces has always been of great importance to humans. Hunters depended on game that migrated into the fertile terrain exposed by retreating glaciers. With the transition to agriculture, communities became dependent on the periodic deposition of nutrients by floodwaters along such rivers as the Euphrates and Nile to sustain soil fertility and civilization. Away from floodplains, farmers have had to manipulate succession to produce crops on infertile sites. Present-day ecologists use lessons from primary succession in many ways (e.g. to create new habitats and to rehabilitate mined lands and pastures). Effective manipulations of our environment to improve fertility, productivity or diversity and an ability to mitigate undesirable conditions all clearly depend on our ability to understand primary succession. The exponential increase in human numbers and declining resource availability lend great urgency to the search for more efficient resource use and habitat rehabilitation. We hope that this book will provide a comprehensive understanding of the mechanisms of primary succession that will facilitate that search.

1.1.1 Humans and disturbance

Humans live in awe of disturbances, particularly catastrophes such as floods, earthquakes, hurricanes, meteor strikes and volcanoes that can create the conditions for primary succession. Of course, the terms catastrophe and disaster place a human bias on these natural phenomena. Therefore we will use the neutral and more encompassing term disturbance to refer to disruptions of the environment that trigger primary succession. However, humans are responsible for an increasing array of disturbances such as acid rain, ozone depletion, nuclear explosions and global warming, all of which can exacerbate natural phenomena such as erosion and flooding. Flooding accounts for nearly half of all human deaths associated with such natural disturbances as earthquakes, volcanoes, windstorms, fires and landslides (Abramovitz, 2001). However, the distinction between natural and human-induced disturbances is increasingly blurred because of the growing extent and impact of human activities.

Historical records demonstrate the impact of natural disturbances on human history. Natural disturbances have altered the balance of power in many parts of the world, often by devastating agricultural production and thereby destabilizing societies (Keys, 2000). The eruption of the island of Thera in the Mediterranean Sea in 1623 B.C. (Oliver-Smith & Hoffman, 1999) destroyed the Minoan civilization, thereby changing the course of Western civilization. Global climates were disrupted by gigantic volcanic explosions such as Taupo in New Zealand (A.D. 186), Krakatoa in Indonesia (c. A.D. 535 and 1883) and Laki in Iceland (A.D. 1783). We are gaining a better historical understanding of climate change through examination of ice cores in Greenland, deposits from meteor collisions and volcanic debris. Many past disturbances are thought to be responsible for catastrophic extinction episodes. The loss of 90% of all marine species about 250 million years ago during the Permian extinction coincided with the lowering of sea levels and the merging of the continents to form a super-continent called Pangaea. The cooling of the oceans about 65 million years ago marked the end of the Cretaceous and was almost certainly the result of a large meteor impact off the Yucatan Peninsula (Mexico). This impact may have caused the loss of 50% of all marine species and terrestrial organisms, including most dinosaurs. More recently, the loss of many large mammal species from North America during the Late Pleistocene was associated not only with human hunting pressures, but also with climate warming and a resultant loss of grassland habitat that was replaced by forests.

With a rapidly growing human population that is increasingly resource-hungry, we have entered a new era of intense human impacts on the planet with no clear outcome in sight. Humans disturb life in ways that are unique in the history of the Earth. Because of growing data on ozone depletion and global warming, concerns about global climate change have gained a new urgency. Global warming will lead to higher ocean levels, extensive coastal flooding and destruction of coastal habitats (e.g. marshes, swamps and mangrove forests), increased hurricane frequency and intensity, accelerated desertification and possible human-triggered disruptions of La Niña and El Niño climatic cycles. To counteract this unprecedented onslaught we must understand the mechanisms of repair. Fortunately, the same communication network that brings us daily reports of disturbances around the globe also has encouraged a globalization of responses to disturbance and an awareness of broader patterns (Holling, 1994; Oliver-Smith & Hoffman, 1999). International responses to disturbances include famine relief and various summits on environmental issues (Stockholm, 1972; Montreal, 1987; Rio de Janeiro, 1992), climate change (Kyoto, 1997; The Hague, 2000), or population issues (Cairo, 1994).

Human responses to disturbance vary with the scale, nature and severity of the disturbance as well as with the cultural context (Barrow, 1999). Abrupt and catastrophic disturbances are best remembered. Having a volcano such as Parícutin emerge in your back yard – as one Mexican farmer experienced in 1943 (Scarth, 1999) – is more memorable than the gradual increase in atmospheric carbon dioxide over the past century. In addition, our ability to perceive ameliorative measures decreases with distance from home, size and duration of the disturbance. Sudden, but somewhat predictable, disturbances such as hurricanes, volcanoes or earthquakes can be planned for. Gradual disturbances (e.g. the loss of Mediterranean forests and soil degradation from centuries of farming, grazing and forestry), develop almost without notice and are accepted as normal – at least until some threshold is crossed (e.g. soils erode in a storm). New types of disturbance (atomic fallout, acid rain) are often poorly understood and unforeseen until the damage is well advanced. Risk assessment is a recent approach that provides early warnings of disturbances that affect humans. Unfortunately, humans are all too willing to settle in areas prone to natural hazards, such as floodplains, hurricane belts and earthquake faults.

Human responses to disturbance are also influenced by the predominant culture. Extreme events displace humans and alter human immigration into the affected area as far into the future as primary succession proceeds. Low-density agrarian cultures are perhaps most resilient (e.g. farmers in Peru after the 1970 earthquake; Oliver-Smith & Hoffman, 1999) because they are relatively self-reliant. If the disturbance is ephemeral (unlike the years of ash fall on Iceland from the volcano Laki that led to mass starvation in 1783), crops can be replanted and cultural necessities re-established. High-density populations may recover rapidly if the people are affluent (e.g. Northridge, California, U.S.A., earthquake in 1994; Bolin & Stanford, 1999), or more slowly when they are poor (e.g. gas explosion in Bhopal, India, in 1984; Rajan, 1999). High-density populations are much more vulnerable than low-density populations because of their dependence on complex physical and cultural infrastructures. Given that more than half of all humans now live in rapidly expanding urban areas, vulnerability to natural or human-induced disturbances will continue to increase.

Humans often migrate to avoid the consequences of natural disturbances. They also may try to endure by changes in diet or behavior. Residents on the slopes of the chronically erupting Japanese volcano Sakurajima all have small concrete shelters by their homes for protection from air-borne volcanic debris (tephra). People may also try to prevent disturbance or ameliorate its effects, as when residents of Heimaey, Iceland, stopped lava from filling their harbor by spraying the lava with sea water (Jónsson & Matthíasson, 1993). Finally, humans may take advantage of the benefits of a disturbance, including the development of new fertile soils. Central American cultures have made use of the ash from Ilopango (A.D. 260) and other volcanoes (depending on the size of the eruption, the depth and acidity of the ash, weathering rates and societal complexity; Sheets, 1999) because volcanic ash is good for soils. Many forms of lava also weather to produce fertile soils suitable for agriculture. The windward slopes of Mount Etna (Sicily) support productive citrus orchards and vineyards, despite millennia of agriculture. Similar benefits come from nutrient-rich sediments that are deposited on floodplains or nutrients released from organic soils by fires. Disturbances can also lead to more governmental aid to build or rebuild infrastructures or increased tourism in areas of dramatic or scenic value (e.g. Mount St. Helens and Kilauea volcanoes, U.S.A.).

Some cultures have therefore adapted to the benefits provided by some disturbances.

1.1.2 Human interest in ecosystem recovery

Humans may be in awe of natural disturbances but we also depend on natural processes that permit recovery of 'ecosystem services' (Daily, 1997). These services include the tangible benefits of croplands, game populations, clean air and a dependable water supply, but also such intangible benefits as esthetic spaces, employment and a familiar landscape that are critical to our feelings of well-being and a sense of place (Gallagher, 1993). These very personal issues are additional reasons for learning about ecosystem recovery.

Humans also have intellectual interests in the recovery of ecosystems. The urge to preserve rare species or habitats is driven both by the desire for stable and familiar surroundings and by esthetic, moral or socially conscious motives. Accelerated rates of species extinction and the loss of species and habitat diversity through the homogenizing influence of invasive species motivate some to preserve the familiar or historical species and landscapes. Habitat rehabilitation (e.g. of abandoned roads for wildlife corridors) usually requires knowledge about the process of primary succession. Large-scale, futuristic habitat modifications such as mariculture (e.g. growing crops of algae on glass plates suspended in the ocean) or terraforming the moon or Mars will certainly be based on the principles of primary succession. Such activities – at any scale – will become an increasing part of our global economy.

Preservation or rehabilitation of ecosystem services following a disturbance can have positive influences. For example, a hurricane may directly reduce crop production through flooding or indirectly reduce visits by tourists. Both consequences disrupt the local economy but efforts to restore crops and revive tourism can stimulate economic development. In the growing field of ecological economics, resources are assigned values, the maintenance of which is dependent on the intricate interactions of ecological processes, including recovery following disturbance.

1.2 Definitions

We define several important terms in this section. Other terms are introduced as they arise or are included in the glossary. Succession is most simply defined as species change over time (turnover). Such a process is readily observed in many ecosystems, but may be less obvious or even non-existent in stressful environments such as deserts or tundra. We adopt this broad definition for its wide applicability. Yet there is value in contrasting the rich variety of types and trajectories of succession (see Chapter 7).

Temporal changes in the characteristics of an ecosystem (nutrients, biomass, productivity), a community (species diversity, vegetation structure, herbivory) or a population (sex ratios, age distributions, life history patterns) are closely associated with species change. Therefore, succession is sometimes measured by these variables (Glenn-Lewin & van der Maarel, 1992). Time intervals are best measured in relation to the species involved, often one to ten times the life span of the species. Thus, succession of microbes occurs in hours, fruit flies in weeks, grasses in decades and trees in centuries. Historical reconstructions of very long-term changes (paleoecology) and fluctuations around a relatively stable community (seasonal patterns) are generally excluded from studies of succession.

Succession also occurs over a wide range of spatial scales, again linked to the relative size of the organisms of interest. Microsite variations (e.g. in pH or nutrient availability) can be vital for soil organisms (see section 4.4 on soil biota) or annual plants, whereas some processes such as long-distance seed dispersal (see section 5.3 on dispersal) are best measured on landscape scales. Patch dynamics (Pickett & White, 1985; Pickett et al., 1999; White & Jentsch, 2001) incorporate spatial patterns and shifting patterns of areas with similar characteristics across the landscape and have many implications for succession (see section 2.1.4).

Studies of succession have followed both holistic and reductionistic approaches (see Chapter 3). The former approach, espoused by Clements (1916) and some modern-day ecosystem scientists (Odum, 1969, 1992; Margalef, 1968a,b) views succession as linear and directional, ending in a final species equilibrium or climax. This innately satisfying concept – that succession has a direction with a predictable endpoint – has been widely challenged, beginning with Gleason (1917, 1926) and further elaborated by Egler (1954), Drury & Nisbet (1973) and Whittaker (1974). The more commonly held view now (and the one we espouse in this book) is that succession is a process of change that is not always linear and rarely reaches equilibrium. Directionality occurs only in the sense that there is a turnover in the species present, not in the sense of heading toward a known or predictable endpoint. Succession can therefore incorporate multiple types of trajectory including ones that are cyclic, convergent, divergent, parallel or reticulate (see Chapter 7). Disturbance often redirects or resets successional trajectories, leading to the observation that stable endpoints are rarely achieved. However, disturbances are not evenly distributed in time or space, so the landscape becomes a mosaic of patches, each at a different stage of successional development (Pickett & White, 1985). Rates of successional change also can vary among patches, depending on local site factors such as soil fertility and the mechanisms that regulate the change. To confound the issue, there are multiple mechanisms that drive species change within any given sere (a sere is a particular successional sequence) (Walker & Chapin, 1987). Such complexity has been appreciated for over a century and led Cowles (1901) to suggest that succession was a 'variable approaching a variable rather than a constant.' It is not surprising, then, that no agreement has been reached on the definition of succession. Some plant ecologists have suggested that because the term succession is loaded with connotations of directionality and equilibrium, the term vegetation dynamics is preferable (Miles, 1979, 1987; Burrows, 1990). However, we have no trouble using the term succession, because it has a broadly accepted general meaning of species change over time. Further, vegetation dynamics does not always refer to processes of species turnover. We caution, however, that the term should be carefully defined for each usage by specifying both temporal and spatial scales of interest. For example, dividing types of succession into increasingly longer time scales (Glenn-Lewin & van der Maarel, 1992), one might distinguish among fluctuations, fine-scale gap dynamics, patch dynamics, cyclic succession, secondary succession, primary succession and secular succession (i.e. changes due to climatic change or those measured by paleoecology).

Disturbance starts, directs and may stop or redirect succession (see Chapters 2 and 7), so an understanding of its complexities is critical to interpreting successional pathways. Disturbance and succession are both broad terms, difficult to define precisely, yet unavoidably crucial when interpreting temporal change. Disturbance and succession are also strongly interactive, although Grubb (1988) points out that succession and disturbance are not invariably coupled. Disturbance is an event that is relatively discrete in time and space and one that alters population, community or ecosystem structure. As with succession, the exact temporal and spatial scales of the components of interest need to be defined. Disturbance is characterized by its frequency, extent and magnitude (see Chapter 2).

Primary succession involves species change on substrates with little or no biological legacy, that is, substrates that have no surviving plants, animals or soil microbes. Secondary succession begins with some biological legacy (Franklin *et al.*, 1985) following an initial disturbance (e.g. intact soils). Primary and secondary succession are not always clearly distinguishable, but are points on a continuum (Vitousek & Walker, 1987; White & Jentsch, 2001) from ecosystem development on sterile substrates (lava) or near-sterile substrates (glacial moraines) to development on well-established soils following fire, forest clear-cutting or abandonment of agricultural lands. Regeneration dynamics is a term used to describe one endpoint of the continuum, where there is damage but no mortality (e.g. only loss of leaves and branches of trees during a hurricane).

Secondary succession on low-nutrient substrates can resemble primary succession on fertile substrates (Gleeson & Tilman, 1990). There are numerous examples that are difficult to categorize as either primary or secondary succession. For example, volcanic lava clearly buries soil and biota, but the effects of volcanic tephra depend on the depth of deposition following the eruption. Shallow layers of tephra may cause only minor damage to leaves or understory plants (Grishin *et al.*, 1996; Zobel & Antos, 1997). Similarly, differential depths of burial by sand, floods or landslides can result in secondary or primary succession. And what about burial by plant matter such as wrack (seaweed or other sea life cast onto the shore; Pennings & Richards, 1998)? We take an inclusive approach in this book, utilizing examples that illustrate our themes whether or not they are primary succession in the strictest sense. Further, we de-emphasize the source of the disturbance because different disturbances can have similar results. For example, many disturbances (e.g. glacial melt waters, volcanoes, earthquakes, road construction, heavy rains or mining activities) can trigger landslides. Yet succession results from the conditions created by many, often interacting disturbances, so an understanding of the disturbance regime is essential. We emphasize the actual site conditions that are the net result of the disturbance regime, including the initial status of soil development. Carefully following soil development over time makes a fascinating study in primary succession where most processes have to start *de novo* (see Chapter 4). Yet too little is known about soil development in primary succession to make robust predictions about successional trajectories.

Autogenic succession occurs when the mechanisms that drive species change are derived from the organisms within the community (e.g. nurse plant effects or competition). Allogenic succession occurs when abiotic forces (generally outside the system) determine species change (e.g. sediment transport from a flood, ash deposition from a volcano). Species change in the early stages of primary succession is due mostly to allogenic mechanisms, but internal, autogenic mechanisms become more important as succession proceeds (Matthews, 1992). However, as with most dichotomies (cf. primary and secondary succession), autogenic and allogenic succession represent endpoints of a continuum. Any given sere is the consequence of a mixture of both mechanisms and it is unwise and naïve to label an entire sere as either autogenic or allogenic (Glenn-Lewin & van der Maarel, 1992).

1.3 Methods

The study of succession involves various techniques adapted to a wide range of temporal and spatial scales (Fig. 1.1). Direct observation of temporal change on permanent plots is best (Fig. 1.2), especially when combined with experimental manipulations that have un-manipulated controls (Austin, 1981; Prach et al., 1993). Repeat photography is useful for identification of changes in populations of long-lived plants (Hastings & Turner, 1980; Wright & Bunting, 1994; Webb, 1996) and it does not involve maintenance of permanent plots. However, where change is on the order of decades and centuries (and photographic records are not available), scientists must use the indirect chronosequence approach, where plots of different ages are presumed to represent different stages of development in the actual succession. The main disadvantage of this space-for-time substitution is that the older sites have different histories than the younger ones (Pickett, 1989; Fastie, 1995). The differences among sites can have multiple causes including stochastic events, changes in landscape context or climate over time, or initiation of succession during different seasons. All of these causes can affect both species invasions and species interactions, resulting in different successional outcomes (see Chapter 7).

A century of observations of seres has provided a wealth of information about the patterns of species change. However, the mechanisms that drive succession remain poorly understood. Typical mechanisms include such aspects as the accumulation of organic matter, competition and grazing (see Pickett et al., 1987). Experimental manipulations are usually necessary to determine mechanisms. Common approaches in succession experiments include alteration of such resources as nutrients through additions (fertilizer, leaf litter) or removals (sawdust, sugar or other carbon sources that immobilize nutrients). Species can also be manipulated through additions (transplants, seeding) or removals (cutting, uprooting, girdling, herbivore exclosures). However, any manipulation can have

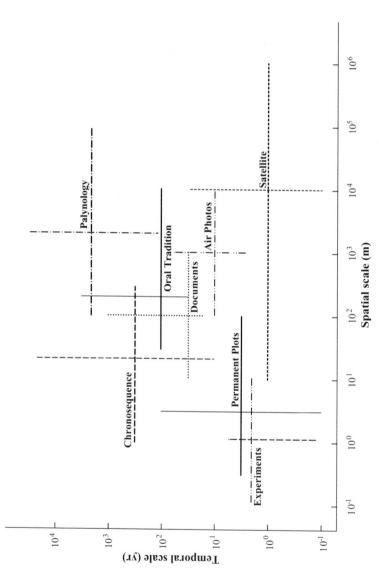

Fig. 1.1. Spatial and temporal scales at which primary succession may be studied. The range of each line is from the finest resolution to the broadest extent normally encountered. The lines cross where the method is most commonly used. Modified from Prentice (1992), with kind permission from Kluwer Academic Publishers.

Fig. 1.2. Primary succession on a landslide in Puerto Rico. (A) Initial seedling invasion at 6 mo; (B) *Cecropia* saplings 4 m tall at 18 mo; (C) near total destruction of *Cecropia* saplings from Hurricane Hugo at 22 mo; (D) shift in dominance to *Cyathea* tree ferns that were 10 m tall after 80 mo. Note the unpredictable shift in species composition from a secondary disturbance.

undesirable side effects (Canill *et al.*, 2001). For example, the removal of the aboveground portion of a small tree results in decaying roots in the soil and a vacant spot in the canopy. Ideally, species effects are measured by excluding a target species from the beginning of a sere, rather than abruptly removing (or adding) it in mid-succession. Alternatively, removals and additions in various combinations (and at various stages) can help to clarify which aspects of the target species are influencing succession.

Multivariate analyses (e.g. gradient analyses) or various modeling approaches (see Chapters 3 and 7) can be used to study succession. Further evidence can be obtained from oral histories, historical observations by travelers such as John C. Fremont and geologists and paleoecologists using fossils (Bhiry & Filion, 1996) or pollen cores (Birks, 1980a). Combining several approaches to studying succession is valuable. Studies of primary succession at Glacier Bay, Alaska, have used Native American legends, reports from the British explorer George Vancouver, detailed observations by geologists, photographs by naturalists, permanent vegetation plots and tree cores (Engstrom, 1995). Most recently, lake-bottom deposits and water chemistry (Engstrom *et al.*, 2000), soil samples and experimental manipulations (Chapin *et al.*, 1994) have helped to understand species change.

1.4 Questions that still remain

It is unclear when primary succession actually ends or is replaced by secondary succession (e.g. succession following a forest fire on a glacial moraine hundreds of years after initial forest establishment). If vegetation change (or any other parameter of interest) is slowed and a relative equilibrium exists for an extended period of time, primary succession can be considered complete (see Chapter 9). However, soil changes may continue long after vegetation change is no longer apparent (see Chapters 4 and 8). We will focus mostly on the dynamic phases of early primary succession but address some of the longer-term changes where data are available from paleoecology (e.g. Jackson *et al.*, 1988) or repeated visits by a series of investigators with long-term plots (e.g. Glacier Bay, Alaska; Cooper, 1931, 1939; Lawrence, 1958; Reiners *et al.*, 1971; Chapin *et al.*, 1994) or without long-term plots (e.g. Krakatau, Indonesia; Treub, 1888; Ernst, 1908; Docters van Leeuwen, 1936; Tagawa *et al.*, 1985; Whittaker *et al.*, 1989).

Another question that remains is the degree to which we can predict successional change. Despite much formal study of succession, no general theory of succession has emerged that can predict the dynamics of species replacements in many different habitats. The enormous variety of the natural world has confounded attempts to generalize, and successional trajectories are very sensitive to site-specific conditions (see Chapter 7). Comparisons of trajectories at different sites within a habitat or across habitats suggest that pathways may converge when biotic controls are strong and diverge when abiotic factors predominate (Matthews, 1992). Several basic mechanisms have been identified (e.g. competitive inhibition, facilitation; Connell & Slatyer, 1977) that appear to be important drivers of most successional change. We discuss how such species interactions affect succession in Chapter 6. Determining the relative importance of each mechanism and how each is modified by local factors is a first step toward a general theory of succession, even if there will never be a 'grand underlying scheme' (GUS) that explains all successional change. We address such theoretical issues about succession in Chapter 3.

Can studies of primary succession lead to practical generalizations about rehabilitation? Environmental regulations increasingly require polluters and excavators to restore severely damaged habitats such as mined lands or roads to some degree of vegetative cover or ecosystem service (e.g. water retention, erosion control, sediment filtering). Site-specific studies examine ways to manipulate primary succession (either by accelerating it or by holding it at some desirable stage). Comparisons of the results of many such studies can lead to predictions about the rates and trajectories of primary succession on anthropogenic disturbances, provided that the studies are done with proper controls and use repeatable methods. Any valid linkage between specific studies and broader theoretical frameworks can then provide generalizations that can be extremely useful to land managers at other sites to address solutions to their particular problems (Luken, 1990). We argue in Chapter 8 that managers and ecologists have much to learn from each other about rehabilitation.

In writing this book we have three objectives. We want to convey our enthusiasm about the dramatic consequences of disturbance and the excitement in observing and studying how plants and animals return following severe disturbances. We also want to contribute to a practical understanding of how to accelerate reconstruction of damaged systems through the process of identifying the spatial and environmental constraints to ecosystem development. Finally, we present the first organized synthesis of what is known about the patterns and processes of the broad topic of primary succession. We address the frontiers for further research in Chapter 9.

2 · *Denudation: the creation of a barren substrate*

2.1 Concepts

Clements (1916, 1928) proposed that six basic processes drive succession: nudation (denudation), migration (dispersal), ecesis (establishment), competition, reaction (site modification by organisms) and stabilization (development of a stable endpoint). With some modifications (see Chapter 3), these concepts still provide a useful framework to study succession (Glenn-Lewin *et al.*, 1992). Denudation is the process of disturbance that creates a barren substrate and is the only process largely independent of the biota. If the site is not completely new (e.g. a recent lava flow, a newly emerged island or a recently formed delta), denudation removes existing plants and animals before succession begins. The degree of removal of the biota determines whether primary or secondary succession ensues on these latter sites (see Chapter 1). In this chapter we examine the often dramatic disturbances that create the initial conditions for primary succession.

2.1.1 Physical environment and disturbance

Our understanding and acceptance of change in the natural world has changed slowly. Early humans often attributed natural disturbances to supernatural forces, and they recorded them as catastrophic, inexplicable or unusual events (Scarth, 1999). Nineteenth century geologists established that landscape changes (e.g. uplift and erosion of landforms) were usually gradual (Lyell, 1850; Davis, 1899), leading to a dynamic view of both landforms and the origin and extinction of organisms as formulated in 1859 in Darwin's *Origin of Species* (Leakey, 1979). Studies in the Swiss Alps, initiated in 1842 and continuing to the present, quantified the movements of glaciers (Aellen, 1981). Finally, the movement of continents or plate tectonics, first proposed by Wegener in 1915 as 'continental drift' (Wegener, 1922), was resurrected in the 1960s (Kearey & Vine, 1996) when a plausible mechanism was demonstrated. All of these discoveries set the stage for the acceptance of changes in the composition of biological communities. We now realize that landscapes are altered by both gradual and cataclysmic changes (cf. Bretz, 1932). Yet these changes are still seen as a dynamic balance between uplift and erosion (Scheidegger, 1997).

Disturbance is a regular part of our lives. We are besieged with reports of catastrophic events at local, regional and global scales. These disturbances have both natural (e.g. hurricanes, El Niño) and human (e.g. ozone depletion, carbon dioxide increase, global warming) causes. Greater attention to disturbances has led to advances in understanding them. Attempts to save lives and reduce property damage has led to one current research emphasis on the prediction of severe disturbances such as landslides, hurricanes and volcanic eruptions. Another, less common emphasis is on how damaged biological communities recover (e.g. the Alaskan coast after the Valdez oil spill; Wheelwright, 1994). Recognition is growing that interactions among disturbances can spawn further disturbances (see section 2.2.6 on disturbance interactions).

Each disturbance determines the nature of the new surface upon which primary succession begins. The type (texture) and shape (slope) of each surface can impact such variables as site moisture and fertility by their effects on substrate stability and accumulation of water, litter, plants, animals and decomposers (Fig. 2.1) (Turner et al., 1997). Clements (1928) noted that the balance of the opposing forces of erosion and sedimentation, subsidence and uplift, flooding and drainage determines the rate of formation of new surfaces for succession. An increase in erosion in one area generally results in a corresponding increase in sedimentation somewhere else, but the effects may not be comparable. Seen from this perspective, most forms of disturbance result from the movement of material across the landscape.

2.1.2 Definitions

A disturbance is an event that is relatively discrete in time and space and that alters the structure of populations, communities and ecosystems. It can alter the density, biomass or spatial distribution of the biota, affect resource availability and change the physical environment (White & Pickett, 1985; Walker & Willig, 1999). Disturbance is a relative term that requires explanation of the spatial and temporal scale of interest for the system under consideration (White & Jentsch, 2001). For example, an animal trail through a forest might disrupt local ant populations but only a larger disturbance such as a road would be likely to have a negative impact on local deer populations. Similarly, large trees might survive a

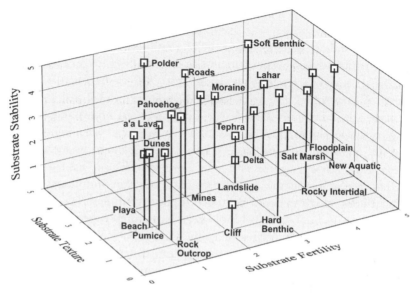

Fig. 2.1. Substrate stability, texture and fertility in different types of primary succession. A score of 5 indicates substrates we consider to be either very stable (substrate stability axis), fine-textured (substrate texture axis) or of high fertility (substrate fertility axis). See also Table 2.1.

short-term flood that killed herbs. Note that the spatial and temporal scales are usually positively correlated. The consequences of small (or short-term) disturbances on large (or long-lived) organisms are not our primary concern, although the impact of large disturbances on small localized processes can be critical (e.g. the consequences of widespread flooding on soil recovery if soil microbes are killed).

Other aspects of disturbance include its origin, frequency, extent and magnitude. Disturbances may originate outside (exogenous) or within (endogenous) the system of interest. The former causes allogenic succession (see Chapter 1) where the biota respond to alterations of the physical environment that originate outside the system (e.g. a glacial retreat or a flood). Endogenous disturbances originate from biotic activity within the system (a gopher mound or competition) and cause autogenic succession. Any given succession is probably driven by a combination of both processes (Miles, 1987), and the relative importance of each will change with time (Matthews, 1992).

Disturbance frequency measures the number of occurrences at a given location per unit of time and/or the probability of occurrence. Residents on the island of Hawaii have their land categorized by the probability that volcanic activity will impact their land. Extent is the amount of area affected. When extensive portions of a landscape such as Hawaii experience frequent disturbances, ecosystem recovery is slow (White & Jentsch, 2001). Magnitude includes intensity (the actual physical force, such as the position of an earthquake on the Richter scale), whereas severity is a measure of the damage caused to the biota. The disturbance regime is the cumulative effect of all the various disturbances that affect a certain area. Disturbance interactions (see section 2.2.6) may be synergistic and enhance the frequency, extent or magnitude of the other disturbances.

2.1.3 Plants and animals as agents of disturbance

Not all disturbances are abiotic in origin. Plants and animals can also create or exacerbate disturbances in many ways. Invasive plants can alter disturbance regimes by increasing or decreasing fire frequency, erosion or herbivore resistance compared to pre-invasion plant communities (Walker & Smith, 1997; Ehrenfeld & Scott, 2001). For example, if an invasive tree species is more shallow-rooted than the species it replaces, erosion may be promoted (Versfeld & van Wilgen, 1986) and the possibility of primary succession increased. Animals create disturbances that include burrowing (e.g. prairie dogs and marmots), wallowing (e.g. mountain goats), building (e.g. beaver dams) and dying (Willig & McGinley, 1999). Such alterations of the physical habitat have been termed ecosystem engineering (Jones et al., 1997) and may have widespread repercussions on biodiversity, energy flow and nutrient cycling. Exposed subsoils from burrows can provide a possible surface for the acceleration of primary succession (del Moral, 1984; Andersen & MacMahon, 1985). Differential browsing may favor either pioneer plants (Ramsey & Wilson, 1997) or later invaders (Kielland & Bryant, 1998). D'Antonio et al. (1999) determined that invaders whose behavior differed qualitatively from native species (e.g. the introduction of feral pigs to Hawaii (Stone, 1984) or sub-Antarctic islands (Challies, 1975)) were more likely to introduce a new disturbance regime than species that only differed quantitatively from natives (e.g. invasive *Pinus* and *Eucalyptus* trees in the South African fynbos that increased fire frequency; Le Maitre et al., 1996).

2.1.4 Patch dynamics

Disturbances are not homogeneous in their effects on the landscape, regardless of the spatial scale of the disturbance (from plate tectonics to

Table 2.1. *Summary of disturbance characteristics*

See also Fig. 2.1.

Disturbance type	Return interval (yr)	Maximum extent (m^2)	Intensity[a]	Severity[a]	Substrate[b] Stability	Texture	Fertility
EARTH							
Volcano							
a'a	10^0–10^5	10^8	5	5	4	2	1
pahoehoe	10^0–10^5	10^8	5	5	5	1	1
tephra	10^0–10^5	10^{10}	4	4	2	3	3
lahar	10^0–10^5	10^8	5	5	3	4	4
Earthquake	10^0–10^5	10^7	4	2	2	variable	variable
Landslide							
terrestrial	10^0–10^4	10^4	3	3	1	3	3
sea floor	10^0–10^4	10^8	3	3	1	3	3
Uprooting	10^0–10^3	10^1	1	2	2	1	4
Rock outcrops	10^0–10^5	10^5	3	5	5	1	1
AIR							
Dune	10^0–10^4	10^7	3	3	2	3	2
WATER							
Floodplain	10^1–10^3	10^8	2	2	3	4	5
Glacial moraine	10^4–10^5	10^7	4	5	3	4	3
HUMAN							
Mineral extraction	10^0–10^1	10^6	5	5	4	3	2
Urban	10^0–10^2	10^6	5	5	5	4	2
Roads	10^0–10^1	10^6	5	5	5	3	2

[a] 1–5 scale where 1 = least; 5 = most (intensity is the force of the disturbance, as in temperature of lava; severity is the degree of damage to the site).

[b] 1–5 scale where 1 = least stable, coarsest texture, least fertile; 5 = most stable, finest texture, most fertile.

local treefall gaps). Instead, after denudation there is a mosaic of patches with differing levels of disturbance and often some partially or entirely undisturbed patches (relicts or refugia; Pickett & White, 1985). Succession may proceed along alternative trajectories and at different rates in the various patches. The initial disturbance severity, its landscape position along resource gradients, the prior disturbance history and many other variables will affect trajectories and rates (see Chapter 7). Alternative trajectories moving at variable rates enhance patchiness in the landscape, with adjacent patches at different stages of succession. For example, a river floodplain is usually a shifting mosaic of younger and older stands of vegetation that are in various stages of succession following erosion and deposition by the meandering river (Drury, 1956). Recovering vegetation of the same age developing along strong environmental gradients can appear to be at different successional stages, depending on such factors as the length of the growing season or effective moisture levels (Dlugosch & del Moral, 1999).

2.2 Types of disturbance that initiate primary succession

In this section we define and describe the types of disturbance that are most important for providing substrates on which primary succession can occur. The disturbances are grouped by the four classical elements, earth, air, water and fire (Walker & Willig, 1999). Disturbances linked to the earth result from tectonic forces. Disturbances involving air, water and fire are products of the interplay of climatic, topographic and soil factors. An additional category is disturbances created by human activities. We discuss the disturbance characteristics (frequency, extent, magnitude) and how multiple disturbances contribute to the disturbance regime. We also examine substrate characteristics found after the disturbance (stability, texture, fertility; Table 2.1, Fig. 2.1) and suggest how these characteristics will influence primary succession. Later chapters expand on soil development and plant and animal succession. At the end of this section we summarize disturbance interactions.

2.2.1 Earth

Earth movements that expose barren substrates form this category. Disturbances of the earth vary in extent from the global impacts large volcanoes can have on climate by the dispersion of tephra and aerosols into the

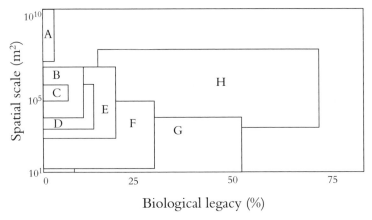

Fig. 2.2. Spatial scales and biological legacies of different types of primary succession. (A) Volcanoes; (B) glacial moraines; (C) mines; (D) transportation; (E) dunes; (F) rock outcrops; (G) landslides; (H) floodplains. Modified from Walker (1999b), with permission from Elsevier Science.

atmosphere (Thornton, 1996) to the small-scale effects of an uprooted tree on soil (Fig. 2.2; Walker, 2000). Similar variation occurs in the severity of the disturbance, as measured by the percentage of biological material remaining (biological legacy; Fig. 2.2). Lava lacks a biological legacy but landslides often contain islands of intact plants, animals and soils. The frequency of tectonic disturbances is also highly variable but is generally inversely related to the extent and severity of the disturbance. Strong earthquakes are much less common than weak tremors; major volcanic eruptions are episodic; landslides are often followed by periods of relative stability until uplift or erosion, for example, result in renewed destabilization. However, steep slopes may remain unstable for centuries until erosion reaches to the ridge top and the slopes stabilize at the angle of repose. Surface textures that result from tectonic disturbances also vary from very fine volcanic ash or fine silts on uplifted seabeds to boulder-strewn avalanches. The ability of a substrate to sustain plant growth is normally greatest on fine-textured particles, although moisture and temperature are also important.

Volcanoes
Volcanic eruptions present one of the most extreme disturbances on the Earth (Table 2.1). They are widespread throughout the world, generally but not exclusively associated with subduction zones of the Earth's crust. Of the 3000 volcanoes that are considered active, 60 to 70

Table 2.2. *Thickness, aerial extent and temperature of the various volcanic deposits at Mount St. Helens in order of declining thickness*

Deposit type	Thickness (m)	Aerial extent (km^2)	Temperature (°C)
Debris avalanche	10^2	80	<100
Pyroclastic flow	10^1	25	350–850
Mudflow	10^0	50	<100
Tephra	10^{-2}	400	<100

From Franklin *et al.* (1985).

erupt annually, directly affecting 500 million people. Volcanic surfaces are so extreme because they begin with a complete absence of nitrogen or organic matter. However, many other minerals are present (e.g. Si, Al, Ca, Fe, Mg, Na, K and P in descending order for basaltic lava), so with adequate additions of N, P and moisture, nutrient-rich soils can develop. Dense human populations in Indonesia and the Philippines are actually supported by intense volcanism and rapid soil development. Many volcanoes erupt infrequently, but some have nearly continuous, low-volume eruptions that continue over decades. Stromboli, off the coast of Italy, has erupted regularly for over 3,000 yr (Scarth, 1999) and nearby Mount Etna has regularly buried villages and farms for at least 1,200 yr. The effects of volcanoes can be global, as when a stratovolcano, which is usually explosive (phreatic; Fig. 2.3A), sends ash particles into the atmosphere that cool the Earth's temperature and cause colorful sunsets. This occurred for several years after Krakatau (1883; Indonesia; Thornton, 1996) and Pinatubo (1991; Philippines). Krakatau also created tidal waves (tsunamis) that were detected half way around the world in the English Channel and severely damaged regional shoreline communities, killing over 36,000 people. More typically, volcanoes directly influence only their immediate surroundings. Shield volcanoes (Fig. 2.3B) are not generally explosive, and their lava builds large, flattened domes, as found in Hawaii and Sicily. Although thick lava is among the most severe disturbances imaginable because it generally leaves no biological legacy, varying depths of volcanic tephra can create conditions that result in secondary succession (where tephra deposits are too shallow to cause mortality; Zobel & Antos, 1992) to primary succession where all biological components are destroyed. On Mount St. Helens there was a wide range of disturbance types, varying in the thickness, temperature and extent of the deposits (Table 2.2). Subsequent eruptions can deflect successional

Fig. 2.3. Two types of volcano. (A) Explosive (Mount St. Helens); (B) shield (Mauna Loa – over 4,000 m tall despite appearing nearly flat on the horizon).

trajectories on volcanoes (Bush *et al.*, 1992). Thus, the disturbance regime is highly variable, both spatially and temporally, with continuous to episodic eruptions creating conditions of variable, but usually extreme, severity.

Volcanic substrates are also highly variable. Lava that is smooth in texture and ropy in appearance is termed pahoehoe lava (Fig. 2.4A). In contrast, a'a lava (Fig. 2.4B) has a crinkly or blocky appearance,

Fig. 2.4. Close-up views of (A) pahoehoe lava; (B) a'a lava; and (C) cinder from Hawaii, to show texture differences.

with a much rougher surface. There is no chemical difference between the two, so the textural differences result from slight variations in viscosity and rates of flow and cooling (a'a is generally more viscous, moves faster and cools more quickly than pahoehoe lava). The explosive eruptions of phreatic volcanoes result in a variety of textures of tephra, or solid materials first thrown into the air (del Moral & Grishin, 1999). Tephra can include lava bombs that are generally smooth-sided spheres weighing several to many kilos, to fist- or pea-sized cinders or lapillae (Fig. 2.4C), to very fine ash or glassy strings called Pele's hair. Again, the chemical composition of each of these is nearly identical, but the shapes (and subsequent surfaces for primary succession) are very different. Volcanoes can also trigger pyroclastic flows, lahars, debris flows and avalanches. Pyroclastic flows are incandescent clouds of gas and solids produced by explosive eruptions that move at speeds of > 100 km h^{-1} down the slopes of a volcano (Francis, 1993). Such flows killed over 28,000 people when Pelée erupted on the Caribbean island of Martinique in 1902. Lahars (mudflows; > 50% water), debris flows (< 50% water) and avalanches (nearly dry) all descend volcanic slopes under the force of gravity. They often result from snow and ice melt during an eruption. Lahars killed 23,000 people in Colombia in 1985 (Nevado del Ruiz eruption) and over 200,000 people are at risk from the huge lahars that would be created by a collapse of the upper reaches of Mount Rainier in Washington, U.S.A. Such a variety of substrates obviously result in many different successional trajectories on volcanic surfaces. For example, *Castanopsis* trees grew only on lava but *Persea* trees grew on both lava and tephra on Miyake Jima Island in Japan (Kamijo & Okutomi, 1995). Surface textures influence erosion, organic matter accumulation and soil development in addition to primary succession. Volcanoes often trigger secondary disturbances including mudflows, landslides and floods by increasing slopes and altering drainage patterns. Mudflows can dam rivers to form lakes (Mount St. Helens, U.S.A.). Earthquakes and volcanoes are products of the same tectonic forces and cluster along active faults (Fig. 2.5).

Earthquakes
Earthquakes can also be a very severe disturbance, with moderate to high frequency along active faults throughout many parts of the globe. Earthquakes have consistently caused the death of millions of people (Table 2.3). They can create surfaces for primary succession by triggering

Table 2.3. *Earthquakes that have caused the deaths of ≥100,000 people in recorded history*

nd, No data available.

Location	Date	Deaths	Magnitude (Richter)	Secondary disturbances
Shansi, China	23 Jan 1556	830,000	nd	–
Tangshan, China	27 Jul 1976	255,000–655,000[a]	8.0	–
Aleppo, Syria	9 Aug 1138	230,000	nd	–
Gansu, China	16 Dec 1920	200,000	8.6	landslides
Xining, China	22 May 1927	200,000	8.3	fractures
Damghan, Iran	22 Dec 856	200,000	nd	–
Ardabil, Iran	23 Mar 893	150,000	nd	–
Kwanto, Japan	1 Sep 1923	143,000	8.3	fire
Chihli, China	Sep 1290	100,000	nd	–
Messina, Italy	28 Dec 1908	70,000–100,000	7.5	tsunami

[a] The official death toll is 255,000, but casualties may have been as high as 655,000. Modified from *http://neic.usgs.gov/neis/eqlists/eqsmosde.html*

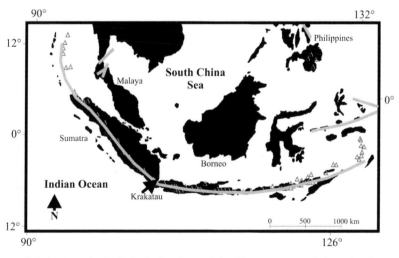

Fig. 2.5. Major seismic faults in Southeast Asia. Lines represent faults; triangles represent volcanoes. Note the location of the volcano Krakatau. Modified from Thornton (1996), after Ninkovich (1976), with permission from Elsevier Science.

landslides (Reid & Taber, 1919; Simonett, 1967; Garwood et al., 1979; R. Allen et al., 1999; Wells et al., 2001) or other forms of erosion, causing uplift or collapse of shorelines, causing floods, uprooting trees, disrupting roads and demolishing buildings. Earthquakes can cause tsunamis (tidal waves) or trigger fires by breaking natural gas lines (e.g. San Francisco, U.S.A., 1989). Tsunamis and fires generally do not initiate primary succession (Lawrence, 1979). Damage is proportional not only to the magnitude of the earthquake, but also to the density of buildings and humans in the affected area. The most severe earthquake ever recorded occurred in Chile in 1960 and registered 9.5 on the Richter scale. It produced massive landslides (Veblen & Ashton, 1978) and a tsunami 10 m tall that killed hundreds of people around the Pacific Ocean. Two other earthquakes have been recorded at 9.0 or above (Alaska, U.S.A., 1964, 9.2; Kamchatka, Russia, 1952, 9.0). Historically, severe earthquakes appear to occur approximately every 1,000 years in Puget Sound, Washington, USA. The last earthquake above 9.0 occurred there in about A.D. 1000 (Bucknam et al., 1992). Worldwide, one earthquake greater than 8.0 and 20 greater than 7.0 typically occur each year. The most destructive earthquake in recorded history killed around 830,000 people (Shansi, China, 1556; Table 2.3). The most destructive one in the twentieth century killed an undetermined number between 255,000 and 655,000 people (Tangshan, China, 1976).

Erosion
Erosion is a natural process along waterways (by water), on steep slopes (by gravity and precipitation) and in arid lands (by wind). Large areas of the Earth are being shaped by erosion (e.g. river valleys, deltas, canyons, alluvial plains, sea floors). Areas of active geological uplift associated with correspondingly high erosion rates include the Himalayas (Shroder et al., 1999), the Andes (Harden, 1996; Horton, 1999) and the Southern Alps of New Zealand (Shi et al., 1996; Koons et al., 1998). On any of these surfaces where soil structure and perhaps nutrients and organic matter have been lost, primary succession can occur. Deposition of sediments can also provide surfaces for primary succession, once the new surfaces stabilize. Other natural disturbances promote erosion when surfaces are destabilized (e.g. by volcanoes, glacial melt water, dune formation or geologic uplift; see section 2.2.6 on disturbance interactions).

Humans have greatly accelerated erosion throughout the world. By the removal of trees and shrubs for fuel and shelter (Wilmshurst, 1997), the creation of croplands and reliance on disruptive agricultural techniques (e.g. plowing, intensive grazing and widespread and simultaneous crop removals) we have increased natural rates of erosion by more than 100-fold (Pimentel & Harvey, 1999). Most human-induced erosion is related to loss of soil from agricultural lands. For example, nearly 50% of grain-producing farmland in Kazakhstan was abandoned as a result of soil erosion between 1980 and 1998 (Brown, 1999). Many other human activities also cause erosion (e.g. construction sites, roadbeds that cross slopes, and paved surfaces that accelerate and channel floodwaters). Over-grazing of more than half the world's pasturelands (World Resources Institute, 1994) has also dramatically increased erosion, particularly along river-banks. Consequences of drastic increases in global erosion include loss of arable land and subsequent increases in forest removal to create more cropland, or increases in fertilizer rates or grazing pressures to increase primary productivity (Pimentel *et al.*, 1995). This downward spiral can only be stopped by reducing the harmful land uses that cause erosion and by rehabilitating eroded lands by using the processes of primary succession.

Landslides
Landslides (or avalanches) are a type of rapid erosion by gravity that results from slope destabilization due to volcanoes, earthquakes, rain, road cuts, forest clear cuts, construction, mining or loss of vegetative cover (Sharpe, 1960; Tang *et al.*, 1997). We use the term landslide to include all downward and outward movement from falls, slides, spreads, slumps and flows (Larsen & Torres-Sánchez, 1996). Avalanches are the result of the rapid destabilization of snow or ice owing to such factors as melting, sonic triggers or earthquakes. When soil and/or rocks move, primary succession may result. Landslides may shape landscape-level features such as treeline in Chile (Veblen *et al.*, 1977) and affect forest re-generation in New Zealand (Vittoz *et al.*, 2001). Landslides vary in size from several square meters to many square kilometers. In earthquake-prone areas of New Zealand, individual landslides have eroded between 400 and 4,000 m^3 ha^{-1} (Sidle *et al.*, 1985). One large earthquake in New Zealand in 1929 created over 1,850 landslides that eroded an av-erage of 210,000 m^3 km^{-2} in a 1,200 km^2 area (Pearce & O'Loughlin, 1985). Mudflows and dirty snow avalanches (ones with rocks and soil) can transport hundreds of cubic meters of debris per year in montane, maritime environments (Keylock, 1997). Landslides are sometimes pre-dictable (e.g. in Puerto Rico based on duration and amount of rainfall; Larsen & Simon, 1993), but those associated with volcanoes (lahars, de-bris flows, mudflows) are sudden. Huge, underwater landslides off the

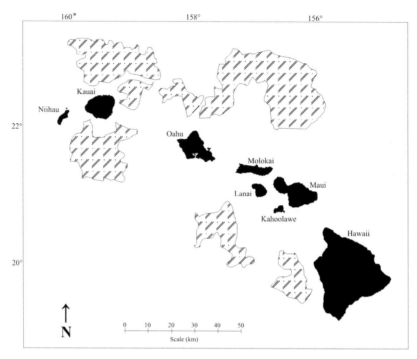

Fig. 2.6. Underwater avalanches (crosshatched) in the ocean surrounding the islands of Hawaii (solid). Modified from G. Walker (1994). ©University of Hawaii Press.

Hawaiian Islands probably caused the tsunamis that deposited lime-stone debris more than 300 m above sea level on the island of Lanai (Fig. 2.6) (G. Walker, 1994). A landslide resulting from a directed blast of Harimkotan Volcano in the Kuril Islands, Russia, in 1933 produced a tsunami 20 m tall. Owing to low population densities on the surrounding islands, loss of life was minimal (Belousov & Belousova, 1996). Landslides can cause flooding by damming rivers (Zion National Park, U.S.A.). Landslides (Haigh *et al.*, 1993) and avalanches (De Scally & Gardner, 1994) are major threats to the lives and livelihoods of residents of the Himalaya Mountains and the Andes.

Landslides often present a very heterogeneous substrate with patches of surviving or rafted soils and plants mixed with areas of nearly sterile subsoil or rock and areas of deposition of jumbled organic matter and mineral soil from above (Fig. 2.7; see also Figs. 1.2 and 5.7B). This mosaic of substrates can lead to various patterns of soil development and plant succession. Key factors determining successional trajectories on Puerto Rican landslides were initial soil stability and organic matter content (Fig. 2.8) (Walker

Fig. 2.7. A landslide in the Luquillo Experimental Forest, Puerto Rico.

et al., 1996). Where rafted or deposited forest soil is stable, return to pre-disturbance forest is an order of magnitude faster than on exposed mineral soil.

Rocks

Rock outcrops (Table 2.1) are areas of exposed bedrock that are caused by tectonic activities (uplift, earthquakes, fault zones, landslides or volcanism), erosion by wind or water (particularly along shorelines or river valleys) or human activities (mining, construction). Cliffs (from several to 700 m high) occur along 80% of the world's coastlines (May, 1997). Cliffs

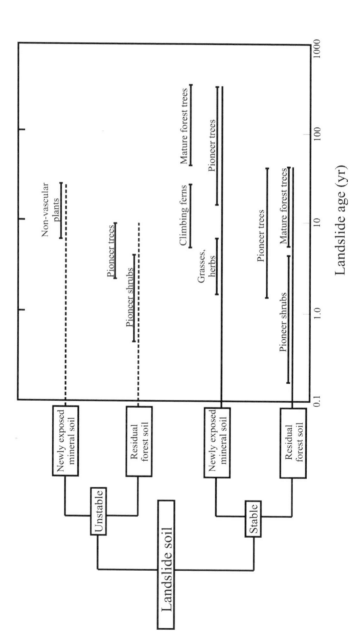

Fig. 2.8. Model of landslide succession from Puerto Rico. Four successional pathways originate from the degree of slope stability and presence/absence of residual organically rich forest soil. Dotted lines indicate that erosion continuously resets succession. From Walker *et al.* (1996). ©1996 by the Association for Tropical Biology, P.O. Box 1897, Lawrence, KS 66044–8897. Reprinted by permission.

can be ephemeral and rapidly covered with soil and vegetation or essentially permanent features of the landscape (e.g. the Niagara Escarpment in Canada; Larson *et al.*, 2000), maintained by active erosion, lack of soil formation, steep slopes or harsh microclimates that inhibit colonization by plants. Stability is a function of both the slope and the frequency and severity of disturbance. Steep cliffs might be very stable in the short term, but are subject to long-term erosion; talus slopes actively erode. The rocks can be intact, as on a cliff face, or jumbled, as on a talus slope. The surface texture has implications for the rate and nature of early succession.

Rock outcrop surfaces range from vertical (cliffs), to sloped (talus), or horizontal (granite domes, lava flows, alvars, tepuis). A typical cliff has a variety of physical characteristics (Fig. 2.9A) that provide microsites where plants can colonize (Fig. 2.9B) (Larson *et al.*, 2000). As a result, plant diversity can be higher on a cliff than in adjacent non-cliff areas (Camp & Knight, 1997). The surface texture varies from relatively smooth (marble), to rough (granite), to pocked by wind and water erosion (sandstone,

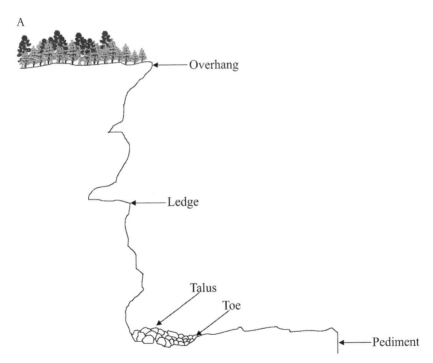

Fig. 2.9. Drawings of cliffs with (A) physical features; and (B) microhabitats for plants. Modified from Larson *et al.* (2000). Reprinted with the permission of Cambridge University Press.

B

Fig. 2.9. (cont.)

limestone). Where soils have formed they tend to be moderately fertile, with nutrient input from birds and water input from seeps. Physical gradients of soil water and soil depth may or may not represent stages in succession on granite rock outcrops in the Appalachian Mountains of the eastern U.S.A. (Burbanck & Phillips, 1983; Collins *et al.*, 1989;

Houle, 1990; Shure, 1999; see section 5.3.2 on dispersal mechanisms) or on flat-topped South American mountains called tepuis (Michelangeli, 2000). Frequent erosion or shallow soils can maintain rock outcrop communities at early stages of succession, promote endemism and reduce the likelihood that adjacent forest vegetation establishes (Wiser & White, 1999). Sometimes the erosion is triggered by biomass too heavy to support itself (Larson et al., 2000).

2.2.2 Air

Wind can be the primary agent of disturbance when it uproots trees (see above) or transports particles that form new surfaces such as dunes, or when it has secondary impacts such as creating waves that damage coral reefs. Several types of primary successional surface are ultimately formed by wind.

Hurricanes
Hurricanes (sustained winds faster than 119 km h^{-1}) uproot trees but also cause floods that have substantial impacts on coastal areas in the tropics and subtropics. Sandy beaches and bays can be reshaped by hurricane-driven waves (Birkemeier et al., 1991) and salty ocean water can be forced up streams and sprayed into coastal vegetation (Gardner et al., 1991). Primary succession on a newly constructed beach in Florida (U.S.A.) was slowed when inundation from storm surge of Hurricane Andrew reduced populations of three dominant species (Gibson et al., 1995). Changes in water turbidity or salinity and wave action can damage algae, invertebrates, fish and coral (Bénito-Espinal & Bénito-Espinal, 1991; Finkl & Pilkey, 1991). Primary succession is thus often initiated by the reshaping of the coastline, and by the formation of newly exposed surfaces (e.g. broken corals, sand deposition, benthic scour). Hurricanes are annual events in tropical and subtropical regions where warm ocean waters sustain the development of these massive vortices, which can be hundreds of kilometers wide. Hurricane return intervals for a given location vary greatly. Because tropical and subtropical coastal areas are often densely settled, hurricanes kill thousands of people annually, destroy homes and damage agricultural land. Bangladesh is a low-lying country that has lost over 800,000 people to periodic cyclones (the term for hurricanes in the Pacific and Indian Oceans) within the last 35 yr. On 12 November 1970, 300,000 people died in a cyclone in Bangladesh (Berz, 1988). Yet former residents return to

farm the fertile land left by successive floods despite the dangers because of land scarcity, population pressures and acute poverty (Zaman, 1999). In wealthier countries, people return to their coastal resort homes, confident that rebuilding costs will be borne by governments and insurers.

Tornadoes can have winds of over 400 km h^{-1}, but are much smaller in scale (less than 1 km wide) than hurricanes. The debris left behind after a tornado can include many surfaces that will undergo primary succession (e.g. exposed mineral soils, damaged roadbeds) if left unmodified by humans. Forest composition affects the degree of uprooting for both tornadoes (Peterson & Rebertus, 1997) and hurricanes (Brokaw & Walker, 1991), which is one way existing vegetation can influence the subsequent course of primary succession.

Uprooted trees
Uprooted trees expose the soil surrounding the roots and create both a pit where the roots were and a mound where the roots remain. The soil associated with deep roots is often less fertile than the surface soils, and plant colonization of pits and mounds can be considered primary succession. The principal cause of uprooting of trees is wind. Hurricanes, tornadoes or other abrupt increases in wind speed are more likely to uproot trees than are the steady, predictable winds found on exposed ridges. Occasionally, catastrophic damage may equal (Frelich & Lorimer, 1991) or exceed (Brokaw & Walker, 1991; Lugo & Scatena, 1999) background levels of mortality. Treefall gaps in both tropical (Whigham *et al.*, 1999) and temperate (Cogbill, 1996) forests can impact 1–2% of a forest annually. Uprooted trees cause about 50% of these gaps and are the only type of disturbance providing exposed mineral soil for primary succession. The remainder comes from snapped tree trunks (Brokaw & Walker, 1991; Whigham *et al.*, 1999). The percentage of gaps caused by uprooting varies widely, depending on the windstorm's intensity and windstorm intervals. The time for the whole area to be affected by all types of tree fall (return interval) is about ten times longer in temperate (1,400–3,000 yr; Cogbill, 1996; Webb, 1999) than in tropical forests (*c.* 100–300 yr; Whigham *et al.*, 1999), suggesting that successional recovery is ten times faster in tropical than in temperate gaps (Brokaw & Walker, 1991). These estimates are tentative as they assume uniformity of both the disturbance and vulnerability to wind-throw across the landscape (Webb, 1999).

Fig. 2.10. Multiple dune ridges on the eastern shore of Lake Michigan (U.S.A.).

Substrates exposed by uprooted trees reflect the variety of forest environments throughout the world. Shallow-rooted trees do not expose the deeper, less organic soils. Uprooting in saturated soils can lead to drying only on the mounds, with pools of water collecting in the pits (Beatty & Stone, 1986). Uprooting in hot climates can lead to desiccation of mound soils (Walker, 2000). Soil fertility is generally lower in the exposed mineral soils of the pit and mound than in undisturbed forest soils where there is more organic matter (Vitousek & Denslow, 1986; Walker, 2000). Forces that uproot trees therefore can create heterogeneity and variable surfaces for succession.

Dunes
Dunes are widespread features of both coastlines and arid interior lands and are principally formed by wind or wind–plant interactions (Fig. 2.10). Dunes are especially abundant around 30° N and S (Rice, 1988). At low latitudes, dunes are less common, owing in part to stabilizing vegetation, low average wind speeds and salt-tolerant vegetation along shorelines (Doing, 1985). Dunes tend to be least common at high latitudes, owing in part to raised shorelines and younger surfaces from recent deglaciation combined with shorter seasons when the soils are not frozen. Many factors can impact the amount and quality of sediments that are deposited or eroded from shorelines, including tides, tsunamis, coastal advance or

retreat, tectonic uplift or subsidence, riparian sediment deposition, land-slides, volcanoes, earthquakes and human activities. Interior dunes can be affected by all but the first three factors. Given a source of wind-blown sand, dunes begin to form when surface roughness disrupts laminar flow and impedes sand movement. This roughness can initially be abiotic, but organisms such as the grass *Ammophila* (Olson, 1997) and mycor-rhizae (Jehne & Thompson, 1981) are effective in trapping sand and, at least initially, accelerating dune formation. Dunes can be ephemeral or mobile, moving across the landscape at rates of 3–15 m yr^{-1}. Such highly unstable surfaces rarely provide a suitable substrate for succes-sion. Alternatively, dune surfaces can remain stable for thousands of years (e.g. the Cooloola Dunes in Queensland, Australia; Walker *et al.*, 1981; Specht, 1997). Either surface instability or lack of available water can de-ter colonization. Many dunes with large sand grains are surprisingly moist because water collects but cannot subsequently evaporate owing to the large pore spaces (Danin, 1991). Ponds often form in interdunal swales and these wetlands undergo primary succession (Lammerts & Grootjans, 1998).

2.2.3 Water

Floodplains
Floodplains (Fig. 2.11) occur wherever river channels are wide enough and flow velocities are slow enough for the deposition of sediments (e.g. meanders, deltas). Floodplain surfaces are often ephemeral because floods frequently redeposit the sediments downstream. As a result, any given floodplain can be either erosional or depositional. However, dur-ing their short duration, they can host pioneer vegetation and provide colonists for other sites. Floods occur when water is no longer con-tained within the primary channel and overflows into the wider flood-plain, an area that often contains farmlands, homes and cities. Floods may simply deposit a layer of silt or mud when they retreat, or they may cause severe erosion. Floodplains can be created or altered by glacial retreat or volcanic activity. Glacially fed rivers or others with high sed-iment loads can act as sources for dune formation. Human alterations of the environment (e.g. deforestation or urbanization upstream) can in-crease the severity and frequency of floods. For example, deforestation in the Himalayas aggravates the already severe flooding in Bangladesh (Zaman, 1999) and India. Extensive annual dredging of the river chan-nels has been suggested to increase channel capacity and elevate the

Fig. 2.11. The Tanana River floodplain in central Alaska (U.S.A.). Note the bands of vegetation (*Salix–Alnus–Populus–Picea*) representing successional development away from the river's edge (the darker bands are *Alnus* and *Picea*).

land, thereby reducing flood frequency and severity (Khalequzzaman, 1994).

Floodplain substrates vary in stability, texture and fertility, each of which will impact primary succession. Interactions of local topography, flooding regime and channel dynamics create a variety of habitats in a floodplain (e.g. point or scroll bars, levees at cut banks, backswamps or oxbow lakes; Malanson, 1993). Flow velocity determines the particle sorting that occurs on a floodplain and texture can vary greatly over short distances. The mosaic of boulders, rocks, silt and clay has broad implications for plant colonization (Grubb, 1987).

Plant succession on floodplains can be rapid and helps to stabilize the riverbanks (Hughes, 1997) (Fig. 2.12). Deposits from upstream plant communities often include substantial amounts of nutrients, organic matter and propagules (Luken & Fonda, 1983; Walker, 1989). However, soil stability, texture or water-holding capacity are often more important determinants of succession than soil nutrients (Kalliola *et al.*, 1991; Shaffer *et al.*, 1992). Human alterations of flood regimes from channelization, dams (Busch & Smith, 1995) or pollution (Muzika *et al.*, 1987) have altered plant community dynamics in many areas. Drops in water level

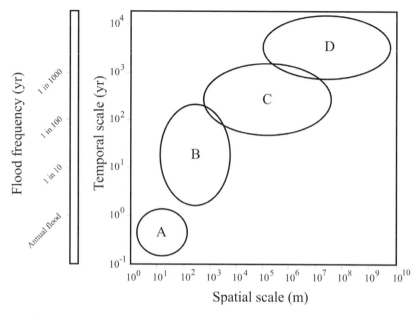

Fig. 2.12. Floodplain components and processes on spatial and temporal scales.
(A) Primary succession with annual floods; (B) primary and secondary succession
with floods of medium magnitude and frequency; (C) long-term floodplain
succession and widespread erosion with high-magnitude, low-frequency floods;
(D) extensive landscape-level floods during post-glacial melting. Modified from
Hughes (1997), with permission.

from natural droughts can cause the emergence of islands, but human activities such as draining lakes (Rydin & Borgegård, 1991) and salt marshes
(Holter, 1984; Andersen, 1995) or creating polders (by building dikes)
also create opportunities for primary succession.

Glaciers
Glaciers (Fig. 2.13) currently cover about 10% of the Earth's surface (90%
of that coverage is in Greenland and Antarctica; Matthews, 1999). During
recent glacial maxima, glaciers covered as much as 32% of the Earth.
Glaciers therefore have a dramatic influence on sea levels, ocean currents,
global climate and fisheries. Worldwide, most glaciers have retreated from
their most recent advances during the Little Ice Age (200–400 yr BP).
Current estimates suggest that 75% of all glaciers are now in retreat
(Haeberli *et al.*, 1999), probably owing to global warming. Tidewater
glaciers are susceptible to rapid retreats, most notably Glacier Bay where
over 100 km have been exposed due to glacial melt during the past

Fig. 2.13. The Franz Josef Glacier in Westland (New Zealand).

200 yr (Chapin *et al.*, 1994). This retreat averaged 0.4 km yr^{-1}, 15-fold faster than any other tidewater glacier (Lawrence, 1958; Goldthwait, 1966). However, glaciers also advance or remain relatively stable. One 'galloping glacier', the Kutiah Glacier in Pakistan, advanced 12 km in 3 months during 1953 (Hewitt, 1965).

Recently deglaciated surfaces, often rebounding (or rising) from the release of the pressure of the glacier, include many types of surface with various textures, stability and fertility. Outcrops of bedrock that have been abraded by rocks carried in the glacier form one end of a continuum of substrate types (stable, hard, infertile). Cracks, ledges or other irregularities in the rock surface provide a more favorable substrate for primary succession. Glacial sediments or till are deposited in a variety of ways and can come from material carried by the glacial base and redeposited further downhill by the moving ice, by melting of stationary or moving ice with sediments embedded in it, or by movement of melt water from beneath or in front of the glacier (Matthews, 1992, 1999). Till can include particles as coarse as boulders or as fine as rock flour (very fine silts with little clay; Haldorsen, 1981). The fine texture particles are generally least stable and most fertile. Recently deglaciated surfaces may not be entirely sterile (some glaciers even transport intact plant communities on their upper surfaces, leading to an inoculation of the barrens when the glacier retreats), although the soil microbes and soil organic matter that initially exists when a glacier melts may (Wynn–Williams, 1993) or may not (Matthews, 1999) promote primary succession. Secondary disturbances following deglaciation include landslides, floodplains, wind-blown loess that abrades and erodes, and winds that desiccate and cool the environment. Human activities, including tourist impacts, can also affect post-glacial succession.

Cold regions
Disturbances in regions of permafrost provide new substrates for primary succession. Extreme temperature fluctuations, drought, wind, and many surface disturbances related to freezing and thawing cycles (e.g. pingos or ice-cored hills, ice wedge polygons, frost boils and ice-scoured riverbanks) contribute to the disturbance regime (Komárková & Wielgolaski, 1999) and regulate chemical reactions.

Drought
Drought can be an important disturbance in both arid and non-arid areas. Although drought typically induces secondary succession because soils are not generally disrupted, severe drought can lead to loss of vegetative cover and surface disruption from wind erosion (MacMahon, 1999). Extensive agriculture and grazing, especially in arid climates, also can lead to loss of topsoils and exposure of subsoil or bedrock (Pimentel & Harvey, 1999). Human activities interact with climatic shifts to cause expansion of deserts in a poorly understood process called desertification. Dramatic examples of how humans can be impacted by widespread wind erosion include the Dust Bowl in the central U.S.A. in the 1930s (Worster, 1979) and the more recent expansion of the Sahara Desert into the semi-arid

Sahel (Tucker *et al.*, 1991; Buol, 1994). Wind-blown dusts are important sources of nutrients, microbial spores and organic matter in many arid lands (Belnap & Gillette, 1998) and can be transported to distant land surfaces.

Marine
Tsunamis can create new surfaces on land or under water. Earthquakes, volcanic explosions or underwater landslides cause these sea waves but little is known about their role in primary succession. Succession in rocky intertidal areas occurs following the addition, removal, or scouring (by ice or waves) of rocky surfaces (Sousa, 1979, 1984). Benthic succession can occur on soft surfaces created by the addition of sediments or, infrequently, on hard surfaces following sediment removal. Agents of underwater disturbance include methane and brine seeps (Van Dover, 2000), or inputs of sediments from beaches, rivers or currents, earthquakes, landslides and hydrothermal vents. Colonization of new surfaces and metabolism of bacteria and bivalves is generally much slower in the low-temperature, high-pressure and low-nutrient conditions of the deep sea (Van Dover, 2000) than on land. However, additions of nutrients (e.g. from whale carcasses; Jones *et al.*, 1998) can increase colonization rates. These extremely fertile patches may be one factor contributing to the relatively high biodiversity on the ocean floor (Grassle & Morse-Porteous, 1987). Hydrothermal vents are even more diverse. Chemoautotrophic organisms at hydrothermal vents (mostly sulfide reducers) are arranged along gradients of decreasing temperature with distance from the vents, but also appear to have various successional gradients, with both biotic and abiotic processes driving species change (Sarrazin *et al.*, 1997). For example, bivalves generally replace tubeworms, perhaps owing to chemical cues produced by the tubeworms (Mullineaux *et al.*, 2000). However, chronic disturbances from changes in the direction and flow rate of vented gases are the dominant influences on this succession (Van Dover, 2000). Hot springs are terrestrial analogies of hydrothermal vents and also have a specialized biota dominated by archaebacteria.

2.2.4 Fire

Fire does not usually create surfaces for primary succession, but it is an important agent of disturbance associated with volcanic eruptions and earthquakes. Very slow-burning or extremely hot fires in grasslands, chaparral or forests can destroy most soil organic matter and seedbanks, leading to primary succession on a low-nutrient substrate. More commonly, fires remove only the aboveground vegetation and secondary succession ensues (Bond & van Wilgen, 1996). However, the loss of vegetation on slopes can precipitate erosion of topsoil and initiate primary succession, either on the slope or where the sediments accumulate. Fires can also be important in altering successional trajectories, as when the burning of early successional grasses on Taal Volcano (Philippines) facilitated the establishment of a fire-resistant tree (*Acacia*; Brown *et al.*, 1917).

2.2.5 Humans

The impact of humans on the Earth is immeasurable. The number of humans on Earth surpassed six billion in October 1999 and could double by 2050. There is considerable debate about what the carrying capacity of the Earth is (numbers of humans the Earth can sustain indefinitely). Most estimates vary between 2 and 15 billion people, with some agreement around 8–12 billion, but any estimates (of future numbers or ultimate carrying capacity) depend on fertility rates and how humans utilize the remaining resources (Cohen, 1995). These estimates do not appear to account for the ripple effects such high numbers will have on biodiversity or the draw-down of natural resources. Loss of natural resources lowers the carrying capacity. Humans now appropriate, directly or indirectly, at least 40% of the net primary productivity of Earth's terrestrial surfaces (Vitousek *et al.*, 1986). Is there food or fuel or shelter or simply space to sustain twice today's human population? Many humans today live without basic necessities, and doubling our numbers will certainly continue to reduce the average quality of life for humans and increase our impacts on biodiversity and natural resources. Humans have now affected every habitat on Earth (McKibben, 1989), and many surfaces undergoing primary succession are either anthropogenic in origin or influenced to varying degrees by human activities. In this section we describe the major direct impacts humans have on land surfaces, ignoring impacts on air and water quality.

Erosion

Erosion (see section 2.2.1) is a major cause of soil loss and often exposes surfaces that consequently undergo primary succession. Humans have accelerated erosion rates around the globe, particularly in association with agriculture, grazing and urbanization (Gardner, 1997).

Fig. 2.14. Hydraulic mining for gold in a streambed in central Alaska (U.S.A.).

Mining

Surface mineral extraction creates many substrates for primary succession and already covers *c.* 1% of the Earth's land (Fig. 2.14). Mining has always been a part of civilization and is a crucial part of the global economy. Coal is the most widely removed product, currently fueling 40% of world electrical production. Most of this coal is now removed from huge surface pits rather than from underground, but even underground mining results in surface waste piles (tailings) and subsidence of the ground can occur above shallow underground mines. Other mined products include other fuels (e.g. oil in shale, peat and uranium), metals (e.g. Al, Cu, Zn, Fe, Pb, Ni and Ti), construction materials (gravel, sand, slate, stone), and other products (e.g. gypsum, phosphate, potash, salt; Cooke, 1999). Mining removes vegetation and soils and creates mine pits, stockpiles of topsoil, tailings, slurry lagoons and surfaces used for roads and buildings (Majer, 1989a). Stream or terrestrial pollution results from mining activities that release chemical wastes. Hydraulic mining destroys riverbanks and other landforms and creates sterile, homogeneous surfaces.

Mined surfaces vary in stability, texture and fertility (see Fig. 2.1). Vertical rock surfaces in rock quarries are among the most stable, least fertile, and most difficult substrates to reclaim. The slopes of newly formed

Fig. 2.15. Derelict urban habitats often undergo spontaneous primary succession that is often dominated by ruderal species such as legumes, particularly in nitrogen-deficient soils (Liverpool, England).

tailings, in contrast, are unstable. Textures range from large boulders (e.g. overburden) to gravel piles rinsed with cyanide (heap leach) to toxic sediments left in slurry lagoons. Stockpiles of topsoil (Munshower, 1993) or compacted access roads (Walker & Powell, 2001) are among the most fertile of mine-related disturbances, but most mined lands are deficient in N and P, reflecting the general lack of organic matter. Toxic levels of Al, Fe, and Mn, very low pH values (e.g. pH = 2.6 on coal mine tailings in Kentucky, U.S.A.) and very high pH values (e.g. pH = 9.0 on soda ash wastes in Britain; Ash *et al.*, 1994) further hinder succession on these substrates (Leopold & Wali, 1992). We discuss rehabilitation of mined lands in Chapter 8.

Urban
Urban habitats (Fig. 2.15) now cover over 3% of the Earth's land surface and house over 50% of all humans, and the percentages continue to increase. Each year between 1982 and 1992 the United States paved over 168,000 ha (the equivalent area of two New York Cities; Gardner, 1997). Studies of urban ecology (Bornkamm *et al.*, 1982; Gilbert, 1989; Sukopp *et al.*, 1990) have become critical as urban centers expand both geographically and in terms of resource use. A wide variety of urban

disturbances reflect the historical development of cities, with severity usually increasing as one approaches the center of a city from its edges (Sukopp & Starfinger, 1999). Many city centers (e.g., Rome, Sao Paulo) have few trees or gardens. Yet primary succession does occur on such urban surfaces as compacted or graveled footpaths, paved sidewalks or gravestones (Bradshaw & Chadwick, 1980; Gilbert, 1989). Urban streets are either a densely compacted dirt surface or composed of crushed rock, flagstone, lava, brick, tar, asphalt or concrete. These surfaces provide stable, low-fertility substrates that can resist invasion for several decades after abandonment. Road edges (see discussion of transportation below) provide open soil, gravel or sidewalks where primary succession can occur, but chronic disturbance from herbicides or continuing pedestrian traffic may preclude significant development. Vacant lots and dumpsites provide other urban surfaces for primary succession and may be more (soil, garden wastes) or less (glass, steel, plastic) fertile than roads or pathways. However, fewer potential colonizers surround vacant lots than are found adjacent to urban parks. This is one reason why natural succession in vacant lots is often arrested in an early phase dominated by weedy species. It also suggests that more subtle effects of landscape context affect primary succession. Any urban habitat can be subject to pollution from toxic waste materials such as battery acid or oil. Urban habitats are also prone to erosion or flooding from surface flow of waters altered, enhanced or redirected by disruption of traditional drainages.

Military
Military activities (especially during wartime) create ample opportunities for primary succession by explosions of bombs, by soil disruption and compaction (from tracked and wheeled military vehicles, marching soldiers, military vehicles, foxholes and trenches; Becher, 1985; Demarais *et al.*, 1999), by metal, petroleum and plastic wastes (including abandoned vehicles) left at battle or training sites and by subsequent soil erosion (especially following use of defoliants such as Agent Orange, which was used in Vietnam; Southwick, 1996). The frequency and severity of peacetime military activities are monitored carefully only in countries with strict environmental legislation. Surfaces created by military activities (except abandoned vehicles) typically remain fertile, but compacted, with or without heavy metal, radioactive or oil contaminants. Stability varies, but where extensive vegetative cover is lost, soil erosion (by water and wind) is common (El-Baz, 1992; Demarais *et al.*, 1999) and can lead to dune formation in arid environments.

Transportation

Transportation of humans has created many surfaces that undergo primary succession when abandoned. Trails or vehicle tracks may have enough intact soil that, if abandoned before the soil is too compacted or eroded, can undergo secondary succession (Ullman & Heindl, 1989). However, any paved surface (roads, runways, parking lots) or otherwise treated surfaces that have been oiled, sprayed, bulldozed, tarred or mounded (as for a railroad bed) are likely to undergo primary succession. Road (and railroad) edges may also undergo primary or secondary succession during the life of the transportation corridor (Klimes, 1987). However, edges are subjected to continual disturbances such as dust, herbicide applications, mowing or lead from burning gasoline and asbestos from brake linings (Hansen & Jensen, 1972; Walker & Everett, 1987; Barrow, 1991; Forbes, 1995). Road construction often creates additional disturbances such as gravel pads (Bishop & Chapin, 1989b) and gravel pits (Kershaw & Kershaw, 1987) that often have unique trajectories of recovery. For example, the recovery from human disturbances in soils affected by permafrost (Komárková & Wielgolaski, 1999) depends on how stable the soil is after thawing (Lawson, 1986; McKendrick, 1987).

Abandonment of transportation corridors typically occurs when newer, straighter thoroughfares are constructed or when towns are depopulated owing to causes such as warfare, disease, fire or economic hardships. Transportation corridors affect approximately 5% of the Earth's terrestrial surface (Walker & Willig, 1999), a surface area that is increasing as human populations expand and travel becomes more common (Spellerberg, 1998). As much as 35% of urban areas can be occupied by roads and related surfaces such as parking lots (Gilbert, 1989).

Transportation corridors provide diverse substrates, from asphalt and concrete, which resemble lava in their disturbance severity, to gravel or packed dirt surfaces where succession may proceed more rapidly than on paved surfaces. Primary succession may also occur on bridges and walls associated with roads and railways (Gilbert, 1989). However, compaction slows succession on unpaved vehicle corridors (Webb & Wilshire, 1983; Bolling & Walker, 2000). Impermeable surfaces (e.g. asphalt) often trap water and organic matter. Succession can be rapid in these relatively favorable microsites (Sader, 1995), especially in the tropics. Heyne (2000) found that plant communities growing on abandoned paved roads in a Puerto Rican rain forest (Fig. 2.16) resembled adjacent forests within only 60 yr. In contrast, in cold (Roxburgh *et al.*, 1988; Auerbach *et al.*, 1997) or arid (Bolling & Walker, 2000) climates, recovery is much slower.

Fig. 2.16. A paved road in Puerto Rico (U.S.A.) that had been abandoned for 5 yr when the photo was taken in 1998.

Roads can destabilize slopes and cause landslides (Guariguata & Larsen, 1990; Haigh *et al.*, 1993) or obstruct drainage and cause flooding. Roads also provide invasion corridors for alien species that can, in turn, alter successional trajectories (D'Antonio *et al.*, 1999; see Chapter 7), particularly if the aliens invade off-road communities. Common alien plants along road edges in southern New Zealand have equilibrated with their apparent environmental tolerances since their introduction from Europe 100–150 yr ago (Ullman *et al.*, 1995). However, the alien species are most common in agricultural landscapes and do not appear to alter vegetation dynamics beyond road edges in the more natural landscapes.

Other surfaces

Other new human-created surfaces that can be considered as opportunities for primary succession include land that has been temporarily flooded, oil spills, and atomic bomb craters (e.g. the Nevada, U.S.A., Test Site). Landfills, usually unconsolidated mixtures of inorganic and organic wastes, may also be nutrient-rich surfaces for primary succession, unless they are capped with soil, which would then initiate secondary succession. New aquatic surfaces include some landfills and waste dumps, artificial reefs, hurricane or human (fishing, snorkeling, poisoning, blasting) damage to natural reefs, benthic areas that are subjected to dredging or blasting (for construction of docks, harbors, cableways or canals), discarded fishing gear, sunken ships, piers and jetties. Nutrient inputs from fish farms (Lu & Wu, 1998) provide an anthropogenic analogy to naturally occurring methane seeps. Building of dikes can lead to primary succession on former seashores and polders (van Noordwijk-Puijk *et al.*, 1979). All of these human disturbances vary in their extent, severity and fertility. Nuclear waste sites and bomb blasts are very extensive and sterilize everything at the center of the disturbance, delaying succession. Initial colonization of oil spills by bacteria or piers by algae can be quite rapid. The newest frontier for fungi and bacteria is in outer space. Russian scientists experienced fungal damage to windows and equipment on the space station Mir (Svitil, 2000).

2.2.6 Disturbance interactions

Primary succession may be initiated by a particular disturbance, but that disturbance is often only part of a set of interacting disturbances or a disturbance regime that influences the initial conditions for succession and subsequent successional trajectories. As we have noted, some disturbances are more often linked than others (Fig. 2.17). Landslides, for example, can result from any disturbance that destabilizes slopes (e.g. volcanoes, dunes, glaciers, mines or roads). Prior landslides (Lundgren, 1978) or deforestation from logging or herbivory (e.g. by possums in New Zealand; James, 1973) can also trigger landslides. Landslides can transport debris onto glacial ice, developing 'in-transit' moraines. Succession on such moraines in Chile is similar to succession on more conventional recessional or lateral moraines (Veblen *et al.*, 1989). Drainage patterns of rivers and successional trajectories on associated floodplains also result from the interactions of many types of disturbance (Kiilsgaard *et al.*, 1986). In a tropical forest in Puerto Rico, half of all possible

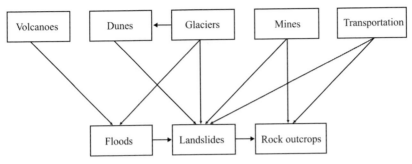

Fig. 2.17. Disturbance interactions. Disturbances in the upper row can cause disturbances in the lower row. Other causal relationships are indicated by horizontal arrows. Upper row disturbances are generally more severe than those in the lower row. Modified from Walker (1999b), with permission from Elsevier Science.

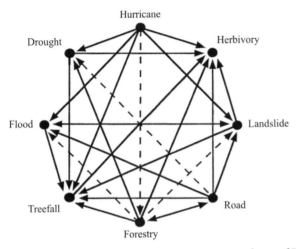

Fig. 2.18. Disturbance interactions in a tabonuco forest of Puerto Rico. Arrows indicate when the occurrence of one disturbance enhances the probability of another disturbance. Weaker links are shown as dotted lines. From Willig & Walker (1999), with permission from Elsevier Science.

linkages between eight disturbance types were considered to be important (Fig. 2.18) (Willig & Walker, 1999) and many of these interactions were reciprocal (e.g. landslides promote floods and vice versa). Two dramatic examples of disturbance interactions include massive flooding in Iceland when a volcano melted a glacier (Monastersky, 1996) and an earthquake in Peru in 1970 that led to huge landslides. Sediments from these landslides were washed into the ocean, increased local sediment loads and created new beaches. These beaches then were blown inland and ruined croplands (Moseley, 1999).

2.2.7 Summary of disturbance types

A wide variety of disturbances create denuded patches of landscape where primary succession can begin. Their return intervals range from continuous eruptions (e.g. Sakurajima, Japan; Mount Etna, Italy) or movement of dune sands to glaciers that only retreat during interglacial periods (see Fig. 2.1, Table 2.1). The size of the disturbances also varies broadly from localized exposure of bedrock or the uprooting of a tree to volcanic particles and gases blasted into the upper atmosphere that alter climates and light interception on a global scale. The most intense disturbances include high wind speeds during hurricanes, lavas, and the shaking of high-magnitude earthquakes. Pyroclastic flows are particularly intense, combining intense heat with rapid air and soil movement. Damage severity reaches its highest values with lava and pavement, where no biological legacy remains and the surface is completely transformed. Disturbances of lowest severity that still create conditions for primary succession include landslides and floodplains where biological legacies remain. However, remnant organisms (seeds, eggs, vegetative parts with the potential to sprout) may not be active for some time. Primary succession is clearly influenced by all these variables. Return intervals determine the predictability of primary substrates and the subsequent neighborhood pool of potential propagules. The size of a given disturbance also influences the likelihood that propagules will reach the disturbed site (see Chapter 5). Disturbance intensity and severity determine what biological legacy remains and how long recovery to pre-disturbance conditions will take.

Surface characteristics determine the rates and trajectories of succession on a local scale (Fig. 2.1, Table 2.1). Soil development (see Chapter 4) is more rapid where surfaces are stable, fine-textured and fertile. Water retention is best in cracks in impermeable pahoehoe lava, but even coarse dunes may retain subsurface moisture. Fertility is determined mostly by the degree of biological legacy, although intrinsic characteristics of the substrate are also important. On landslides, floodplains and in uprooted patches of forest, the presence of some residual organic matter (and/or proximity to patches of undisturbed soil) greatly increases substrate fertility and potentially accelerates primary succession. In contrast, pahoehoe lava and rock outcrops are least fertile. In subsequent chapters we will explore how soils develop (Chapter 4), how plants and animals invade new sites (Chapter 5), and how they interact (Chapter 6) to form different patterns of primary succession (Chapter 7). We will then explore how the lessons learned from studying primary succession can be applied to more effectively rehabilitate or restore profoundly damaged ecosystems (Chapter 8).

3 · *Successional theory*

3.1 Introduction

Succession is as central a concept for ecology as evolution is for biology (Margalef, 1968a). Studies of succession integrate concepts and tools from ecosystem, community, population and organismal ecology, soil science, geology, meteorology, conservation biology and other disciplines. Succession is at once an easily observable phenomenon and an unpredictable puzzle. It has intrigued humans for centuries and has become increasingly relevant as a model for habitat rehabilitation. In this chapter we provide both a history of ideas about succession and an overview of current perspectives, including attempts to describe and predict successional trajectories with conceptual and mathematical models. We trace the development of several sharply contrasting views of succession that have persisted in several forms for the past 100 yr and are still represented in the conflicting modern approaches to succession. This variety of perspectives has considerably enlivened studies of succession and emphasizes that we have yet to understand the complexity that underlies ecosystem development (Anderson, 1986; Weiher & Keddy, 1999).

Most theoretical considerations of succession have addressed secondary succession, but some of these also are applicable to primary succession. Primary succession has played a pivotal role in the study of succession because it represents visually dramatic development following the most extreme disturbances, was the subject of most early studies of succession and is now increasingly relevant for rehabilitation of disturbed lands. In this chapter we assess the relevance of succession to studies of primary succession (cf. Matthews, 1992).

Several seminal studies provided a highly visible profile for primary succession (Fig. 3.1). These included studies of coastal dunes in and near Denmark (Warming, 1895), the Cooloola Dunes in Queensland (Australia; Coaldrake, 1962) and dunes along the shores of Lake Michigan (U.S.A.; Cowles, 1901; Olson, 1958). Classic studies of glacial moraines

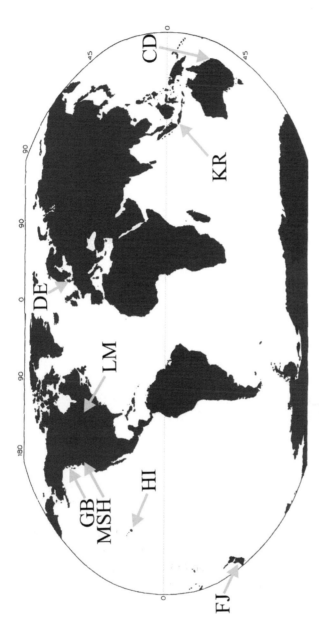

Fig. 3.1. Locations of some classical studies of primary succession (dates of early publications in parentheses). Dunes: DE, Denmark (1895); LM, Lake Michigan, Michigan (U.S.A.; 1901); CD, Cooloola, Queensland (Australia, 1962). Glacial moraines: GB, Glacier Bay, Alaska (U.S.A.; 1923); FJ, Franz Josef, Westland (New Zealand; 1968). Volcanoes: HI, Hawaii (U.S.A.; 1971); KR, Krakatau (Indonesia; 1908); MSH, Mount St. Helens, Washington (U.S.A.; 1980).

have included those at Franz Josef Glacier in Westland (New Zealand; Stevens & Walker, 1970; Wardle, 1980) and Glacier Bay, Alaska (U.S.A.; Cooper, 1923, 1939; Crocker & Major, 1955). Notable volcanic studies include Krakatau Volcano (Indonesia; Ernst, 1908) and various volcanoes in Hawaii (Eggler, 1971; Smathers & Mueller-Dombois, 1974). More recent studies in the same systems reinforce the importance of long-term studies and the continued importance of primary seres to the understanding of vegetation dynamics. These recent studies include Jensen (1993; Denmark), Petersen (2000; Denmark), J. Walker *et al.* (1981, 2001; Cooloola), Lichter (1998, 2000; Lake Michigan), Wardle & Ghani (1995; Franz Josef), Chapin *et al.* (1994; Glacier Bay), Whittaker *et al.* (1989, 1999; Krakatau) and Kitayama *et al.* (1995; Hawaii).

3.2 Early observations

Humans have evolved in environments prone to repeated and interacting disturbances (see section 1.1.1 on humans and disturbance). The process of ecosystem recovery, or succession, was essential in determining immigration routes and agricultural and hunting activities. As humans became more sophisticated, they consciously manipulated the landscape to expand or improve croplands by clearing forests, draining wetlands, and flooding or burning pastures (see multiple historical references in Clements, 1928). These activities implied an intuitive awareness of invasions, regeneration, nutrient cycling and succession. For example, crop rotations involved an understanding of seedbanks and germination requirements. Clements (1928) credited De Luc with the first use of the term succession in 1806, and quoted him describing how 'lakes and pools are converted into meadows and mosses'. Before about 1850, most writings focused on practical land management issues. Subsequent observations began also to describe natural phenomena. Thoreau (1860) documented the succession of farms to forests in Massachusetts (U.S.A.). Successional changes on German floodplains (Reissek, 1856) and Danish dunes (Warming, 1895) contributed to a growing interest in the dynamic nature of the landscape. At the end of the nineteenth century, a fusion of natural history (including observations of succession) and the study of causation merged with the study of function (physiology) to become what McIntosh (1985) termed self-conscious ecology, with succession as its central theme. A century later, succession remains one of the most important themes in ecology (McIntosh, 1999; Wali, 1999a). Members of the British Ecological Society ranked succession as the second most important concept in ecology (after the ecosystem concept) (Cherrett, 1989), perhaps because of the widely variable yet easily observable changes that occur during succession and the increasing relevance of successional concepts to the process of ecosystem rehabilitation (Wali, 1999a). Table 3.1 highlights milestones in the early development of ideas about succession. Table 3.2 provides a lexical chronology of terms about succession.

Warming (1895, 1909) compiled the first synthesis of succession in a book widely considered as the foundation of modern ecology (Taylor, 1912). Warming considered primary succession on sand, volcanoes, landslides and other newly exposed surfaces in a chapter entitled *The Peopling of New Soil*. His coverage included a discussion of pioneer, transitional and final communities and the concepts of species richness, life forms, migration and light reduction (Warming, 1895; Clements, 1928). Warming viewed plant communities as being formed by an unceasing struggle among organisms and considered equilibrium with the environment to be a rare event owing to continual physical disturbances (McIntosh, 1985). These concepts of species interactions and disruption from external disturbances still permeate discussions of succession.

Cowles (1899, 1901), like Warming, studied dune succession. He emphasized the importance of the physical landscape, or physiography in determining succession, distinguishing local causes (water, wind, ice, gravity, volcanoes) from regional ones (e.g. extensive changes induced by melting glaciers or changing climates; Golley, 1977). Cowles also distinguished biotic causes of succession, but recognized that a mixture of physiographic and biotic variables drives most seres (McIntosh, 1999). Cowles also recognized the variability inherent in successional trajectories and the possibility of retrogression, or repetition of earlier successional stages, describing succession as a 'variable approaching a variable' (Cowles, 1901). Thus the two most comprehensive studies of succession by the end of the nineteenth century dealt with primary succession on dunes and presaged modern reductionistic concepts of succession (see section 3.6).

3.3 Holism

Clements (1916, 1928, 1936) gave successional studies a conceptual framework. He also provided an historical summary of prior work, offered his own syntheses and placed the dynamic view of the natural world into a sharp focus. Clements' ideas were so compelling and logical, if dogmatic, and his detailed examples so extensive, that his ideas

Table 3.1. *Milestones in the development of the concept of ecological succession prior to 1970*

Date	Observer	Contribution
300 BC	Theophrastus	Changes in vegetation of river floodplains
1685	King	Changes in bog vegetation; inferences about past changes from plant remains in peat
1714	Lancisi	Successional changes of seashore vegetation
1729	Degner	Trenches filled with water and aquatic plants then to bog
1735	Linnaeus	Succession in bog communities
1742	Buffon	To reclaim forests, plant shrubs; recognition of importance of different life forms, longevities
1749	Beiberg	Sequence on rocks from lichens to mosses to herbs to shrubs
1806	DeLuc	Sequence from lake to aquatic plants to solid ground
1820	de Candolle	Vegetation changes
1825	de la Malle	Used 'succession' in its ecological sense and long-term observations
1830	Hundeshagen	Replacement of broadleaved trees by evergreens
1850	Humboldt	Vegetation changes
1850	Lyell (geologist)	Nature is orderly
1856	Reissek	Vegetation changes are dynamic
1857	Vaupell	Reconstruction of post-glacial vegetation changes by examining macrofossils in peat deposits
1860	Thoreau	Coined 'forest succession'; succession after logging
1876	Gremblich	Cycles of vegetation
1895	Warming	Vegetation is dynamic. Dune grasses bind sand and affect later colonists
1896	Meigen	Succession tends toward equilibrium
1899–1911	Cowles	Succession on sand dunes, importance of physical factors
1904–1936	Clements	Succession deterministic, directional to climax
1908	Ernst	Succession on Krakatau Volcano
1917–1939	Gleason	Every species a law unto itself, communities are random assemblages of individuals
1920–1930	Tansley	Succession and community theory
1923–1939	Cooper	Permanent plot data from post-glacial succession
1924	Ramensky	Individualistic nature of plant communities
1939	Cain	Analysis of climax concept
1940–1948	Oosting	Old-field succession, seeds, climax communities
1947	Watt	Pattern and process in the plant community
1950	Olson	Succession on sand dunes
1951–1954	Egler	Initial floristic composition model
1950–1979	Keever	Experimental studies in secondary succession

Table 3.1. (*cont.*)

Date	Observer	Contribution
1951–1973	Whittaker	Critique of climax theory, gradient analysis, ordination
1955	Crocker & Major	Ecosystem changes during post-glacial succession
1955–1972	MacArthur	Stability, invasions, island biogeography
1956	Drury	Vegetation cycles on floodplains
1967–	McIntosh	Plant community theory and history, continuum concept
1969–	Odum	Generalizations about ecosystem succession
1969–	Bormann & Likens	Ecosystem succession in watersheds

Entries prior to 1900 are from Clements (1928) and Wali (1999b).

dominated successional theory for 50 yr (Glenn-Lewin *et al.*, 1992). He was a keen observer of natural phenomena and both codified disturbance types and linked them clearly to succession. Clements (1916, 1928) proposed that six processes drive succession. Nudation creates bare land (see Chapter 2). Migration and ecesis deal, respectively, with the arrival and establishment of organisms at the site (see Chapter 5). Competition encapsulates interactions of the species, and reaction is the modification of the site by the organisms (see Chapter 6). Finally, Clements viewed stabilization as the development of a stable endpoint or climax stage, in a sense the result of the first five processes (Pickett *et al.*, 1987). Clements explicitly recognized succession on completely barren and less damaged areas, originating the concept of primary and secondary succession. He saw primary succession as presenting extreme conditions with respect to water content and lack of propagules where reaction would take a long time to prepare the soils for the climax community. Secondary succession was less extreme, with viable propagules and shorter, simpler seres. Clements discussed many examples of primary succession, including volcanoes, dunes, glacial moraines, floodplains and emergent land surfaces. His work was also well received because of his attention to practical issues such as range improvement and fire ecology of the prairies of North America, a practical implication of successional studies that continues to the present (Cook, 1996).

Clements incorporated his six processes into a concept of succession that was highly deterministic and directional. Emphasizing the importance of autogenic processes, and particularly of reaction, he believed that

Table 3.2. *A lexical chronology of terms related to succession, arranged by decades*

See Glossary for definitions of most terms.

1900
Association, climax, chronosequence, convergence, disturbance, dynamic, equilibrium, eutrophication, pioneer, progressive, retrogressive, stability, zonation
1910
Biome, competition, ecesis, hydrarch, individualistic, migration, mosaic patch, nudation, primary, reaction, secondary, xerarch
1920
Allogenic, autogenic, biogeochemistry, gap dynamics, holism
1930
Ecosystem
1940
Energy, trophic dynamics
1950
Continuum, cybernetics, gradient analysis, holistic, initial floristics, thermodynamics
1960
Computer models, keystone species, r- and K-selection, strategy
1970
Assembly rule, facilitation, inhibition, intermediate disturbance, nutrient retention, resilience, resistance, tolerance
1980
Resource ratio, river continuum
1990
Complex adaptive systems, complex ecology, ecological law of thermodynamics

---

Modified from McIntosh (1999).

despite the multiple starting points of succession, species influences on a site would result in the convergence of all seres within a given climate. This endpoint would be a stable climax, capable of self-reproduction and no longer subject to succession. In his words (Clements, 1928, pp. 105, 107):

Every complete sere ends in a climax. This point is reached when the occupation and reaction of a dominant are such as to exclude the invasion of another dominant . . . The climax is thus a product of reaction operating within the limits of the climatic factors of the region concerned. The latter determine the dominants that can be present in the region, and the reaction decides the relative sequence of these and the selection of one or more as the final dominant, that is, as the adult organism.

Clements (1928) noted that although the climax 'marks the close of the general development' and a point of equilibrium, much of the landscape would be at various stages or relays approaching the climax and that a climatic climax could take a long time to reach and often be hard to identify. He also stated his belief that the climax concept was universally applicable. Finally, he linked the development of plant communities to the development of organisms. Thus, Clements' writings began with an emphasis on disturbance and variability, but later emphasized equilibrium and organismal analogies. Although Clements provided the first global perspective of succession and a conceptually satisfying order to the unruly natural world, his rigidly structured views fostered the beginnings of dissent (Egler, 1951; Colinvaux, 1973).

The organismal analogy that Clements used for succession was influenced by the post-Darwinian view of gradualism and organismal metaphors for growth and evolution (Glenn-Lewin et al., 1992) that was shared by several of his contemporaries (McIntosh, 1985). For example, Davis's geographical cycle (Davis, 1899) was a view of an orderly landscape that was uplifted and then gradually eroded in stages to a rolling peneplain, similar to the stages in the development of an organism. Clements (1928) developed the organismal analogy in detail. He likened succession to reproduction, and conceived of the climax plant community as a sort of superorganism, analogous to the adult stage of an organism. Obstacles to climax development resulted from abnormal conditions. Although a climax was supposed to take a long time to develop, Clements (1928) provided explanations for each absence of a climax, introducing a number of alternative climax types such as potential, sub-, dis-, pre- or post-climaxes. Each of these had some reasonable and often ecologically interesting explanation, such as local soil factors overriding regional climatic factors, intervention by humans or periodic disturbances such as fire. But with each qualification, the argument for an overarching climatic climax was weakened (Tansley, 1935; Whittaker, 1953). Interesting parallels can be drawn to the recent tendency to over-use the prefix 'eco' and the word 'ecology' (Wali, 1999b) and thus dilute the sense of any clear definition of climax on the one hand or ecology on the other. The plethora of recent descriptors of facilitation is another example of the expanding use of terms and the attempts to explain the variety of nature by using one general concept (cf. Table 6.1). Certainly,

Table 3.3. *Clementsian processes that drive succession, and modern analogs*

Clementsian processes	Modern analogs
Nudation	Allogenic disturbances, stochastic events
Migration	Life history characteristics: dispersal
Ecesis	Life history characteristics: establishment, growth, longevity
Competition	Competition, allelopathy, herbivory, disease
Reaction	Site modification by organisms, facilitation
Stabilization	Development of climax

recent terminology on facilitation has gone well beyond what Clements envisioned.

3.4 Neo-holism

Post-Clementsian studies of succession have not reached such levels of generalization again, perhaps because of the gradual disillusionment with Clementsian paradigms. The organismal analogy found little support (but see Phillips, 1934–35), perhaps because it was carried to such a detailed extreme. McIntosh (1985) pointed out that critics thought it odd that a successional superorganism could have multiple embryonic starting points and no genetic basis, but, as he succinctly observed, 'superorganisms are not easily killed by logic alone'! Although the emphasis Clements placed on reaction (site modification) has been criticized, it remains a widely recognized process. He correctly emphasized that reaction was more important in primary than in secondary seres. The climax concept has retained its hold on the imagination of some ecologists and the public and remains as a basic tenet of ecology by applied disciplines such as forestry and wildlife management. Yet because it must be qualified so much to be applicable, its usefulness is questionable.

Despite all the critiques of his organismal analogies and emphasis on a climatic climax, Clements' emphasis on the dynamics of vegetation that responds to disturbance is a major contribution and has remained a focus of modern studies of succession (McIntosh, 1999). Many of his successional processes, with a few modifications, remain valid (Table 3.3) and provide the basic organizational framework for this book. Nudation now encompasses studies of allogenic disturbance (Tansley, 1920) and stochastic events. Migration is studied in the context of how dispersal, a life history characteristic, modifies succession. Ecesis includes examination of the life history characteristics of establishment, growth and longevity. Competition has been broadened to encompass allelopathy, herbivory and other inhibitory species interactions but is still considered pivotal for succession. Finally, reaction is now subsumed under site amelioration, and is often considered a type of facilitation, the importance of which is under renewed scrutiny (Callaway, 1995). In summary, Clements made many contributions. Some of these have shaped studies of succession for the last century but others have provoked strong negative reactions and fomented a reductionistic approach (see section 3.6 on reductionism).

Neo-holists have continued to view succession as a deterministic phenomenon, modifying Clements' views in several ways, but retaining the holistic perspective on vegetation dynamics. Textbooks by Oosting (1948) and Odum (1953, 1971) accepted the notion of a climax without providing a critical examination. Odum (1969) suggested that during succession, change in many ecosystem-level processes is predictable. With little data to support them, and a clear basis in Clementsian thinking (Christensen & Peet, 1981; Hagen, 1992), Odum's predictions none the less were popular, perhaps because we continually seek generalizations to explain natural complexity (Odum, 1992). However, with the suggestion of links between the laws of thermodynamics and ecosystem function (Watt, 1968; McIntosh, 1999) and an international effort to study ecosystems in the 1960s and 1970s (The International Biological Program), Odum's principles provided partial support for the emerging systems ecology (McIntosh, 1985). Systems ecology, equipped with new tools such as isotopic tracers, examined the roles of biogeochemical cycles in succession and renewed the holistic approach of Clements (McIntosh, 1999). Margalef (1968a) suggested cybernetic analogies for succession where the system limits outside input as it develops and climax is measured by maximum (internal) energy flow. A book entitled *Complex Ecology* (Patten & Jorgenson, 1995) provides the latest adaptation of the holistic philosophy to ecology (McIntosh, 1999). Species' impacts on ecosystem function were seen as part of a complex adaptive system by Brown (1995) and an ecological law of thermodynamics was even proposed as an update of Clements' organismic analogy (Jorgenson, 1997). It appears that both Clements' and Odum's holistic views remain alive in the realm of theory, but most experimental work in succession remains in the reductionist camp.

Not all ecosystem studies are predicated on holistic philosophies. At best, measurements of energy and nutrient fluxes through ecosystems are made independently of how one believes the ecosystem is organized.

Watershed-level studies of nutrient cycles at Hubbard Brook Forest in New Hampshire (U.S.A.; Bormann & Likens, 1979) found some support for, but also some clear contradictions of, Odum's ecosystem-level predictions for successional change in community properties. Bormann and Likens attributed some of the contradictions to stochastic events and local disturbance leading to a shifting mosaic of species and a higher degree of spatial heterogeneity than predicted by Odum (Hagen, 1992). Marks and Bormann (1972) noted the critical role that even a single shrub species (*Prunus pensylvanica*) can have on nutrient cycles at Hubbard Brook. Stream ecologists also found mixed support for Odum's predictions and emphasized the importance of disturbance (Fisher *et al.*, 1982). These and similar studies set the tone for modern ecosystem ecology and the analysis of the dynamics of nutrient cycles in response to disturbance and through succession (cf. Vitousek & Reiners, 1975). These integrative approaches that combined disturbance, nutrient cycling and population biology presaged the current uneasy merger of holistic and reductionistic philosophies.

Clementsian deductive approaches to complexity were built on the belief that communities have emergent properties that one must address in order to understand succession. Holists argued that approaching nature's complexity without a set of first principles was to lose oneself in large amounts of trivial observations (Forbes, 1887) in the seas of experimental reductionism (Cody, 1981). In a successional context, these first principles included, if not a climax, at least a steady state between the vegetation and the environment.

3.5 Phytosociology

The ideas of the American Clements were popular in the U.S.A. and to a lesser extent in Britain, but not in the rest of Europe, where the Zurich–Montpellier school of phytosociology developed a more taxonomic approach to plant communities under the leadership of Braun-Blanquet (1932, 1964) and Ellenberg (1956). Plant communities are evaluated by the use of standardized plots called relevés and the emphasis is on detailed analysis of floristic composition rather than vegetation change across temporal gradients. This approach reflected the smaller physical scales, the sharp discontinuities induced by humans and the better understood floras of Europe. Clements' ideas were more suited to larger landscapes with more unknowns, less intense human influence and a greater emphasis on natural disturbances. The applicability of

Braun-Blanquet's system is therefore most relevant to primary succession when species change is rapid and qualitative changes are important (Mueller-Dombois & Ellenberg, 1974; Poli Marchese & Grillo, 2000). Phytosociology provided the conceptual basis for many modern analytical tools, such as detrended correspondence analysis, that are useful in floristically based studies of primary succession (see section 7.2.1 on definitions).

3.6 Reductionism

Clements' ideas were based on a Germanic idealism and philosophy developed by Kant, Goethe and Humboldt (McIntosh, 1985) that supported a top-down deductive approach to problem solving. A sharply contrasting approach to plant ecology and succession followed from a Darwinian emphasis on individuals and inductive reasoning. Warming (1909) was hesitant to accept emergent properties such as Clements' climax communities. Cowles (1901) also conducted his studies of the Lake Michigan sand dunes from this reductionist approach.

Gleason (1917, 1926, 1939) carried the reductionistic focus on the individual species to its logical extreme. He refuted Clements' organismal analogy and described succession as an indeterminate and often unpredictable process dependent on the properties of individual species. Gleason, like Clements, used quadrats or small plots to study vegetation, but focused on distributions of individual species rather than properties of communities, detail rather than generalization, pattern rather than process. He noted that, temporally and spatially, each assemblage of species was independent of other assemblages. Given that each species was often independently dispersed across a variable landscape (Fig. 3.2), Gleason saw no reason to expect predictable or permanent communities (Matthews, 1996). He discussed some examples of primary succession such as dunes and rock outcrops, but did not emphasize areas of severe disturbance as much as Clements. Neither the ideas of the American Gleason nor those of his like-minded Russian contemporary Ramensky were popular during most of their lifetimes (McIntosh, 1985). Matthews (1996) attributed this lack of acceptance to several causes. Gleason's views countered a deeply set belief that plant communities were somehow more than the sum of their parts, if not actually analogous to organisms. Ecologists were also busy classifying communities and Gleason's views made those efforts problematical. The reductionistic approach was also misconstrued to imply random collections of species that did not interact

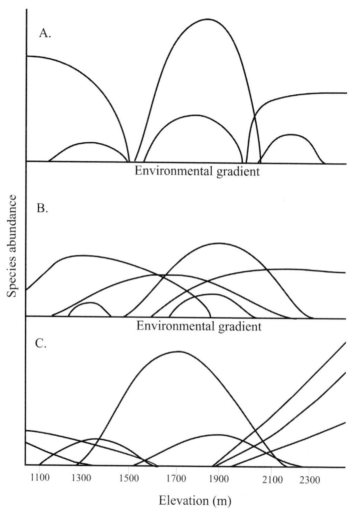

Fig. 3.2. Patterns of species abundance along an environmental gradient.
(A) Distinct clusters of species that have similar distributions defined by sharp
ecotones: a pattern compatible with a Clementsian view of community
organization. (B) No distinct clustering: compatible with a Gleasonian view with a
continuum of species distributions. (C) Patterns of species abundance along an
elevational gradient in southern Nevada that partly support both views. Data for
(C) from L. Walker, unpublished data. See also Lei & Walker (1997).

competitively. Gleason promoted neither random assembly nor a lack of competition (Nicolson & McIntosh, 2002).

In a dramatic turnaround in 1959, the Ecological Society of America awarded the self-described 'ecological outlaw' (Gleason, 1953) the title of eminent ecologist (Cain, 1959), effectively validating Gleason's contributions. This action set the stage for a gradual increase in the emphasis on reductionism and decline in the Clementsian influence during the next half century.

3.7 Neo-reductionism

The goal to construct mechanistic explanations of succession from data collected on individuals in particular habitats has fueled ecological and successional studies since the 1950s and could be termed neo-reductionism for the continuation of Gleason's reductionistic approach. The lack of emphasis on integration or generalization stems, in part, from the failure of Clementsian or other views to provide a global structure for succession. Attempts to revive top-down general principles of succession (Odum, 1969) or ecology (Odum, 1992) have not altered the reductionist momentum. Instead, there has been a growing collection of conceptual models that are based on observational and experimental data, but no replacement for Clements' framework of successional ideas. Instead, a collective realism has accumulated that any grand underlying scheme (GUS) for describing succession in broad, predictive terms will remain elusive (Glenn-Lewin *et al.*, 1992; McIntosh, 1999). This has not stopped the search for generalization at smaller scales or for the purposes of modeling or prediction. Approaching successional processes at various spatial and temporal scales has proved to be a useful organizing tool (O'Neill *et al.*, 1986; van der Maarel, 1988) but most studies that contrast various scales do so within a sere. Comparisons across types of primary sere are rare (Messer, 1988; Walker, 1995; Schipper *et al.*, 2001) yet a necessary first step toward generalization. Thus the pendulum swings from generalizations based on overarching concepts to ones based on empirical data.

An emphasis on the stochasticity of species establishment, disturbance and spatial patterns has characterized the approach of neo-reductionists to succession (Glenn-Lewin, 1980). Recognizing, as Cowles did decades earlier, that variability (e.g. of trajectories and disturbance regimes) is innate to most systems, ecologists now are more likely to focus on components than on generalities. Measurements of characteristics of individual

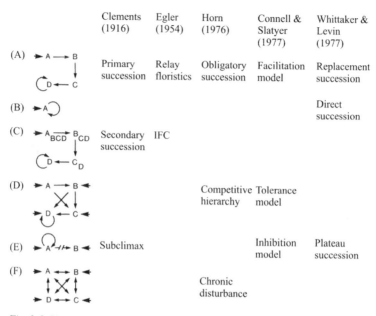

	Clements (1916)	Egler (1954)	Horn (1976)	Connell & Slatyer (1977)	Whittaker & Levin (1977)
(A)	Primary succession	Relay floristics	Obligatory succession	Facilitation model	Replacement succession
(B)					Direct succession
(C)	Secondary succession	IFC			
(D)				Competitive hierarchy	Tolerance model
(E)	Subclimax			Inhibition model	Plateau succession
(F)			Chronic disturbance		

Fig. 3.3. Various models of succession. From Noble (1981) as modified by Miles (1987). Reprinted by permission of Blackwell Science, Inc.

plants (e.g. biomass, seed production) and species (e.g. stem density) have become the currency for data analyses and modeling, replacing assessments of community properties (e.g. diversity, stability). Succession has become, in this approach, less an interpretation of community development and more a population-level process where life history characteristics such as establishment are important until a forest canopy develops, after which mortality processes predominate (Peet & Christensen, 1980).

Egler (1954) proposed an alternative to Clements' view of waves of species colonizing after the habitat had been altered by previous species (relay floristics). Egler suggested that succession might be simply described as the sequential conspicuousness of species that all arrive at the beginning (initial floristic composition, IFC) but grow at different rates and mature at different heights (Fig. 3.3). This scenario was actually first proposed by Clements (1916), who noted that seeds and fruits of all dominants could be present initially in secondary succession. IFC was first applied to abandoned old fields and is less likely in primary than in secondary succession because of the harshness of initial conditions. However, some individuals of later stages may be among the initial colonists on floodplains (Walker *et al.*, 1986) or glacial moraines (Cooper, 1923; Chapin *et al.*,

1994; Vetaas, 1994), even if they do not dominate the vegetation for many decades. Therefore, the IFC concept is applicable to some primary seres.

Harper (1977) provided further emphasis on the individualistic approach by subdividing the individual plant into a population of modules, recognizing the conflicting requirements of seeds, clonal structures and vegetative parts. Harper also noted the evolutionary dilemma a species faces in succession: to stay and adapt to environmental change (generally in mid to late succession) or reproduce quickly and move to a similar (often pioneer) habitat. This dichotomy, formulated as the r-selected or K-selected strategy, respectively (Pianka, 1970), helped focus attention on different traits and their evolution. Pioneers of the early stages of primary succession often fit the r-selected strategy, although invasion can come from local, mature, K-selected vegetation where initial conditions are severe or the potential flora is limited. Raup (1981) interpreted succession as the process of recovery from disturbance rather than progression toward a climax. His studies in the Arctic underscored his interest in the impacts of chronic or recurrent disturbance (Raup, 1971). Pickett (1976) also emphasized disturbance but drew a different lesson about the relationship between disturbance and succession than did either Clements or Raup. Declaring ecologists had abandoned the Clementsian views of an ordered progression of stages to a climax, Pickett proposed an evolutionary sorting of populations of species along changing environmental gradients as the mechanism driving succession. Despite variously emphasizing plant parts (Harper, 1977), plant reproduction (Pianka, 1970), disturbance (Raup, 1971) or evolution (Pickett, 1976), these four approaches to succession are all based on the reductionistic viewpoint.

A further development that emerged from Gleason's individualistic concept was the study of gradients or the continuum concept (Curtis, 1959; McIntosh, 1967; Austin, 1985; Matthews, 1996). Unique distributions of species across environmental gradients such as moisture, temperature or elevation support Gleason's view. Areas of congruent distributions combined with abrupt shifts to a new set of species support a Clementsian view of discrete communities. Community boundaries may be internally caused or a response to some sharp environmental gradient. Ordination analyses developed as a method of measuring species distributions along environmental gradients (Whittaker, 1973) and they can be applied to describe temporal change, but they do not address the mechanisms of that change.

The integration of disturbance and spatial patterns has long been a theme of adherents of Gleasonian views. Whittaker (1953) attempted an integration of the spatial emphasis of Gleason and Cowles and the temporal emphasis of Clements. He proposed a climax pattern hypothesis, suggesting that multiple climaxes were possible across landscape-scale gradients. Whittaker recognized that there is a continuously changing pattern of vegetation across the landscape, using the primary succession analogy of a braided stream meandering across a floodplain. Cooper (1913) and later Watt (1947) and Drury (1956) interpreted the changing patterns of vegetation across a landscape as a shifting mosaic where each patch was at a different successional stage. Pickett (1976) and colleagues (Pickett & White, 1985) have advanced these ideas with the concept of patch dynamics, or a landscape differentially impacted by disturbance and succession (see section 2.1.4 on patch dynamics). Several contributors to Pickett & White (1985) addressed primary seres. Veblen (1985) demonstrated the critical role of disturbances (volcanoes, earthquakes, landslides, mudflows, tree falls and drought) in shaping the primary successional dynamics of Chilean forests. Intertidal (Sousa, 1985) and subtidal (Connell & Keough, 1985) substrates are also subject to primary succession following disturbances ranging from volcanoes and landslides to coral breakage or exposure of rocks or shells. The concepts of patch dynamics (Pickett & White, 1985; Pickett, 1999), metapopulations (Levins, 1969) and landscape ecology (Naveh & Lieberman, 1984) are natural outgrowths of the earlier dynamic ecology of Warming and Cowles. However, this geographically explicit approach to disturbance (modernized with geographic information systems or GIS) has not been adequately integrated with the temporal changes needed for successional models.

The philosophical and practical divide between holism and reductionism continues to permeate ecology and successional studies (Matthews, 1996; McIntosh, 1999), but this divide also offers an opportunity to select from a wealth of perspectives. Indeed, even Odum and Margalef recognized the value of both the holistic and reductionistic approaches (McIntosh, 1985), although they clearly preferred the former. The dichotomy boils down to whether one focuses on the forest or the trees. Both approaches are useful for studies of primary succession (Finegan, 1984).

3.8 Ecosystem assembly

Ecosystem assembly refers to the coalescing of groups of species into communities during succession (Drake, 1990). A subset of those species that arrives at a site establishes and may later reproduce. Succession (the shifting pattern of species abundance over time) may be a completely

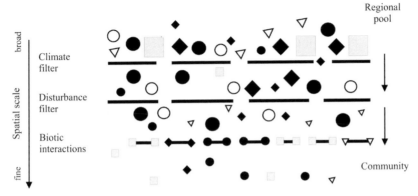

Fig. 3.4. Environmental conditions (climate, disturbance, biotic interactions) that serve as filters for species. From Díaz et al. (1999). Reprinted with the permission of Cambridge University Press.

haphazard confluence of individuals with little predictability (Gleason) or a tightly choreographed, highly predictable assemblage (Clements). If the latter case is at least partly true, are there any decipherable assembly rules that dictate the permissible combinations of species at a site? Diamond (1975), following Clements, suggested that assemblages of birds on islands in New Guinea were determined by intraguild competition. In this context, assembly rules are restrictions on the potential coexistence of species based on interactions of those species (Wilson et al., 1996). Keddy (1992) expanded Diamond's concept of assembly rules to include any environmental filter that limited colonization. J. Wilson (1999) decried this dilution of the more autogenic nature of assembly rules, and considered assembly rules distinct from environmentally mediated patterns. Several decades of debate (Wilson, 1994; Weiher & Keddy, 1999) have not resolved whether assembly rules exist or not, in part because of unclear meanings of the terms 'assembly' and 'rules.' Semantics aside, we suggest that Keddy's approach recognizes the importance of interactions at various scales (see section 5.6.4 on priority effects) and is readily adaptable to studies of primary succession.

Although largely framed in an evolutionary and biogeographical context, several aspects of the discussion about assembly rules are relevant to primary succession, particularly the emphasis on disturbance (see section 7.1.1 on converging trajectories). Climate and disturbance provide broad-scale filters, followed by biotic interactions at finer scales (Fig. 3.4). These filters might selectively remove plants or animals with certain traits or functions (Díaz et al., 1998, 1999) or prevent their initial establishment (see section 5.6.4 on priority effects). In a sense, plant functional types are analogous to animal guilds. Successional rates and trajectories are strongly influenced by the types of functional group represented. For example, the presence of a dominant, shrubby, N-fixing plant in primary succession alters soil nutrients, the light regime, and opportunities for colonization (see sections 4.5.1 on N fixation and 6.3.2 on inhibition). Subsequent establishment of other N-fixers or shade intolerant species is less likely than establishment of shade-tolerant, high-N-adapted species. Colonizing species, particularly non-indigenous invasive plants, alter many other ecosystem functions. Examples include species that are highly flammable (e.g. *Bromus* grasses on disturbed patches of the Mojave Desert, U.S.A.; Smith *et al.*, 2000), that increase soil salts (e.g. *Tamarix* trees on southwestern U.S.A. and Australian floodplains; Walker & Smith, 1997) or that add nutrient-poor, recalcitrant litter, thereby reducing nutrient availability (e.g. *Pinus* trees in South Africa; Versfeld & van Wilgen, 1986). Determining what functional groups are present can then help predict how succession will develop.

A second link between concepts of ecosystem assembly and primary succession is the process of rehabilitation that we discuss in Chapter 8. Successful rehabilitation relies on an understanding of how to re-establish the structure and function of an ecosystem as well as species replacement dynamics over time (Lockwood & Pimm, 1999). Structure and function are readily replaceable. However, the re-establishment of successional trajectories that continue once management ends appear to be much more difficult to achieve. Colonization from undisturbed natural areas appears to be the key factor promoting succession. We discuss factors controlling colonization in primary succession in Chapter 5. Our inadequate understanding of how communities assemble and change over time means we have not been able to deliberately reconstruct damaged systems. Therefore we have, for now, flunked the acid test of ecology (Bradshaw, 1987).

3.9 Models

Models of succession attempt both to explain past trajectories and to predict future trajectories and are flexible tools that can be applied to any given sere. There are many ways to define and classify models of succession (Table 3.4), but classification is arbitrary. The clearest approach is to define the parameters of the model of concern. We first describe verbal models that address characteristics of species, communities or ecosystems

Table 3.4. *Methods of classifying models about succession*

Usher:
    Verbal (descriptive explanations using words or diagrams)
    Compartment (computer-based projection of simple entities such as a parcel of
        land)
    Mathematical
        Population dynamic (focus on each species, emphasis on organisms)
        Statistical (each subprocess within succession assigned a probability)

van Hulst:
    Population
    Ecosystem
    Phenomenological approach (Markov)
    First-principles approach
    Factor of interest (model)
        a. present on-site vegetation (Markov, linear differential; least relevant to
           present primary succession)
        b. present surrounding vegetation (island biogeography, descriptive, dispersal
           dynamics)
        c. past vegetation (differential equation)
        d. present resource levels (resource ratio)
        e. disturbance level (C–S–R model by Grime, 1977; lottery)
        f. stochastic factors (Markov)
        g. invasion model (integrates a–f)

From Usher (1992) and van Hulst (1992).

without assigning values to individual components. We subdivide verbal models into ones that focus on successional change that is driven mostly by species interactions (autogenic models) and ones that focus on a combination of autogenic and allogenic processes (process models). We then discuss mathematical models where probabilities can be assigned to such variables as relative species cover or biomass (a population approach), species richness (a community approach), decomposition rates (an ecosystem approach) or the spatial distribution of species (van Hulst, 1992). All models are vulnerable to over-simplification because unknowns are either ignored or assigned values of varying degrees of arbitrariness. For example, to model succession one must define the sequential replacement of species (Usher, 1992). Plant species composition may have little correlation with plant cover, microbial activity, herbivore levels or other potentially important driver variables. The set of variables used to define succession limits the applicability of the model. Attempts to model entire seres are mostly unsatisfactory (McIntosh, 1980; Franklin *et al.*, 1985).

Modeling is therefore a process of trying to fit the simplest model to observed behavior (a phenomenological approach) or a test of some conceptual idea (first principles approach; van Hulst, 1992). Both approaches are hampered by the quantity of unknowns in succession, yet modeling can help guide empirical studies. In turn, extensive data sets from well-studied seres (see section 3.1) serve as empirical models.

3.9.1 Verbal models

Successional theory advanced significantly with the widespread adoption of experimental approaches in the 1970s and 1980s. Experiments began to test such paradigms as whether facilitation was essential. Keever (1950, 1979) and McCormick (1968) conducted experiments with old-field plants (secondary succession) and found little evidence for the necessity of early colonists improving the environment for later species (see section 6.3.1 on facilitation) because the second wave of colonists (perennials) did better when the first wave of colonists (annuals) was experimentally removed. The results of these seminal experiments were widely promulgated and followed by a challenge from Drury & Nisbet (1973) to address the paucity of data with further experiments, particularly in primary succession. They suggested that the importance of soil development via organic matter accumulation for primary succession was overstated, because of rapid colonization of glacial moraines (Cooper, 1923) and dunes (Richards, 1952). Drury & Nisbet also emphasized that differences in life history characteristics of species could largely explain succession. With the growing number of experimental tests of old paradigms and the publication by Drury & Nisbet (1973), the stage was set for the next round of experiments and conceptual models that began to directly incorporate life history parameters.

Autogenic models
Connell & Slatyer (1977) presented three different succession models (Figs. 3.3, 3.5) that considered only autogenic successional changes (in the absence of allogenic disturbance) and which emphasized secondary succession. In other words, their models attributed species change to species interactions (see Chapter 6). The facilitation model only slightly revised Clements' concept that Egler (1954) termed relay floristics. In this scenario, succession is seen as a series of sequential invasions, each dependent on the site amelioration provided by earlier colonists. Species die out because the changes in the environment make it more suitable

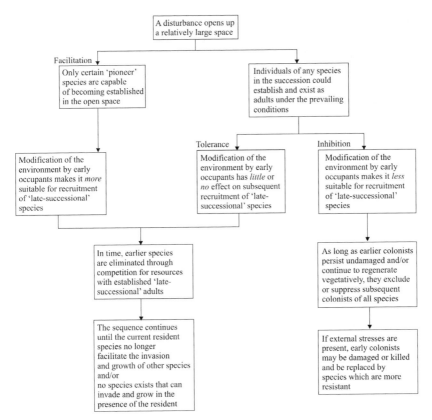

Fig. 3.5. Facilitation, tolerance and inhibition models. From Connell & Slatyer (1977) as modified by Begon *et al.* (1990). Reprinted by permission of Blackwell Science, Inc.

for later colonists and less suitable for their own survival or regeneration. In Connell & Slatyer's facilitation model the emphasis was on the facilitation of new arrivals, yet earlier species must be actively displaced by competition for resources or germination sites (Walker & Chapin, 1987). Connell & Slatyer's facilitation model is further over-simplified because facilitative interactions can affect arrival, establishment, growth or longevity of a species (Walker, 1999b; see Table 6.2 and section 6.3.1 on facilitation). Connell & Slatyer suggested that most evidence for this facilitation model came from primary seres including the studies of glacial moraines at Glacier Bay (Crocker & Major, 1955) and the dunes of Lake Michigan (Cowles, 1899; Olson, 1958).

The tolerance model of Connell & Slatyer (1977) resembles the facilitation model in the later stages, when species are competitively displaced by species that are more tolerant of declining resource conditions (active tolerance, *sensu* Pickett *et al.*, 1987). However, in the tolerance model, successions simply depend on arrival times and growth rates and earlier species do not facilitate or inhibit later ones (passive tolerance). If all key species arrive early in succession, this scenario resembles the IFC described above. The tolerance model may explain basic patterns of species replacements in primary succession (Walker & Chapin, 1986; Chapin *et al.*, 1994) but is perhaps most applicable in low-nutrient secondary seres (Partridge, 1992), where establishment does not require substantial habitat modification.

The inhibition model suggests that early arrivals competitively inhibit establishment of later ones. These inhibitors often form thickets and monopolize resources until they die (see section 6.3.2 on inhibition). Connell & Slatyer (1977) found much support for this model in secondary succession. Subsequent studies have also found evidence for inhibition in many types of primary succession (Iverson & Wali, 1982; Wilson & Agnew, 1992; del Moral & Bliss, 1993; Walker, 1993; Chapin *et al.*, 1994; Russell *et al.*, 1998). However, as we explore in Chapter 6, the complexities of species interactions and the many types and results of competition make a simple model difficult to test.

The Connell & Slatyer (1977) models have been widely adopted as testable hypotheses about how species interactions drive succession. The original intent of the models was to focus on the net effect (positive, neutral or negative) of the pioneer species on the establishment of the next colonists (Connell *et al.*, 1987). However, the models are easily interpreted as attempts to explain whole seres. Numerous papers have provided data that suggest support for one or several of the models. Yet most seres are a combination of positive, neutral and negative interactions at all stages of succession (Huston & Smith, 1987; Pickett *et al.*, 1987; Walker & Chapin, 1987) so net effects must be compiled across each sere. These net effects vary with successional type, stage, disturbance regime and resource availability, so progress on this task has been slow. Indeed, shifting conditions can make even a relatively simple interaction between two species vary from mutually beneficial to mutually antagonistic, with many other possible outcomes as well (see Table 6.1). Nevertheless, Connell & Slatyer (1977) appropriately focused researchers on the important influence of species interactions that range from positive to negative to neutral. Yet the presence of one type of interaction does not rule out the presence of any other, and the variation in the relative importance of each interaction over time may be the most important determinant of succession.

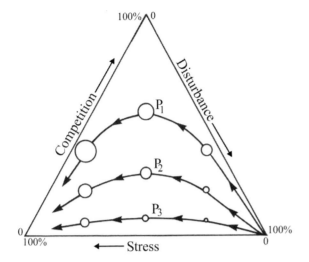

Fig. 3.6. Successional pathways in relation to Grime's three plant strategies (species adapted to competition, disturbance or stress) at high (P₁), intermediate (P₂) and low (P₃) levels of potential primary productivity. Circle size represents biomass. From Grime (1979). © John Wiley & Sons Limited. Reproduced with permission.

Grime (1977, 1979) proposed a second type of autogenic model. He suggested that plant species have evolved different strategies in response to disturbance, stress and competition and that these differences help to explain successional change. His ideas were an extension of the dichotomy of r and K selection (see section 3.7 on neo-reductionism) into a three-way model. Grime suggested that succession begins with disturbance-adapted species, goes through good competitors and ends with species adapted to stress. This model may be too over-simplified to offer any predictive value to community theory (Wilson & Lee, 2000). It does not explicitly address primary succession, but Grime suggested that in sites of low productivity good competitors would not be abundant (Fig. 3.6). Applications of this model to primary succession are complicated by species longevities. A species that is short-lived may experience a flood as a disturbance, but one that is long-lived endures the flood as a stress (Menges & Waller, 1983).

A final autogenic model was developed based on three species characteristics called vital attributes (Noble & Slatyer, 1980). The first was the method of arrival of a species at a site after a disturbance or its method of persistence during the disturbance. The second was the ability to establish and grow to maturity. Finally, they included the time to reach reproductive maturity combined with the duration of the species population and its propagule pool. After reducing all possible combinations of attributes to biologically realistic ones, Noble & Slatyer explored the relevance of these attributes to two fire-prone forest seres. The integration of the disturbance regime with life history characteristics and the simplification of the complexities of species life histories are the strengths of this approach. However, it has not been widely applied to studies of primary succession (Matthews, 1992) and appears best suited for secondary seres responding to recurrent fires (Noble & Slatyer, 1980).

Process models

Pickett *et al.* (1987) compiled a hierarchical list of processes that might affect succession (Table 3.5) that grouped processes by site factors and the availability of species and species performance. Each process was further defined by more specific factors (for example, herbivory is impacted by climate, consumer cycles, plant condition, community composition and patchiness). This checklist clarified the large number of processes, and many are relevant to primary succession. However, no guidance is given as to how the relative importance of these processes might vary during succession.

Walker & Chapin (1987) proposed a generalized model of how the relative importance of several processes might vary across succession (Fig. 3.7). This model provided an idealized contrast between severe and favorable environments that could easily represent primary and secondary succession. It further proposed that seed dispersal, rather than vegetative propagules, is most important for dispersal in primary succession. This model has been a useful heuristic tool that presents testable patterns but few tests have been made of these predictions, except for the most commonly studied processes, facilitation and competition (see Chapter 6). This model may promote the comparison of processes both within and between seres (Glenn-Lewin *et al.*, 1992). Walker found support for the model by determining that the relative importance of facilitation increased with decreasing levels of soil nutrients in a comparison of two Alaskan seres (Walker, 1995). A detailed comparison between primary succession on the glacial moraines at Glacier Bay (Chapin *et al.*, 1994) and on the floodplain of the Tanana River (Walker *et al.*, 1986; Walker & Chapin, 1986, 1987; Walker, 1989, 1999b) indicated that facilitation was more important to succession at Glacier Bay, where initial soil nitrogen was lower. Very similar vegetation at both sites improved the comparison. Seedlings of the tree *Picea*, grown in successional soils (in both the

Table 3.5. *A hierarchy of successional causes*

General causes of succession	Contributing processes or conditions	Factors that modify the processes
Site availability	Coarse-scale disturbance	Size, severity, time, dispersion
Differential	Dispersal	Landscape configuration
species availability	Propagule pool	Disperal agents, time since disturbance, land use
Differential species performance	Resource availability	Soil conditions, topography, microclimate, site history
	Ecophysiology	Germination requirements, assimilation rates, growth rates, population differentiation
	Life history strategy	Allocation pattern, reproductive timing reproductive mode
	Stochastic environmental stress	Climate cycles, site history, prior occupants
	Competition	Presence and identity of competitors, within-community disturbance, predators, herbivores, resource base
	Allelopathy	Soil characteristics, microbes, neighbors
	Herbivory, disease, predation	Climate and consumer cycles, plant vigor and defense, community composition, patchiness

From Pickett *et al.* (1987), with kind permission from Kluwer Academic Publishers.

field and greenhouse experiments), responded more favorably to inputs of nitrogen from the N-fixing shrub *Alnus* at Glacier Bay than at the Tanana River. Competition was more important in the favorable environment on the Tanana River. Comparisons of primary seres across gradients (e.g. of soil fertility) are essential tests of the generality of successional models (see section 6.6 on net effects of interactions).

Matthews (1992) stated that much of the Walker & Chapin model was useful in his global evaluation of succession on glacial moraines. Particularly relevant were the importance of competition (but not necessarily facilitation) and the emphasis on environmental severity. Matthews also pointed out several ways in which none of the previous models

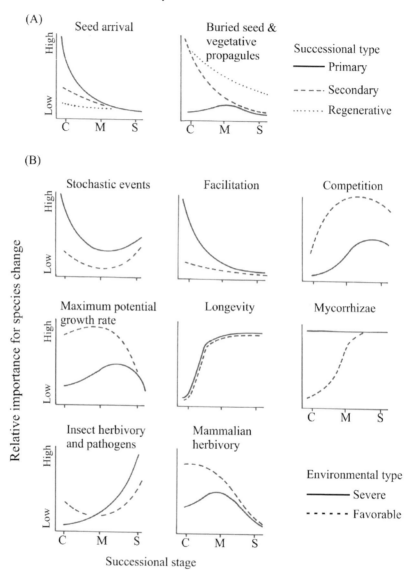

Fig. 3.7. Influence of type of succession (top) and environmental severity (bottom) upon major successional processes that determine change in species composition during colonization (C), maturation (M) or senescence (S) stages of succession. From Walker & Chapin (1987).

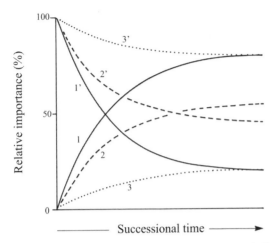

Fig. 3.8. The relative importance of autogenic and allogenic factors in primary succession on glacial moraines. With greater environmental severity, increasing from solid to broken to dotted curves, the relative importance of allogenic processes (1' to 3') increases and autogenic processes (1 to 3) declines. From Matthews (1992). Reprinted with the permission of Cambridge University Press.

address certain characteristics of glacial moraines. For example, he noted that moraines are rarely sterile and propagule dispersal is generally not as limiting as establishment for early successional species. Establishment on a moraine is subject to predictable, gradual allogenic change (e.g. decreases in frost-sorting or wind erosion and acidification of soils from cation leaching). Matthews argues that these changes are not similar to the stochastic events of Walker & Chapin (1987) and proposes that the relative importance of allogenic changes decreases during succession (Fig. 3.8). However, he concurs with Walker & Chapin and others (White, 1979; Pickett & White, 1985) that those smaller-scale disturbances that follow the destabilization in a glacial valley when the ice retreats (e.g. land-slides) are critical in directing primary succession on moraines (Matthews, 1999). Matthews (1992) joins the reductionists in stating that moraines are subject to continual disturbances at many scales and are therefore in permanent disequilibrium. Clearly, such a conclusion is meant to apply to large temporal and spatial scales.

Extensive studies of primary succession following the 1980 eruption of Mount St. Helens in Washington (U.S.A.) (Halpern & Harmon, 1983; Wood & del Moral, 1987; Morris & Wood, 1989; Halvorson *et al.*, 1992; del Moral 1993b, 2000a; del Moral & Grishin, 1999) have found general support for the initial importance of facilitation, followed by increasing

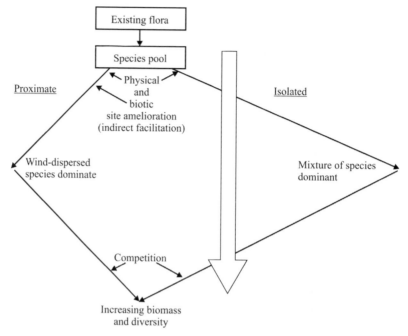

Fig. 3.9. A summary of two successional trajectories on Mount St. Helens, depending on the proximity of the barren sites to existing vegetation. The broad arrow represents successional time.

competition (Fig. 3.9). In addition, these studies highlight the importance of dispersal and support Matthews' claim that allogenic factors become less critical as succession proceeds. The local flora and the degree of isolation of the disturbed ground from that flora determined potential colonists. All newly barren sites on Mount St. Helens became less stressful for plants through physical site amelioration (indirect facilitation; see Table 6.1), but additional facilitative biotic influences accelerated the process at sites near surviving flora. The composition of the more isolated sites was dominated by wind-dispersed species, whereas proximate sites were dominated by species representing a mixture of dispersal types. Competition became more important than facilitation at both isolated and proximate sites as additional species invaded and biomass and diversity increased. These general patterns parallel findings found on other volcanoes (Grishin et al., 1996; Poli & Grillo, 2000). Generalizations about the relative importance of successional processes during primary succession on volcanoes (del Moral & Grishin, 1999), as for other types of primary sere, are still based on relatively few studies.

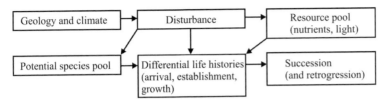

Fig. 3.10. General model of factors affecting succession.

Burrows (1990) presented another generalized model of succession that also focused on processes. He noted that most successional models only address linear replacement of species, ignoring other possible changes such as fluctuations, cycles, direct replacement and responses to gradual climate change. His model condenses all of these patterns into three basic modes of temporal change. His first mode covers increases in numbers of individuals on previously unoccupied ground (primary succession on new terrain or gaps in secondary succession). His second mode is one of replacement of species in stands after gradual habitat change due to either allogenic or autogenic influences. The final mode involves replacement after death of individuals through senescence. Obviously, his first mode is most appropriate to primary succession, but his model, like that of Matthews, is too general to have much practical application. Models that indicate how to contrast the effects of actual vegetation parameters during primary succession will be most productive. These may have to be more specific about species life history characteristics, successional stage and type, resource availability and disturbance regime.

Disturbance not only initiates primary succession but can also influence its rate and trajectory (Walker, 1999b). Yet models that integrate disturbance and successional processes are still very general and tend to emphasize one variable at the cost of the other. For example, the intermediate disturbance hypothesis (Connell, 1978; see section 6.6 on net effects of interactions; Fig. 6.13) focused on disturbance and species coexistence but not explicitly on succession. Only a system in which the disturbance regime is intermediate in frequency, extent or intensity will approach the maximum potential diversity of the system. Some models (see, for example, Figs. 3.7, 3.8) present the impacts of several levels of environmental severity, but environmental severity and disturbance are not always correlated, particularly in later stages of succession.

Disturbance affects succession in at least three ways (Fig. 3.10). It affects the available species pool, the available resources such as light and nutrients, and species performance (arrival, growth, survival) at a site.

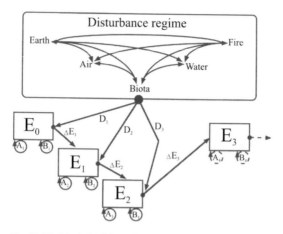

Fig. 3.11. Model of disturbance effects on succession. Elements of the disturbance regime derived from earth, air, fire, water or biota interact as a composite forcing function (D) on succession. At any given time, this function is the product of both past and current influences of disturbance. The successional trajectory of an ecosystem (E) is also modified by abiotic (A) and biotic (B) feedbacks at the site. From Willig & Walker (1999), with permission from Elsevier Science.

Disturbance also comes in many forms and from many sources (earth, air, water, fire, biota) as described in Chapter 2. Further, the likelihood of a particular disturbance affecting succession is dependent on the cumulative impact of past disturbances, abiotic and biotic feedbacks, and the current ecological space that an ecosystem occupies (Fig. 3.11). Models such as these that emphasize disturbance are relevant to primary succession but could develop more explicit links to biological processes. Successional trajectories are affected by the interplay of disturbance and the sum of all abiotic and biotic conditions (Willig & Walker, 1999).

Peet (1992) proposed two process models for secondary succession that have some relevance to understanding primary succession. The gradient-in-time model suggests that species characteristics determine species replacements along a temporal gradient of environmental change. This model seems to apply to primary succession, as long as interactions between species and their environments are included. The competitive-sorting model suggests that succession proceeds from an initially random collection of species to a predictable convergence on a community composed of the best competitors. This model predicts that the mean breadth of habitats occupied by the species within the community becomes reduced because specialists usually exclude or reduce the distribution of generalist species. This second model appears less applicable to primary succession where allogenic factors cannot be ignored. These models, like the positive-feedback switches of Wilson & Agnew (1992) provide further perspectives on the mechanisms that drive succession by focusing on the interactions of species and the environment (Gitay & Wilson, 1995).

We develop additional process models in Chapter 7 that explore how species interactions and environmental stress impact rates and trajectories of succession.

3.9.2 Mathematical models

Most mathematical models address secondary succession, presumably reflecting the preponderance of literature on secondary versus primary succession rather than any inherent difficulties in modeling primary succession. Yet there are many elements that are also applicable to at least the later stages of primary succession and some mathematical models that are ideally suited for studying the unique early stages of primary succession. Primary succession differs from secondary succession by the initial instability and sterility of the substrate (Miles & Walton, 1993) and the unoccupied surface. Models that address dispersal and establishment dynamics of plants and animals as well as increases in resource levels therefore are most applicable to primary succession. Studies that focus on current or past vegetation are least relevant to primary succession (van Hulst, 1992) (Table 3.4). Invasion models incorporate many of these factors in predictions of the successional influences of both naturally occurring pioneers and invasive aliens.

Models that address gap dynamics in forests such as the individual-based JABOWA family of models (Shugart & West, 1980; Urban & Shugart, 1992) and the spatially explicit SORTIE model (Pacala et al., 1993) are less useful to early primary succession. However, forest dynamics in later stages of primary succession have been successfully modeled. For example, in bottomland hardwood forests along a floodplain in South Carolina an individual-based forest model named FORFLO successfully predicted current species composition based on seed germination, tree growth and tree mortality (Pearlstine et al., 1985). The model was then extended to predict the effects of increased flooding from a diversion of the river. Another forest gap model was derived based on five life history characteristics (sapling establishment rate, maximum growth rate and size, shade tolerance and longevity) (Huston & Smith, 1987). Simulations were conducted with pairs of species that had contrasting characteristics (shade tolerance and intolerance) and many successional trajectories were found; however, competition for light emerged as a more influential factor than life history characteristics. Nevertheless, this model suggests that succession is generally predictable. If succession is seen as a game, the rules (outcomes of species interactions) are set, but any combination of players can take part in any arena (Austin & Smith, 1989). There are always limits to players (because of dispersal) and arenas (because of limited safe-sites), so succession in a given location should be predictable (Urban & Shugart, 1992). However, the number of combinations is so large that specific predictability remains elusive.

The arrival and successful establishment of plants in safe-sites in primary succession can be represented as a series of chance events or lottery. The death of colonists, their immediate replacement and arrival of new ones may all be governed by unpredictable events in what has been termed a carousel model (van der Maarel & Sykes, 1993; see sections 5.4.1 on germination, 5.3.2 on dispersal and 7.3.2 on stability). If one assumes that safe-site colonization is independent of previous inhabitants (a good assumption for primary succession) and that a successful colonist stays alive long enough to reproduce, some level of predictability is introduced and the lottery process can be modeled (van Hulst, 1992). Species with the highest relative abundance in the seed rain and lowest seedling mortality can be expected to succeed. The carousel model is thus a highly appropriate model for studies of primary succession (see del Moral, 2000a).

Markov models have been useful for predicting successional change when successional states are clearly defined (Usher, 1981, 1992). Like other transition models, Markov models are stochastic because at any time the transition from one state to the next may or may not occur. Markov models assume that the transition probabilities are constant and that change is independent of initial conditions. Neither of these assumptions is often true in either secondary or primary succession (Facelli & Pickett, 1990). For example, when Gibson et al. (1997) used a Markov model to predict the development of artificial barrier islands in Florida (U.S.A.), they found that the model predicted that seres would develop more rapidly than was the case. This led to an analysis that revealed that local disturbance chronically reset the clock. In this case, a fault of the Markov model (that transition probabilities were assumed constant) was used to learn about mechanisms in the sere. Childress et al. (1998) attempted to use Markov models to predict change in dense populations of *Lupinus lepidus*. Their analysis also failed because trajectories were subject not only to the initial conditions, but also to fluctuating environmental conditions and the composition of adjacent plots. The dependence of

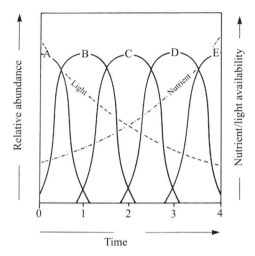

Fig. 3.12. The resource-ratio model of succession, showing how adaptations of species to unique combinations of light or nutrient levels may drive species change. From Tilman, D.; *Plant Strategies and the Dynamics and Structure of Plant Communities.* Copyright ©1988 by Princeton University Press. Reprinted by permission of Princeton University Press.

vegetation change on initial conditions and its continuing effect on transitions throughout succession suggest that more complex models are needed. Starting points do matter, but a given state also depends on continuing processes and spatial constraints. Equilibrium states (another prediction of Markov models) are rarely reached (van Hulst, 1992). Yet Markov models illustrate that autogenic processes do affect successional trajectories.

Tilman (1985, 1988) developed the resource ratio model (Fig. 3.12), which suggests that vegetation responses to changing resource levels drive succession. This model has been useful in focusing on the ecophysiological aspects of succession, and may have certain lessons for primary succession. However, its several assumptions may not be realistic (van Hulst, 1992). For example, the model assumes that resources limit plant species in complementary ways (the resource that limits species A does not limit species B and vice versa), that resource availability limits and determines plant growth and that each successional state reaches an equilibrium between resource levels and plant growth. The emphasis has typically been on light levels (that decrease with succession) and N levels (that increase). Tilman suggested (1988) that early colonists in primary succession would be adapted to high light and low nutrients and later species to the reverse conditions. He used the succession on glacial moraines in Glacier Bay (Alaska) as an example and stated (Tilman, 1988, p. 217):

... many of the features of primary succession might be explained as a slowly shifting trajectory of equilibrial plant communities, with the composition at any point mainly determined by the relative availability of the limiting soil resource [N] and light.

Tilman (1988) also recognized that early stages of succession could be dominated by transient species for a relatively short period before an equilibrial state is reached. An update of this concept incorporates dispersal dynamics and neighbor effects (Tilman, 1994). Transients in primary seres should be species with low maximal (aboveground) growth rates because of high allocation to nutrient acquisition. As a partial explanation of major shifts in species composition and vegetation structure, the resource ratio model is useful. It also points out the paucity of data on light and nutrient requirements of species in succession. Perhaps the biggest problem in applying the model to primary succession is the assumption of equilibrium between plants and their resources. High levels of disturbance and the prevalence of stochastic colonization events make any equilibrium unlikely (Matthews, 1992).

Invasion models (van Hulst, 1992) serve as good integrators of various demographic and environmental factors. Primary succession is a series of successful invasions. Determining what controls species invasions is key to predicting succession. Characteristics of good invaders are well known (e.g. high seed production, widely dispersed seeds, rapid growth) but characteristics that make a community invasible (or resistant to invasion) are less clear (Crawley, 1987; Luken & Thieret, 1997). Progress in modeling invasion dynamics is essential to prevent the spread of alien plants and animals and to predict the impacts of the removal of naturalized aliens on native species (see section 8.4.8 on alien plants).

3.10 New directions

Successional theory has progressed in both content and sophistication since the early 1900s. However, the complexity of succession has become increasingly obvious. While that complexity has required numerous alternative approaches to succession, it has also created a pronounced contrast between holists and reductionists. This contrast of viewpoints has its roots in the origins of the formal study of succession and ecology and has served as a catalyst to many ecologists to define carefully their biases and assumptions. In Chapter 9 we revisit successional theory and project its future development, after discussing the roles of life histories, species interactions and successional trajectories in Chapters 5, 6 and 7, respectively.

The holists see the forest but often neglect the trees. They defend the integrity of communities, arguing that the sum is more than the parts. Strong interactions sharpen the contrasts between communities and help to unify a community. Holists also are quite willing to generalize across ecosystems and types of sere, and lean toward verbal and mathematical models that apply across broad temporal and spatial scales. Holists argue that there are important and repeatable patterns of change during succession and one should not lose sight of them even if the patterns are weak and there are many exceptions.

Reductionists focus on the trees and do not always see the forest. To them, nature is full of exceptions to every rule and biotic boundaries are diffuse or non-existent (or simply explainable by abiotic gradients). Organisms mix in space and time but only in stochastic ways, and there are few if any emergent properties in the loosely assembled communities. Interactions occur, but they are relatively weak and facultative. For reductionists, successional models must stay focused on the details, and cannot hope to explain the complexity of multiple trajectories, variation in life histories and spatial patterning of resources. A more likely result is that each situation is unique and no useful patterns can be discerned.

Inevitably, there is validity to both the holistic and reductionistic approaches, depending on the questions being asked. The variety of viewpoints about succession and community organization seems a logical fit in a field that has many continua: from landscape-level gradients, to the gradient between primary and secondary succession, to the range of possible positive to negative interactions between two organisms. While speaking in dichotomies, we recognize that reality has more options. This seems proper for studies of change over successional time, a variable that is quintessentially continuous.

4 · *Soil development*

4.1 Background

The absence of soil is the defining characteristic of the first stage of primary succession. Therefore the development of soil is a crucial aspect of primary succession. Because soil development is a product of both physical and biological processes, it links the abiotic and biotic variables that drive primary succession (Matthews, 1992). In Chapter 2 we discussed how various physical disturbances alter substrate stability, texture and fertility. In this chapter we will explore how these and other environmental variables govern soil formation and how soil properties such as water-holding capacity and nutrient content vary with succession. We will then examine how such processes as nitrogen fixation and decomposition influence organic matter accumulation. We conclude with a discussion of the spatial variability in soil development during primary succession. Throughout, we will contrast soil development among different types of primary succession and along physical gradients while looking for the level of specificity at which generalizations are possible. For readers that want a more thorough coverage of pedogenesis (soil formation), we refer them to other sources (Stevens & Walker, 1970; Swift *et al.*, 1979; Birkeland, 1984; Killham, 1994; Paton *et al.*, 1995; Wood, 1995; Coleman & Crossley, 1996; Brady & Weil, 1998). Our intent here is to establish the importance of soil development in the context of primary succession.

The types of soil and the rates at which they form are critical in determining the rates and trajectories of primary seres as well as their community and ecosystem properties. The development of soil is extremely complex and the details are poorly understood. Indeed, even defining soil can be difficult, given the possible range of physical, chemical and biological characteristics. However, broad patterns are recognizable and attributable to a basic set of variables, as outlined most clearly by Jenny (1941, 1961, 1980). Following the lead of earlier scientists, including

Dokuchaev (see citations in Crocker, 1952), Jenny proposed that soils are formed by dynamic or 'active' external inputs (e.g. erosion, immigration of biota) and passive internal factors (e.g. parent material, topography). Time, or the age of the site, is the third important factor. Time incorporates the cumulative effect of all active processes at a given site. The best-known elaboration of these active and passive variables includes climate (cl), organisms (o), topography (r), parent material (p) and time (t) together in Jenny's equation where soil formation (s) is a function of all the variables:

$$s = f(cl, o, r, p, t).$$

Each variable has been considered dependent or independent of the other variables (Matthews, 1992) yet time, having no direct impact on soil formation, is often separated such that all other processes are a function of time:

$$s = f(cl, o, r, p) \, dt.$$

The robustness of this approach is demonstrated by its persistence in nearly identical form for many decades (Amundson & Jenny, 1997). Other useful modifications have explicitly included human land-use practices and fire (Wali, 1999a). The unmet challenge is to adequately test this theory under field conditions. The interdependence of each variable means that removing one while keeping the others can only be approximated, even when permanent plots are monitored over time. Chronosequences (see Chapter 1) are more frequently used to study primary succession than permanent plots, but the use of chronosequences for this purpose requires that factors affecting soil formation at various sites of different ages have remained constant over time. This is obviously unlikely due to the spatial heterogeneity of site variables such as parent material and topography and the temporal heterogeneity of the influx variables of erosion and biota (Stevens & Walker, 1970; Pickett, 1989). Rates of soil formation can be approximated when site ages are carefully established (as for many glacial moraines; Matthews, 1992), but determining the relative importance of each soil-forming variable remains unrealistic. We will now examine both broad environmental parameters and more local physical properties that impact soil formation.

4.2 Environmental controls

In this section we present the passive or endogenous site characteristics (parent material and topography) and the active or exogenous variables (erosion and biota) that determine how soils form in primary succession. First, however, we discuss the broadest of all variables, i.e. the climate, in which soil formation occurs.

4.2.1 Climate

Clements (1928) proposed that, within a regional climate, plant communities would reach a stable endpoint or climax. He also proposed that soil development, altered by plant influences (reaction) and climate, would reach a stable equilibrium. However, Clements recognized that heterogeneity in soils at a finer spatial scale than regional climates could result in particular endpoints on particular soil types (edaphic climaxes). Although the emphasis now is on local variation rather than regional similarity (a dynamic view that emphasizes disturbance and disequilibrium over homeostasis; see Chapter 3), climate does determine the overall rate and direction of soil formation.

Temperature and water availability are perhaps the most important aspects of climate that influence soil formation. Temperature extremes limit plant enzyme function for photosynthesis and the presence of water in liquid form. Too much water reduces soil oxygen levels for root respiration whereas too little water limits decomposition, mineralization and many physiological functions, including photosynthesis. Animals are limited in similar ways. Many organisms have temperature optima between 15 and 35°C. Extreme temperatures are frequently encountered in primary succession, where cold (e.g. glacial) or dry (e.g. desert) conditions limit soil formation and plant establishment. Under such conditions, the biotic influence on soil formation is minimal (Matthews, 1992), although many organisms have adapted to such conditions and have temperature optima of less than 15°C or high levels of drought tolerance. However, the ability of some microbes to survive in nearly all terrestrial habitats leads to the potential for primary succession in most habitats (although it might be limited to microbial succession; see section 4.4.2).

Both the amount and timing of precipitation are very influential in soil formation. High levels of precipitation result in leaching of nutrients, as on tropical volcanoes (Whittaker et al., 1989) that can slow soil development. However, in cooler yet rainy climates such as in river valleys (Tonkin & Basher, 2001) and on glacial moraines (Sommerville et al., 1982) in southern New Zealand, soils develop rapidly. In one river valley that received c. 10,000 mm of precipitation per year, spodosols formed within 500–1500 yr (Tonkin & Basher, 2001). In contrast, low levels of precipitation combined with high temperatures can result in net upward movement of solutes and precipitates that form surface crusts of salts, as in deserts (Smith *et al.*, 1997). Intermediate (mesic) levels of precipitation and moderate to warm temperatures result in high rates of plant productivity and decomposition, which is often correlated with rapid soil organic matter accumulation, as on subtropical landslides (Zarin & Johnson, 1995a,b; Walker *et al.*, 1996). Soils at the relatively warm and moist coastal site of Glacier Bay, Alaska, developed 50–100 yr faster than at a cooler and drier site 160 km inland (Klutlan Glacier, Canada; Jacobsen & Birks, 1980). A comparison of rates of soil formation in primary seres in six habitats throughout the world further supports more rapid formation in warm and wet than in cold and dry climates (Birkeland *et al.*, 1989). Decomposition is most accelerated by changes in temperature and moisture such as occur during wet–dry or freeze–thaw cycles (Taylor & Parkinson, 1988).

Soils in primary seres are initially entisols (azonal or undeveloped). Soils typical of the region (e.g. spodosols in cool, wet climates or aridisols in arid regions) may eventually develop on the primary site over successional time, but any of the environmental controls discussed below can direct soil development in unique directions.

It is at the scale of the microclimate that soil formation is ultimately regulated, because local conditions must be favorable for soil biota to function. However, conditions that promote net primary production (mesic, warm, aerated) also increase decomposition and mineralization rates (see section 4.3.5 on N). Therefore, in warmer regions there is more organic matter loss and less carbon accumulation in the soil. In addition, warm sites are often too dry for optimal functioning of soil biota. These counteracting influences make it difficult to predict the effects of both local and larger (e.g. global warming) influences on soil.

4.2.2 Parent material

The substrate that remains following a severe disturbance is obviously an essential feature that directs the process of soil formation through its physical and chemical properties. Primary surfaces encompass a wide variety of textures and levels of stability and fertility (Fig. 2.1). The surfaces have either been transported to the site or modified *in situ*. Transported substrates result from deposits by wind (aeolian, e.g. dunes), water (e.g. floodplains, deltas, glacial moraines) or gravity (e.g. rock scree, landslides, mine tailings, soft benthic surfaces). These substrates are generally unstable and soil formation may be accelerated when colonizing plants begin

to stabilize the surface. Volcanic activity provides a variety of new substrates of varying stability (pahoehoe lava > a'a lava > lahars > pumice >
ash; del Moral & Grishin, 1999). When the added layer is shallow or per-
meable, surviving organisms in established soil layers below can influence
succession (for example, gophers bring buried soils to the surface of new
tephra on Mount St. Helens; Andersen & MacMahon, 1985). Substrates
that are altered *in situ* can result from scouring (some glacial surfaces,
river beds), compaction or scraping (roads, river beds), draw-down (lake
shores) or erosion (cliff faces). Subsequent stability (see section 4.2.4 on
erosion) is related to texture: rocks are less likely to re-slide than is clay,
sand is less likely to blow away than is silt, gravel is less likely to wash
away than is sand.

The chemical composition of the parent material can affect patterns
of soil formation. Neutral to slightly alkaline surfaces (e.g. limestone)
break down faster into soil than acidic surfaces (e.g. granite) or very alkaline surfaces (e.g. sodium-rich marine clays), owing to conditions that
favor biotic weathering by soil microbes (see section 4.3.4 on pH) or
abiotic weathering. Some surfaces are rich in minerals (e.g. serpentine:
magnesium; gypsum: calcium sulfate) that support unique flora and form
distinctive soils. Physical characteristics that affect soil formation and plant
colonization include particular weathering characteristics (granite exfoli-
ates, slate or mica come off in sheets) and porosity (water drains through
a'a lava but puddles in cracks of pahoehoe lava). Differences attributable
to parent material may be temporarily obscured in primary succession by
the impacts of land use such as cultivation and grazing (Puerto & Rico,
1994).

4.2.3 Topography

Steep slopes are unstable and offer fewer microsites for seed retention than
flat surfaces. On an Illinois (U.S.A.) mine tailing, flatter surfaces supported
higher nodulation rates of the shrub *Alnus*, presumably owing to more
favorable water conditions (Dawson *et al.*, 1983). Concave microsites in
volcanic tephra on Mount St. Helens had higher nutrient levels after 7 yr
than convex surfaces (Zobel & Antos, 1991). Slope and aspect (compass
direction of the slope) combine to affect surface temperature and many
other soil properties. Surfaces will be hotter and drier if they face toward
the equator and may experience three times as much evapotranspiration
as pole-facing slopes (Le Houérou *et al.*, 1993), higher rates of decomposition and mineralization (Gerlach, 1993), and lower plant productivity

(Viereck *et al.*, 1983). For succession on mined lands in North Dakota (U.S.A.), Wali (1999a) demonstrated that south (equator)-facing slopes had lower accumulation rates of both N and C than north-facing slopes. Soil development was equally rapid on gentle and steep slopes in a high rainfall area in New Zealand, suggesting that topography is not always important (Basher, 1986). Similarly, slope had no effect on species distribution on Alaskan landslides (Lewis, 1998). The impacts of local surface textures on succession are discussed in section 4.3.1.

4.2.4 Erosion

Steep surfaces made of fine particles are most likely to erode (see section 2.2.1 on erosion). Eroded surfaces are likely to continue to lose soil until the angle of repose is reached or other forces (e.g. plant colonization) stabilize the surface. However, subsequent forces (e.g. earthquakes, road cuts, floods, vegetation removal) can cause the whole surface to erode anew (secondary erosion). On a landscape scale, secondary erosion is dramatically represented by the lahars that were deposited around Pinatubo volcano in the Philippines when rains eroded initial tephra deposits (Newhall & Punongbayan, 1997). Stabilized pumice deposits have been eroding rapidly owing to recent quarrying on the island of Lipari (Italy; R. del Moral, pers. obs.). Active glacial moraines, recently devoid of the stabilizing ice and subject to melt waters from the retreating glaciers, are very unstable substrates (Matthews, 1999). Instability is also a constant factor on floodplains (Malanson, 1993). Only fast-growing species are likely to colonize unstable surfaces such as floodplains (Johnson *et al.*, 1985; Walker *et al.*, 1986), dunes (Sykes & Wilson, 1990) and volcanic tephra (Chiba & Hirose, 1993). High plant productivity accelerates stabilization through root growth and aboveground interception of wind-blown or water-borne particles.

On local scales, erosion can vary within a disturbance type. Landslide surfaces include the upper slip face that is very unstable and likely to resist any soil formation, the middle chute that is frequently scoured to bedrock or mineral soil, and the lower deposition zone where organic debris mixed with subsoil can support relatively rapid colonization and soil formation (Lundgren, 1978; Smith *et al.*, 1986; Adams & Sidle, 1987; Walker *et al.*, 1996). Similarly, dunes are composed of areas of active erosion and relative stability that influence colonization and succession (Moreno-Casasola, 1986). McLachlan *et al.* (1987) determined that both plant and animal biomass were positively associated with sand stability

Table 4.1. *Plant and animal biomass (g m⁻²) in five vegetation zones on dunes in South Africa*

Zones 1 and 2 (slip face and windward slope) from Fig. 4.1 are combined into dune slopes in this table. The other zones in Fig. 4.1 are 3 (pebble corridor), 4 (*Sporobolus*), 5 (*Gazania*) and 6 (*Psoralea*). Biomass increases with increasing stability from left to right.

Component	Dune slopes	Pebble corridor	*Sporobolus* zone	*Gazania* zone	*Psoralea* zone
Aboveground plants	0.0	0.0	20.6	61.7	394.8
Belowground plants	0.0	0.0	50.3	241.1	752.9
Orthoptera	0.0	0.0	0.3	0.3	0.8
Coleoptera	0.6	0.1	0.8	7.5	5.2
Lepidoptera	0.0	0.0	0.4	2.4	2.3
Homoptera	0.0	0.0	0.0	0.0	1.4

Data from McLachlan *et al.* (1987).

on coastal dunes in South Africa (Table 4.1). Six distinct topographic zones with unique vegetation were identified (Fig. 4.1). Biomass increased with succession for about 6–7 yr, after which the vegetation was buried. Tielbörger (1997) delineated seven plant communities on dunes in the Negev Desert (Israel) that were related to surface stability. She proposed but did not test that increased stabilization paralleled successional development. Highest plant cover and species richness were found on the most stable surfaces (Fig. 4.2). Thus plants require a certain level of surface stabilization in order to establish and further stabilize the surface as they grow and spread.

At microsite scales, small erosive channels can displace seeds and seedlings, resetting succession repeatedly. Roads or trails can act as effective erosion channels, funneling runoff, carving deep incisions in the landscape and delaying soil formation, particularly in cold habitats where water accumulates when the permafrost melts and thermal properties of the surface are altered (Komárková & Wielgolaski, 1999). Because soil removed by wind or water is up to five times richer in organic matter than the soil left behind (Allison, 1973) and loss of developing topsoils reduces water-holding capacity, nutrients and soil organic matter, the effects of erosion on soil formation are especially destructive (Pimentel & Harvey, 1999). Global effects of aeolian transport of wind-eroded materials are discussed in sections 4.3.6 and 4.3.7.

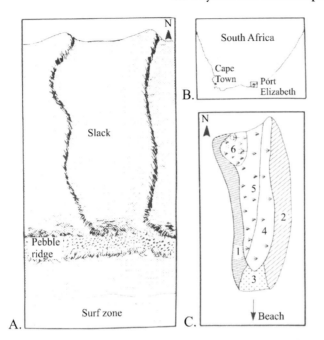

Fig. 4.1. Six zones of vegetation on a South African dune, determined in part from the relative stability of the sand. Dune movement was about 7 m yr^{-1} (west to east) and slack width was 45 m, so succession lasted only 6–7 yr. Vegetation zones in order of increasing stability: 1, slip face (no vegetation); 2, windward face (no vegetation); 3, pebble corridor (no vegetation); 4, *Sporobolus* zone; 5, *Gazania* zone, subject to burial by advancing sand; 6, *Psoralea* zone in wind-shelter. Modified from McLachlan *et al.* (1987). See also Table 4.1.

4.3 Physical and chemical properties

4.3.1 Texture

Surface textures are extremely important in determining which propagules will be trapped, germinate and establish in primary succession. Smooth surfaces are slowest to be colonized, particularly in exposed windy situations such as desert playas (Fort & Richards, 1998) or where seeds are not trapped. Smooth surface crusts (e.g. salts from evaporative surfaces) can impede entrapment (by reducing surface heterogeneity) and reduce establishment when the radicle of the newly germinated seed cannot permeate the crust, and the seedlings dry out (Harper, 1977). However, favorable microsites such as small pockets between salty peaks or gaps in the distribution of a crust often allow germination (Krasny *et al.*, 1988). Alternatively, biotic surface crusts (typically more heterogeneous

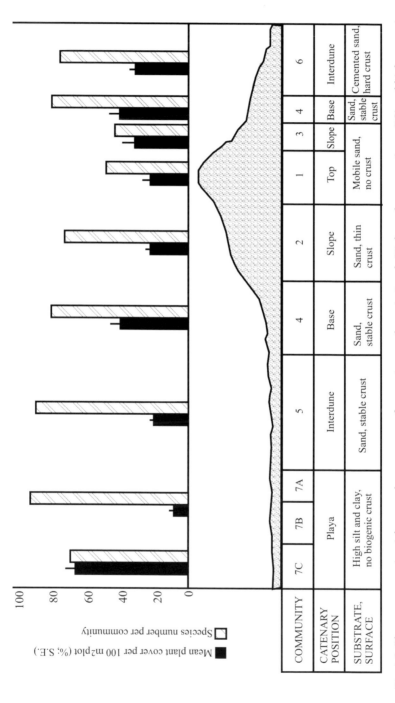

Fig. 4.2. Plant cover, species richness and seven zones of vegetation on dunes in Israel. Highest plant cover (zone 7C) and highest species richness (zone 7B) were on the stable playa between dunes. Cover and richness were both low on unstable dune slopes. Dominant plant genera in each zone were: 1, *Stipagrostis, Heliotropium*; 2, *Moltkiopsis, Convolvulus*; 3, *Echinops*; 4, *Noaea, Artemisia*; 5, *Echiochilon, Thymelaea*; 6, *Cornulaca*; 7, *Anabasis*. Modified from Tielbörger (1997).

than abiotic crusts and composed of various combinations of mosses, lichens, fungi, algae and bacteria) can facilitate entrapment and retention of seeds and spores, collect water, organic debris and nutrients and promote soil development and succession (Walker, 1999b). These biotic crusts, sometimes called cryptogamic crusts, can stabilize dunes (Danin, 1991), glacial moraines (Worley, 1973) and desert soils (Rychert et al., 1978) or invade once other colonizers stabilize a surface. For example, once grasses colonize a dune, finer sediments are trapped, microsite moisture increases and crust-forming cyanobacteria multiply (Forster & Nicolson, 1981; Danin, 1991). The N-fixing cyanobacteria promote further plant growth and subsequent soil formation. Cracks, boulders, existing plants, or shifts in topography also provide windbreaks and promote soil formation (Wright & Bunting, 1994). Volcanic tephra (del Moral & Grishin, 1999) or blowing sands (McLachlan et al., 1987) can reduce this local surface heterogeneity, thereby retarding plant colonization. However, colonizing plants can re-establish variation in the surface (Alpert & Mooney, 1996). Also, tephra can be quite variable in chemistry and depth, with higher levels of nutrients often found in finer particles (Zobel & Antos, 1991).

Surface roughness presents a different set of challenges for plant colonization and soil formation than smoothness. Entrapment is not a problem for seeds and organic matter that land on a'a lava or in a boulder field. However, germination can still be a challenge due to few horizontal surfaces that have enough light and water retention to support the accumulation of organic debris, germination and growth of plants (Walker & Powell, 1999a). Even where small pockets of soil form, root development may be limited by the larger, sterile surfaces. Eggler (1941) suggested that succession was slower on the rough a'a lava than on the smooth pahoehoe lava in Idaho (U.S.A.) because the irregular surface of the a'a required a greater accumulation of fine soil to support plant growth.

Materials deposited at a site clearly impact surface texture and primary succession (Friedman et al., 1996a,b). Floodplains provide good examples of substrates that are created by the sorting and resorting of particles based on their relative size (Malanson, 1993). Grubb (1986, 1987) suggested that substrate particle size largely determined what type of pioneer species would establish (Fig. 4.3). He suggested that grasses would invade sandy and silty areas, herbs invade gravels, trees invade rock crevices, and bryophytes invade exposed rock surfaces. Thus, long-lived species that grow slowly (trees and lichens) invade substrates with larger particle sizes. Grubb further suggested that such long-lived species also colonized sites

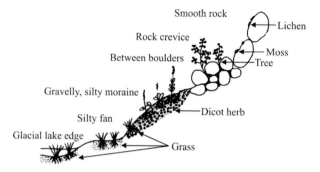

Fig. 4.3. Surface textures characteristic of primary succession, and likely colonists. Modified from Grubb (1986).

with low water and nutrient availability. Trees occur as invaders on lava (Aplet & Vitousek, 1994), glacial moraines (Cooper, 1916), talus slopes (Weaver & Clements, 1938), coarse-textured floodplain deposits (Gill, 1972) and mined lands (Bramble & Ashley, 1955; Bradshaw, 1983b), in part because their roots can reach moisture found in deep cracks on otherwise inhospitable surfaces. Although exceptions to these patterns occur, as when mosses (Smith *et al.*, 1986; Delgadillo & Cárdenas, 1995) and trees (Grubb, 1987; Kalliola *et al.*, 1991; Walker, 1999b) invade fine-grained sands and silts, the conceptual framework is useful and needs to be rigorously tested. A necessary refinement to Grubb's proposal is to recognize the variation of preferred substrates within plant life forms. For example, Helm & Collins (1997) found that three woody genera (*Salix, Populus* and *Dryas*) dominated increasingly coarse substrates from silt to cobble on an Alaskan floodplain. Further discussion of this issue is presented in section 4.5.1 on N-fixers.

4.3.2 Compaction

Many primary seres begin on compacted surfaces. Compaction can occur naturally on riverbeds, glacial moraines or some dunes but more often occurs on artificial surfaces such as agricultural areas (Sojka, 1999), roadbeds (Webb, 1983) or mined lands (Bradshaw & Chadwick, 1980) because of the weight of the vehicles and machinery. Small particle sizes (especially clays and fine sands) compact more easily than large particles. Compaction can alter surface texture and erosiveness of a substrate. By changing infiltration or drainage patterns, compaction also influences movement and retention of water and organic matter, and, indirectly, the chemical composition and decomposition dynamics of the developing soil. For example, Bolling & Walker (2000) found that the soils of abandoned vehicle tracks, twice as dense as adjacent desert soils, had lower available pools of N and P, possibly because of reduced decomposer activity (Belnap et al., 1994). Higher decomposition could also lead to reduction of N and P pools if leaching losses were large, but this scenario was unlikely in the more highly compacted tracks. Bladed roads that were similar in compaction to desert soils were also similar in soil nutrients. However, Walker & Powell (2001) found that desert roads had more organic matter, water content and nutrients than desert mine tailings, even though the roads were more compacted than the tailings. Most studies of compaction in forest soils indicate reductions in plant growth. However, a few studies show increased growth, presumably from increased water-holding capacity in soils with low bulk densities (Greacen & Sands, 1980). Clearly, the relationship between compaction and soil qualities is complex and depends on the degree of initial organic matter removal, soil particle size, drainage and precipitation patterns, among other variables.

Compaction can be alleviated during succession through the action of roots, earthworms and other soil organisms. However, Sojka (1999) noted that many hardpans are not broken up even when earthworms are supplied with abundant organic matter. Bolling & Walker (2000) did not find a reduction in compaction on abandoned dirt roads in the Mojave Desert (U.S.A.) in an 80 yr chronosequence but R. H. Webb (1983; pers. comm., 2000) did find gradual decreases in compaction over 70 yr of recovery of abandoned roads in ghost towns in Nevada (U.S.A.).

Further complications arise from methodological considerations. Compaction is rarely measured directly, but is instead inferred from bulk density (g cm^{-3}). Bulk density is calculated on the mass of fines (the fraction of particles of less than 2 mm diameter) divided by their volume. However, the volume of fines can be calculated either as the total volume minus the rock volume (rocks for this purpose are greater than 2 mm in diameter; unsieved bulk density) or as the volume of fines in a container of known volume following sieving of the soil. The former method incorporates soil air pockets (macropores) into the volume and bulk densities can be lower than when the volume of fines is based on the second measurement, which excludes these air spaces. Bulk densities are used not only as a measure of soil compaction but also as a surrogate for areal estimates of soil nutrient content. Therefore, inconsistencies in measurements (and large variability in the size and presence of air pockets in soils) contribute to difficulties in site comparisons of soil parameters in studies of primary succession (cf. Chapin *et al.*, 1994).

4.3.3 Water content

The presence of adequate water is essential to soil development and has a major influence on the rate and trajectories of primary succession by determining which species can germinate, grow and survive (de Jong *et al.*, 1995; Braatne & Chapin, 1986). Major sources of water include rain, fog drip, dew, runoff, snow or ice melt and ground water. The temporal distribution of water sources can be as important as absolute amounts in determining vegetation and consequent soil development. Water content is affected by all the large-scale environmental controls described above as well as most of the physical and chemical properties described in this section. For example, limitation of primary productivity by water and by N may covary across moisture gradients (Hooper & Johnson, 1999). Fine-textured, clay-rich soils and organic soils hold the most water but that water may be less available to roots and microbes than water in drier, sandier soils with larger pore sizes (Chadwick & Dalke, 1965). Yet dune plants that cannot access stored ground water depend on regular rains, especially in coastal areas where rains reduce the negative effects of salt spray on growth (Barbour *et al.*, 1985). Low water availability on coarse-textured landslides may also limit succession (e.g. in Tanzania; Lundgren, 1978). Depth to water table is another critical variable that determines plant colonization and soil development, particularly on floodplains (Gill, 1973; Menges & Waller, 1983; Oliveira-Filho *et al.*, 1990). However, coarse-textured substrates are drier than fine silts, even when they lie closer to the water table (Gill, 1972; Fonda, 1974). The creation and destruction of distinct landforms (e.g. bars, banks, abandoned channels, swales) may be more important than either depth to water table or substrate texture in floodplain succession (Hupp & Osterkamp, 1985; Kalliola *et al.*, 1991; Prach, 1994a) as each landform has a unique disturbance regime. On surfaces behind advancing dunes (deflation fields) on the coast of Poland, Piotrowska (1988) found two distinct successional paths that differed by depth to the water table. Succession on small (0.5–2.5 m tall) ridges of sand was much slower and the vegetation more stunted than succession in the troughs where the soils were moister (Fig. 4.4).

   Soil organisms vary in their tolerance to water stress, with fungi generally more tolerant than bacteria. Actinomycetes, or bacteria that grow in filaments resembling fungal hyphae, are the most tolerant of both heat

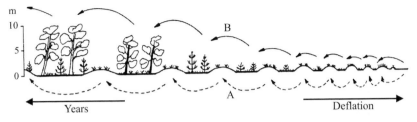

Fig. 4.4. Two contrasting successional trajectories on coastal dunes in Poland. Succession in the wetter troughs (B) was faster than on the drier ridges (A). From Piotrowska (1988), with kind permission from Kluwer Academic Publishers.

and drought (Killham, 1994). Saturated soils may be too anaerobic for roots to grow and mycorrhizal fungi are particularly susceptible to low oxygen levels and are not common in flooded soils.

Soil water infiltration generally increases with soil development in primary succession. Plant roots and burrowing animals increase infiltration and developing organic horizons retain water. However, several soil features can reduce infiltration. Abiotic surface crusts, such as those found on an Alaskan floodplain from evaporative salts (Dyrness & Van Cleve, 1993) can reduce infiltration and delay succession (Chapin & Walker, 1993). Such crusts disappear once a permanent and continuous plant cover is established. Biological surface crusts in arid lands have various influences on water infiltration (Belnap et al., 2000). In hot deserts, smooth mucilaginous cyanobacterial crusts reduce infiltration. In cool deserts subject to frost heaving, cyanobacterial crusts increase infiltration because they increase surface heterogeneity and trap water. Similarly, crusts composed of lichens and mosses (cryptogamic crusts) tend to increase surface texture and infiltration in both hot and cool deserts (Belnap et al., 2000). Recovery of arid land soil crusts can take several centuries, during which time low water infiltration and high erosion reduce the rate of crust formation (Belnap, 1995).

Subsurface layers can restrict water infiltration beyond a certain depth. One example is the caliche layer that forms in arid lands from precipitation of calcium carbonate when surface evaporation exceeds precipitation. Other examples include the layer of ice in tundra soils (permafrost) and the formation of impervious pans in the subsoil through podzolization. These types of subsurface layer often take centuries to develop (for example, spodosols on dunes in eastern Australia continued to develop over a period of 700,000 yr; Walker et al., 1981; Thompson, 1983, 1992), although permafrost can develop within 100 yr in central Canada and is essentially continuous further north (Crampton, 1987). Along Alaskan floodplains, mosses and *Picea glauca* forests may (Drury, 1956; Van Cleve *et al.*, 1991) or may not (Mann *et al.*, 1995) promote development of permafrost and a subsequent shift to *Picea mariana* and *Larix* trees and bog vegetation within 200–1000 yr. Disturbances that disrupt these layers (e.g. roads that expose the caliche or melt the permafrost, floods or winds that erode or deposit sediments or fires that burn off insulating vegetation) alter soil hydrology and soil chemistry, thereby impacting primary succession.

4.3.4 pH and cations

Parent material, soil organic matter, cation exchange capacity (CEC) and organic acids all affect soil pH (a measure of hydrogen ions). Soil pH generally declines in primary succession as organic acids accumulate (see, for example, Jacobsen & Birks, 1980; Sival, 1996), although volcanic ejecta and some types of mined land initially are very acidic. A survey of 28 studies showed significant declines of soil pH for glacial moraines, floodplains and dunes but not for mined lands or volcanoes (Fig. 4.5). Messer (1988) found significant declines in soil pH in 10 of 18 glacial moraines in Norway (Fig. 4.6). On a glacial moraine in Alberta (Canada), pH declined under *Dryas* shrubs but not in bare areas over a 100 yr period (Fitter & Parsons, 1987).

Cation exchange capacity measures the potential of soil to store and supply cations. The values of CEC generally increase through the early stages of primary succession as organic matter accumulates. Micelles, or clay particles, accumulate a negative charge on their surfaces that attract and hold cations (e.g. Ca, Mg, Na and K). Cations held to these surfaces are less readily leached than unattached cations in solution and can be taken up by roots (exchanged for other cations) over time.

Nutrient availability is strongly linked to pH and CEC because elements such as P become unavailable at low pH (fixed as hydrous oxides of Al or Fe) or high pH (fixed as calcium phosphates). The type of soil N source also impacts pH. If plants are taking up ammonium, the release of hydrogen ions increases acidity (lowers pH). When nitrate is the major source, hydroxyl ions decrease acidity (raises pH). Soil acidity also impacts nutrient retention because excess hydrogen ions in acidic soils displace the cations on the micelle surfaces and the cations are leached from the rooting zone. Soil fungi are generally more tolerant of acidic conditions than are soil bacteria. The absence of cations and the high H

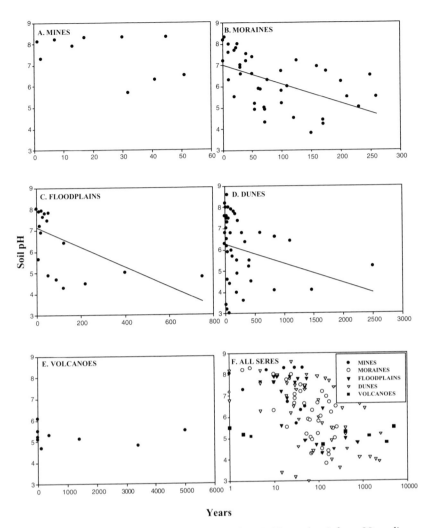

Fig. 4.5. Changes in pH in surface mineral soils (10–30 cm deep) from 28 studies in five types of primary sere. Lines represent significant ($p < 0.05$), first-order linear regressions. (A) Mines ($n = 2$; $r^2 = 0.17$; $p = 0.20$); (B) glacial moraines ($n = 9$ studies; $r^2 = 0.27$; $p = 0.0003$); (C) floodplains ($n = 5$; $r^2 = 0.35$; $p = 0.004$); (D) dunes ($n = 9$; $r^2 = 0.10$; $p = 0.02$); (E) volcanoes ($n = 3$; $r^2 = 0.009$; $p = 0.80$); (F) all seres combined. The following references were used. Mines: Leisman (1957); Wali (1999). Moraines: Crocker & Major (1955); Crocker & Dickson (1957, two moraines); Persson (1964); Viereck (1966); Jacobsen & Birks (1980); Messer (1988); Bormann & Sidle (1990); Helm *et al.* (1996). Floodplains: Viereck (1970); Wilson (1970); Van Cleve & Viereck (1972); Fonda (1974); Walker (1989). Dunes: Olson (1958); Olff *et al.* (1993); Emmer & Sevink (1994); Gerlach *et al.* (1994); Conn & Day (1996); Sival (1996); Berendse *et al.* (1998); Lammerts & Grootjans (1998); Lichter (1998). Volcanoes: del Moral & Clampitt (1985); Kitayama (1996a); Raich *et al.* (1997). Note the increasing temporal scales of the *x* axes and the logarithmic *x* axis in (F).

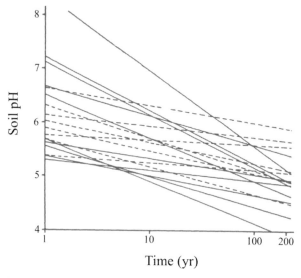

Fig. 4.6. Soil pH changes during primary succession on 18 glacial moraine chronosequences in Norway showing a general pattern of decline. Statistically insignificant chronofunctions ($p \leq 20\%$) are indicated with broken lines. From Messer (1988).

(and often Al) concentrations are toxic to many plant species (Foy, 1984) and may explain species distributions on rock outcrops (Ware & Pinion, 1990; Tyler, 1996) or dune slacks (Sival & Grootjans, 1996).

Temporal trends in cations and anions have been poorly studied in primary succession although nutrient pools normally increase during the early stages. Matthews (1992) found no temporal patterns of change for Al or Fe on glacial moraines. Along a dune toposequence where distance from the beach reflected site age, Henriques & Hay (1992) found decreases in Na, Ca and Mg. Chadwick *et al.* (1999) found that although all cations decreased over the very long Hawaiian chronosequence (4.1 million yr), Al was not leached as quickly as Mg or Ca. Yet Ca never limited growth in the chronosequence. Significant long-distance aeolian inputs of Ca and, by extrapolation, other essential elements such as Mg, offset the gradual declines from leaching.

Because Al and acidity generally increase during pedogenesis and most cations decrease, a base saturation index (BSI) is useful to measure the proportion of cations divided by cations plus Al. Normally, BSI values decline during primary succession. Zarin & Johnson (1995b) found an unusual increase in BSI values over time on landslides in Puerto Rico (Fig. 4.7).

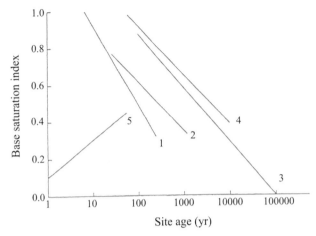

Fig. 4.7. Base saturation indices for five different types of primary sere. 1, Glacier Bay, Alaska; 2, Mount Shasta, California; 3, Baffin Island, Canada; 4, North Island, New Zealand; 5, Luquillo Experimental Forest, Puerto Rico. Reprinted from Zarin & Johnson (1995b), with permission from Elsevier Science.

They suggested that their anomalous data might reflect the early stages of pedogenesis compared with the longer perspective of the other studies and the oligotrophic or extremely low BSI values characteristic of the underlying parent material (saprolite) in Puerto Rico. Rapid increases of organic matter on the landslides may also contribute to the increase in BSI.

4.3.5 Nitrogen

Nitrogen (N) often limits net primary production (Vitousek & Howarth, 1991) and perhaps animal growth as well. Differential responses to N by different species suggest that species will change in response to N level changes over time (Tilman, 1988). Therefore, N availability is critical in defining both the rate and trajectory of most primary seres. For example, on the Tanana River floodplain in central Alaska (U.S.A.), an early successional stage dominated by *Alnus* supplied much of the N for the system (Van Cleve *et al.*, 1991). The transition from *Alnus* to the next stage dominated by *Populus* trees might be due to a decrease in N availability resulting from lower N inputs from *Populus* than from *Alnus* leaves (Clein & Schimel, 1995). The leaves of *Populus* contained tannins that inhibited microbial growth and litter decomposition (Schimel *et al.*, 1996) and perhaps reduced N fixation by *Alnus* with a subsequent increase in the soil C : N ratio (Schimel *et al.*, 1998).

Nitrogen availability to plants usually increases in primary succession (Vitousek & Farrington, 1997; Berendse et al., 1998) and is influenced by many abiotic factors (e.g. water, other nutrients, pH and soil organic matter; Houle, 1997) and some biotic factors (see below). Across a precipitation gradient in arid lands, Hooper & Johnson (1999) found that water and N limitation were tightly coupled. Nitrogen is most reliably available from moist, organically rich soils. Rates of gross and net mineralization (conversion of organic matter to an inorganic state) and nitrification (conversion of ammonia to nitrite and nitrate) determine the release of N from organic matter. Rates of mineralization and nitrification typically remain stable or increase in primary succession (Matthews, 1992; Riley & Vitousek, 1995; Binkley et al., 1997) in parallel with accumulations of organic matter. In forests, mineralization appears to be inversely proportional to the lignin : N ratio (Binkley & Giardina, 1998). However, increasing C : N ratios, increasingly recalcitrant organic matter and poorer soil drainage can make conditions less favorable for N turnover in surface soils and can affect species composition and successional trajectories (Van Cleve et al., 1993; Kitayama, 1996a; Aikio et al., 2000).

Biotic controls over N availability to plants include the activities of microorganisms and the plants themselves. Mineralization results from the breakdown of carbon that can be performed by many soil organisms (Sprent, 1987). Only a few bacteria are capable of nitrification and the fixation of atmospheric N into ammonia (see section 4.5.1 on N fixation). Plant assimilation of ammonium is less energetically expensive than nitrate, yet most plants rely on nitrate, perhaps because ammonium can be toxic (Sprent, 1987). In infertile sites (typical of primary succession), absorption of organic N in the form of amino acids may predominate (Kielland, 1994; F. Chapin, 1995), particularly when mycorrhizae are present. Atmospheric sources of N are discussed in section 4.5.1 on N fixation.

Retention of N is dependent on the many physical, chemical and biotic variables discussed in this chapter. Organic matter retains N on the surfaces of soil clay particles; plants withdraw available forms of N from the soil for growth and reproduction; animals recycle N from live and dead plant parts and from other animals. Nitrate, which is readily leached, can be converted to ammonium via nitrate reduction, thereby preventing N loss from the system.

Losses of N occur through leaching, denitrification or disturbances such as erosion. On several landslides in the Kumaun Himalaya, losses of N (and soil and other nutrients) declined with landslide age

Table 4.2. *Inputs and outputs of nutrients, loss of soil and increase of plant biomass on Kumaun Himalaya landslides*

Soil loss is the average of two monsoon seasons. There was a significant negative relationship between loss of N and plant biomass ($r = -0.946$, $p < 0.01$). C, organic carbon.

Landslide age (yr)	Soil loss (kg ha^{-1})	Runoff loss (kg ha^{-1} yr^{-1})					Plant biomass (t ha^{-1})
		N	P	K	Ca	C	
6	81	0.27	0.05	0.74	1.59	1.43	0.9
13	62	0.21	0.05	0.53	1.28	1.07	1.0
21	42	0.20	0.04	0.44	0.92	0.98	8.2
40	37	0.20	0.04	0.42	0.87	0.98	10.0
Undisturbed forest	26	0.15	0.03	0.31	0.65	0.71	11.0
Rainfall input	–	6.0	1.0	8.4	12.0	7.5	–

From Pandey & Singh, 1985.

(Table 4.2) (Pandey & Singh, 1985). N losses were greatest where plant biomass was minimal, suggesting that plants have a role in stopping erosion and retaining N and other nutrients. Excessive leaching of N into ground water can occur owing to high levels of precipitation or acidic soil conditions. Denitrification, the conversion of nitrite to gaseous N, is performed by several bacterial taxa but only under anaerobic conditions. Therefore, denitrification rates are closely coupled to soil water content (Riley & Vitousek, 1995). Disturbance that causes the removal of plants, animals or soil results in the loss of N. Many studies of N budgets show some imbalance between inputs and outputs, suggesting significant spatial and temporal heterogeneity in N fluxes (Chestnut et al., 1999). Spatial heterogeneity is explored further in section 4.6.

The accumulation of total N has been relatively well studied in primary succession. Consequently, one of the most robust generalizations we can make is that levels of total soil N generally increase during primary succession, in conjunction with soil C (DeKovel et al., 2000). Sources of N are discussed in section 4.5.1 on N fixation. Comparisons of N accumulation across sites both within and among types of primary sere are difficult owing to the variety of measurements used and the lack of data about inputs and outputs from each system. In a 1989 survey of 150 studies covering 141 primary seres, Walker (1993) compared 20 studies that measured total N accumulation (on an areal basis) over three or more dates in surface mineral soils during primary succession. Accumulation rates varied from 27 to 163 kg N ha^{-1} yr^{-1} (Luken & Fonda, 1983; Marrs *et al.*, 1983) and were greatest in the first 50–200 yr, followed by an asymptote between 2,000 and 5,000 kg N ha^{-1} (Walker, 1993). This leveling off of accumulation was reached faster on floodplains (within 100 yr) than on dunes (*c.* 1,000 yr; Olson, 1958; see also Lichter, 1998) but there was large variability in accumulation rates and asymptote levels within each type of sere (Walker, 1993). In Fig. 4.8 we have updated Walker (1993) and summarized 26 studies that measured total soil N. As in the earlier study (and as with soil pH), dunes had the lowest and volcanoes had the highest asymptotes. Temporal increases were significant for all types of sere except volcanoes. Total N reached an asymptote of about 3,000 kg ha^{-1} within 6,000 yr for a volcanic chronosequence in Hawaii (Riley & Vitousek, 1995; Kitayama, 1996a) and, although N concentrations continued to increase, total N remained at similar levels for 4.1 million years. Marrs *et al.* (1983) suggested that a minimum of 1,000–1,800 kg N ha^{-1} was needed to support woody vegetation, a process that took about 100 yr on several mined lands. Forests developed in a dune system with only 400 kg N ha^{-1} (Olson, 1958), but the process also took about 100 yr.

4.3.6 Phosphorus

Phosphorus (P) is provided mainly by weathering of phosphate-rich rocks, and is generally most available immediately after disturbances that uplift or otherwise expose such rocks. Availability of P for plant uptake depends on the type or fraction of P. Water-soluble and resin-extractable forms of P are most available. Next are hydrochloric acid-extractable P, followed by non-occluded and occluded forms of P that are less available and bound with Ca (under alkaline) or Fe and Al (under acidic) conditions. Phosphorus in organic forms is considered to be less available to plants than the other forms of P. Transformations from Ca-bound P found in volcanic ash to organically bound P in surface soils can occur within 100 yr in the tropical climate of Krakatau Volcano (Indonesia) (Schlesinger *et al.*, 1998). Walker & Syers (1976) presented a model suggesting that total and available forms of P generally follow the opposite pattern of N and decline during successional time (Fig. 4.9). The initial rise in less available forms of P is offset over long periods of time by a loss of more than 90% of the original P in the system (Walker & Syers, 1976). Evidence to support Walker & Syers' model comes from studies

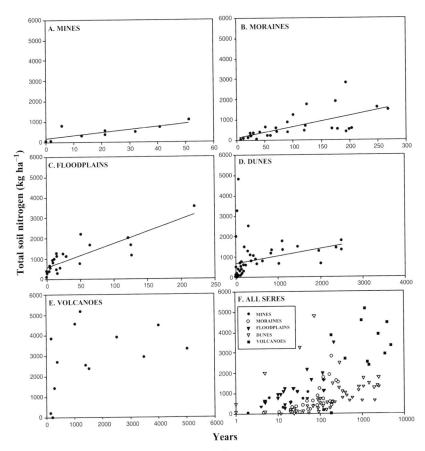

Fig. 4.8. Changes in total nitrogen (kg ha^{-1}) in surface mineral soils (10–30 cm deep) from 26 studies in five types of primary sere. Lines represent significant ($p < 0.05$), first-order linear regressions. (A) Mines ($n = 3$; $r^2 = 0.57$; $p = 0.019$); (B) glacial moraines ($n = 6$ studies; $r^2 = 0.41$; $p = 0.0002$); (C) floodplains ($n = 6$; $r^2 = 0.70$; $p < 0.0001$); (D) dunes ($n = 7$; $r^2 = 0.08$; $p < 0.0001$); (E) volcanoes ($n = 4$; $r^2 = 0.18$; $p = 0.145$); (F) all seres combined. The following references were used. Mines: Leisman (1957); Dancer *et al.* (1977); Palaniappan *et al.* (1979). Moraines: Crocker & Major (1955); Crocker & Dickson (1957, two moraines); Viereck (1966); Jacobsen & Birks (1980); Chapin *et al.* (1994). Floodplains: Viereck (1970); Van Cleve *et al.* (1971); Wilson (1970); Luken & Fonda (1983); Krasny *et al.* (1984); Walker (1989). Dunes: Olson (1958); Ayyad (1973); Robertson & Vitousek (1981); Olff *et al.* (1993); Gerlach *et al.* (1994); Berendse *et al.* (1998); Lichter (1998). Volcanoes: Tezuka (1961); Vitousek *et al.* (1983, two substrate types); Kitayama (1996a). Note the generally increasing temporal scales of the *x* axes and the logarithmic *x* axis in (F).

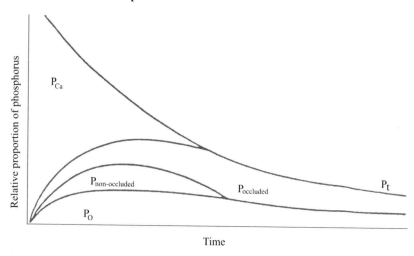

Fig. 4.9. Changes in forms and amounts of soil P with time. P_t, total P; P_{Ca}, HCl-extractable P; P_O, organic P; occluded and non-occluded P are other fractions. Reprinted from Walker & Syers (1976), with permission from Elsevier Science.

of long-term chronosequences over thousands of years on dunes and moraines in New Zealand (Syers & Walker, 1969; Walker & Syers, 1976) and Australia (Walker *et al.*, 1981) and over millions of years on lava flows in Hawaii (Crews *et al.*, 1995). Although Crews *et al.* (1995) found that the pools of P as fractions of total P resembled the model, their results also differed because total P was highest at their mid-successional (150,000 yr old) site and non-occluded P persisted throughout the entire 4.1 million yr chronosequence (Fig. 4.10). Resin-extractable P also peaked at 150,000 yr, reflecting the active cycling of the non-occluded P fraction. Inputs of P (and other elements) through aeolian dust from Asia can be substantial and sustain primary productivity long after the P in local rocks has been weathered or bound in forms unavailable to plants (Chadwick *et al.*, 1999). These inputs may be partially responsible for the unexpectedly high values of P later in the Hawaiian succession (Crews *et al.*, 1995). Matthews (1992) summarizes evidence that substantiates short-term increases in non-occluded P over several centuries on moraines. Decreases in non-occluded P at Glacier Bay (Borman & Sidle, 1990) were presumably from uptake of P into *Picea* foliage. However, P levels in *Picea* foliage subsequently drop in later stages of succession at Glacier Bay (after much more than 200 yr; L. R. Walker, unpublished data).

Bird droppings are another source of P in early primary succession. For example, birds attracted to such open areas as volcanoes in New Zealand

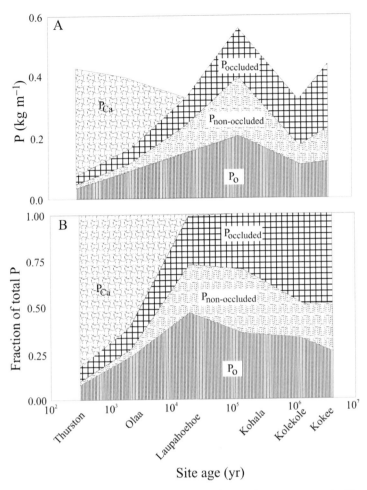

Fig. 4.10. Changes in P fractions along a volcanic chronosequence in Hawaii. From Crews *et al.* (1995). The fractions follow those in Fig. 4.9.

(Clarkson & Clarkson, 1995), rocky offshore islands on many coastlands (Skaggs, 1995), desert soils in Mexico (Anderson & Polis, 1999) or ice-free outwash gravels in Antarctica (Beyer *et al.*, 2000) can add large quantities of nutrients, including N and P. Most vegetation on Surtsey is associated with nutrient-rich sea bird colonies (Fridriksson, 1987). These additions can alter succession by promoting growth of non-indigenous grasses (Clarkson & Clarkson, 1995). The presence of mycorrhizae can increase P uptake for plants growing in low P environments (Allen, 1991).

4.4 Soil biota

The plants and animals that colonize primary seres are an integral part of the process of soil formation. Biotic contributions to primary succession generally begin with wind-blown invertebrates and seeds (the neogeoaeolian inputs; Howarth, 1987; Crawford et al., 1995; Hodkinson et al., 2002; see section 5.3 on dispersal). Smaller openings are often dominated by vegetative expansion from remnant vegetation around the edges of the disturbance. Plants provide the bulk of organic material for soil formation but also impact soil formation in many other ways. Through root growth, root exudates, nutrient cycling and water uptake, plants stabilize substrates, modify surface and subsurface textures, increase mineral weathering of parent material, decrease compaction, contribute to soil aeration and oxygen levels, modify water content, and facilitate the growth of soil organisms (Wood, 1995; Binkley & Giardina, 1998). Mesic climates and high plant productivity result in maximum litter fall and active microbial populations that are associated with high rates of decomposition. Under such circumstances, soil formation can be very rapid (but see section 4.2.1 on climate for contrasting examples). Where biotic input is low, soils will develop slowly as on arid lavas (Wright & Bunting, 1994), in polar deserts (Komárková & Wielgolaski, 1999) or on dunes (Le Houérou, 1986).

The identity and function of the rich diversity of soil biota are poorly described. Estimates of the number of species still unidentified range from one to three orders of magnitude above currently known species for viruses, fungi, bacteria, protozoa and arthropods. Soil biota include (in approximate descending order of biomass) plants, fungi, bacteria, arthropods, annelids, protozoa, algae, nematodes, vertebrates and viruses (Pimentel & Harvey, 1999). In primary succession, soil biota are surprisingly abundant, even in early stages of soil development. Initial sterility is probably limited to new lava because soil biota are widely dispersed, often as wind-blown spores. Soil bacteria have been found under glaciers in the Alps (Matthews, 1992) and in newly exposed glacial moraines in Antarctica (Wynn-Williams, 1993). We now review the various types of soil organism and describe their roles in primary succession.

4.4.1 Plants

Under ideal conditions, plant roots can occupy up to 5% of the surface soil volume. They have many impacts on the physical and biological structure of soils. Compaction of the soil from root growth can decrease water-holding capacity in the rhizosphere (the zone of soil influenced directly by roots), but this reduction is offset by an air space between the root and the soil that facilitates water movement (Killham, 1994). As roots die (turnover can be rapid for some fast-growing plants or plants in harsh environments), the remaining root channels increase the porosity of the soil. This is particularly important in arid regions where surface crusts or subsurface hardpans rich in clays (argillic horizon) or calcium (caliche) inhibit infiltration (Smith et al., 1997) and water availability for plants (Hamerlynck et al., 2000). Primary surfaces that are more porous (e.g. dunes) often support more vegetation than less porous soils because of better root penetration and better infiltration of water, provided the sands are not too deep. Plant root uptake of water further alters soil hydrology, generally drying soils in the rhizosphere.

Plant roots also alter the chemistry of soils. Root respiration can account for 30% of total soil respiration (Killham, 1994; Coleman & Crossley, 1996), as roots take up oxygen. Roots also release exudates full of carbon that increase mineral weathering, alter soil pH, and have a major influence on the microbial populations in the rhizosphere. Some microbes benefit the plant through N fixation, protection from diseases or provision of growth-promoting chemicals. Other microbes bring diseases that destroy root tissue. The net effect of root growth and decay is development of a crumb structure composed of soil aggregates where organic matter is broken up and partially encased in mucilaginous polysaccharides that protect the organic matter from decay by microbes (Killham, 1994). This protection delays the turnover of soil organic matter, thus leading to accumulations that favor successional change in species adapted to more developed soils.

4.4.2 Soil microbes

The composition of the soil microbial community (primarily composed of fungi and bacteria but also including algae) is diverse and poorly known. Direct measurements of changes in soil microbial diversity during primary succession are hampered by the problem that less than 5% of soil bacteria can be cultured (Bakken, 1985). Therefore, measurements have focused on function rather than taxonomy. Functional diversity can be measured by the relative activity of soil microbes on multiple substrates (the BIOLOG system; Garland & Mills, 1991; Degens & Harris, 1997). Microbial respiration is measured from CO_2 production. Microbial biomass can be estimated by direct counts, using various stains, by extraction with chloroform (Brookes *et al.*, 1985) or indirectly through respiration. Recent advances have also led to measurements of the PLFA (phospholipid fatty acid) composition of soil microbes that can distinguish between bacterial and fungal components and between Gram-positive and Gram-negative bacteria (Frostegård & Bååth, 1996; Bardgett *et al.*, 1999). Ritz *et al.* (1994) and Øvreås (2000) provide useful overviews of microbial measurements.

Soil microbes and the soil animals that interact with them largely control the decomposition of organic matter and the recycling of nutrients. Their ubiquity and importance to nearly all soil processes cannot be overstated. Temperature, precipitation, soil organic C, total soil N and soil pH all influence microbial biomass (Wardle, 1992). Soil microbes occur in virtually every soil pore space that is large enough, but occupy less than 0.5% of total pore space (Killham, 1994). Yet the density of soil microbes is patchy, with greatest concentrations in areas of greatest fertility such as the zone of soil near the roots (rhizosphere) or inside the guts of animals (Virginia *et al.*, 1992).

Fungi usually make up the majority of the total microbial biomass, although bacteria may dominate early stages of primary succession. On an Alaskan floodplain, Binkley *et al.* (1997) found that total fungal biomass ($400–1000$ $\mu g\ g^{-1}$ soil) increased across a 220 yr chronosequence and far outweighed bacterial biomass ($2–9$ $\mu g\ g^{-1}$ soil), which declined during succession. The ability of fungi to tolerate higher acidity than soil bacteria results in dominance by fungi in acidic soils. Fungi are the only decomposers of recalcitrant lignin and complex organic acids, so recycling of wood is dependent on fungi. Another critical function of fungi is their interaction with plant roots, particularly in the form of mycorrhizae. Decayed fungal sporocarps facilitated colonization and growth of algae and mosses on Mount St. Helens, presumably by concentrating nutrients and organic matter on recent tephra deposits (Carpenter *et al.*, 1987).

Temporal changes in soil microbes and microbial functions occur in response to changes in the biotic and abiotic properties at a site (Insam & Haselwandter, 1989; M. Allen *et al.*, 1999). Heterotrophic microbes are initially dependent on the carbon supplied by plants and respond to changes in the quantity and quality of the carbon and to changes in species composition of the plant communities. Certain changes in microbial decomposition may also reflect increasing competition among microbes and increased niche heterogeneity in primary succession (Jackson *et al.*, 2001). Herbivores can also alter microbial communities. Fungal succession in experimental microcosms was promoted by the presence of collembolan herbivores that preferentially ate the early colonizers on leaf litter of *Picea* and *Abies* trees, allowing the later colonizers to dominate (Klironomos *et al.*, 1992).

On glacial moraines, microbial communities shift during primary succession from dominance by bacteria to dominance by fungi, perhaps in response to microsite changes and dominance by shrubs such as *Alnus* (Ohtonen *et al.*, 1999). Fungi accumulate C, sequester N and can thereby slow development of primary seres. Bacterial communities were associated with mineralization and transformation of inorganic nutrients in early succession at Glacier Bay, Alaska, when levels of C and N were accumulating rapidly. Mycorrhizal fungi became important as a source of N in stands 100–200 yr old where N cycling was slow and direct plant uptake of nutrients was less than in earlier stages (Hobbie *et al.*, 1999). In later stages, fungi were dominant where recycling of organic forms of nutrients was more important (Bardgett, 2000). In very late stages of succession at Glacier Bay (after more than 2000 yr) microbial biomass declined (Bardgett, 2000). Similar increases in microbial biomass followed by late successional decline were found in primary succession on the Franz Josef moraine in New Zealand (Wardle & Ghani, 1995).

4.4.3 Mycorrhizae

Mycorrhizae, or associations of roots and soil fungi, are found in most plant species and some species are obligately mycorrhizal (Allen, 1991). They facilitate uptake of many nutrients and water primarily by increasing contact between the root and a greater amount of soil volume. Alteration of the chemistry of the rhizosphere by mycorrhizae is also common. Mycorrhizae can modify acidic, basic or toxic soils and improve uptake of many nutrients, including P, N, K, Zn, Cu, Mg and S (M. Allen *et al.*, 1999). Mycorrhizae can mobilize nutrients directly from minerals through excretion of organic acids (Landeweert *et al.*, 2001). All of these processes are highly interactive. For example, on soils from a glacial moraine in Washington (U.S.A.), the P uptake abilities of mycorrhizae were dependent on adequate levels of available N and organic matter (Jumpponen *et al.*, 1998a).

Early stages of succession are often colonized by non-mycorrhizal or facultatively mycorrhizal species (Janos, 1980; Miller, 1987; Allen & Allen, 1990). For instance, winter annual grasses and herbs on dunes in The Netherlands were not infected with mycorrhizae, whereas perennial plants were (Ernst *et al.*, 1984). However, in other dune systems, mycorrhizae, as part of aggregates of surface cryptogamic crusts, appear to be critical in reducing wind erosion, increasing moisture, and promoting plant colonization (Forster & Nicolson, 1981; Corkidi & Rincón, 1997). Mycorrhizae are also important in the early colonization of plants on glacial moraines (Helm *et al.*, 1996; Jumpponen *et al.*, 1998a), many volcanoes (Gemma & Koske, 1990; Allen *et al.*, 1992), disturbed alpine ecosystems (Allen *et al.*, 1987), floodplains (Krasny *et al.*, 1984), and mined lands (Schramm, 1966; Allen & Allen, 1980). Mycorrhizae can survive certain disturbances *in situ*, particularly if the entire soil profile is not removed (Boerner *et al.*, 1996).

Widespread dispersal of mycorrhizal spores and fragments occurs by water, wind, erosion and animals. Only in large, severely disturbed environments such as Mount St. Helens (Allen *et al.*, 1992; Titus *et al.*, 1998) is mycorrhizal colonization significantly delayed. Spores of vesicular–arbuscular mycorrhizae (VAM) have been found to precede plants to isolated dunes or to colonize dunes in association with dispersing plant fragments, suggesting that survival in sea water is common (Koske & Gemma, 1990; Koske *et al.*, 1996). Survival of mycorrhizal fragments in water may be higher than survival of spores in air dispersal (M. Allen *et al.*, 1999) because air-dispersal brings the added problems of drought and high radiation intensities. Animals can bring spores to remote areas but are more likely to be involved in the introduction of fungal populations from undisturbed edges or from soils buried by the disturbance (e.g. tephra on Mount St. Helens; Allen *et al.*, 1992). If soils containing fungal hyphae and plant hosts are not transported to the new colonized site, the symbiosis must establish *de novo*. The renewal of the symbiosis is dependent on contact between both partners and appropriate N : P ratios. Such physical contact can be difficult in unfavorable habitats, such as early primary succession, where host roots are widely spaced (M. Allen *et al.*, 1999).

Types of mycorrhizae change as plant communities develop in primary succession. Helm *et al.* (1996) found 24 types of ectomycorrhizae (EM) associated with *Populus* within the first 100 yr of succession on a glacial moraine in Alaska. High levels of P and N and low pH levels (Killham, 1994), conditions that typically occur in later stages of succession, often suppress VAM. In turn, later successional species are probably more dependent on mycorrhizae (Cázares, 1992; Hobbie *et al.*, 1999). Both EM and VAM are suppressed by too much or too little water.

4.4.4 Animals

Soil invertebrates are typically divided into three size classes: microfauna (< 200 μm in body length), mesofauna (200 μm–2 mm) and macrofauna (2–20 mm). Microfauna include protozoa and nematodes; mesofauna include mites and springtails; macrofauna include earthworms, ants and termites. Arthropods (especially beetles, ants and termites) and annelids (especially earthworms) are critical to the development of soil structure. In arid lands, ants may turn over up to eight times more soil than earthworms (Majer, 1989a). The tunneling activities of ants, earthworms and termites increase soil aeration, drainage, mixing and root penetration (Wali & Kannowski, 1975). Waste products from these animals (notably earthworm casts and dead animals) enrich the soil with nutrients. In turn, soil invertebrates are generally more abundant in nutrient-rich soils. Chan et al. (1997) found three- to five-fold higher densities of soil invertebrates in crushed granite substrates used as caps over landfills than in granite without landfills. The landfills apparently leaked ammonium-rich leachates that can be favorable to growth under some conditions. Animals also initiate the breakdown of organic matter such as leaf litter, increasing the surface area for fungal and bacterial attack.

Nematodes play a pivotal role in soil development, particularly in the decomposition of plant roots. They are extremely diverse and numerous. Nematode species richness appears to be highest in mid-latitudes but can be nearly as high on developing dunes as on more fertile soils (Boag & Yeates, 1998). Increases occurred during the first decade of primary succession on dunes in Poland, but then remained stable (Wasilewska, 1970).

Majer (1989a) estimated that soil invertebrates could resemble predisturbance communities within 5–10 yr on mined lands, if plant cover were re-established immediately. However, earthworm colonization was typically delayed, probably by the lack of adequate organic matter. Springtails, carabid beetles or enchytraeid worms preceded earthworms on various coal mine tailings, but soil changes were rapid once earthworms arrived.

Even though vertebrates have the least biomass of any soil biota except viruses, they still play an important role in soil mixing and nutrient cycling. Moles and rodents tunnel extensively. Large herbivores affect plant productivity and fertilize the soil with their feces and carcasses. Prolonged over-grazing reduces litter production, and this in turn has a negative impact on soil formation by reducing the activity of soil and litter invertebrates that recycle litter (Abbott, 1989). Large herbivores can

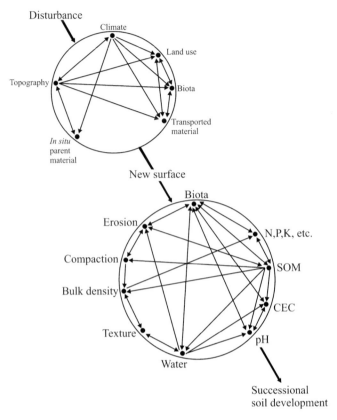

Fig. 4.11. Interaction of factors that affect soil development. Arrows outside the circles indicate time; arrows inside the circles indicate influences.

also compact soils, especially shallow, poorly developed ones. We discuss plant–animal interactions, including herbivory, in Chapter 6.

4.5 Soil processes

In this section we discuss two processes critical to soil development in primary succession, N fixation and the accumulation of soil organic matter. Soil formation is the net result of many interrelated variables (Fig. 4.11), but we select these two to illustrate the interactions among variables that further soil development.

4.5.1 Nitrogen fixation

Vascular N-fixers (vascular plants with symbiotic N-fixing bacteria in root nodules) probably contribute most of the N to developing primary seres (Marrs *et al.*, 1983; Blundon & Dale, 1990) provided the soils are not too acidic (Binggeli *et al.*, 1992). Root nodules of legumes (Fabaceae) are associated with the N-fixing bacterium *Rhizobium* and 23 woody genera from eight other plant families are associated with the N-fixing bacterium *Frankia* (Sprent & Sprent, 1990). However, other important contributions can come from lichens (Vitousek, 1994), free-living fixers (Sprent & Sprent, 1990; Vitousek & Hobbie, 2000), bird droppings (Beyer *et al.*, 2000), wind-blown arthropods (Edwards, 1988) and abiotic sources such as lightning (Sprent & Sprent, 1990), precipitation (Table 4.2; Pandey & Singh, 1985; Vitousek & Walker, 1989), salt spray (Morris *et al.*, 1974; Barbour *et al.*, 1985), or volcanic gases (Eggler, 1963). The cyanobacterium *Nostoc*, associated with colonizing mosses, was an important source of N on the new volcanic island of Surtsey (Henriksson *et al.*, 1987) and in the Canadian Arctic (Dickson, 2000). Nitrogen-fixing bacteria are often associated with lichens and mosses. However, the generalization that lichens and mosses are the essential first stage of soil development in primary succession has been correctly disputed for at least five decades (Cooper & Rudolph, 1953; see section 6.4.1 on mutualisms).

Atmospheric N that is fixed at the surface of active lava flows in Hawaii (Huebert *et al.*, 1999) can be later deposited in cloud water to provide more N to volcanic ecosystems than all other sources combined (Heath & Huebert, 1999). Free-living N-fixers can be important sources of N in both polar (Dickson, 2000) and temperate (Smith *et al.*, 1997) deserts, mesic tundra (Chapin *et al.*, 1991) or glacial moraines (Worley, 1973), particularly when associated with cryptogamic crusts. These crusts can provide all the N used in plant growth in some polar deserts (Bliss & Gold, 1999; Dickson, 2000), although wind-blown debris is also important. In temperate deserts, there is evidence of increased N levels in vascular plants associated with crusts (Belnap & Harper, 1995). Nevertheless, vascular N-fixers provide the largest inputs of N when they are present (Vitousek & Walker, 1989; Blundon & Dale, 1990) and are therefore widely used in reclamation and agriculture (Sprent & Silvester, 1973; Sprent & Sprent, 1990).

Dispersal of *Rhizobium* or *Frankia* to newly disturbed sites does not appear to constrain early successional development along rivers (Huss-Danell *et al.*, 1997) and in new volcanic ash (Vitousek & Walker, 1989). However, *Frankia* does not appear to be widespread in circumpolar Arctic soils (Huss-Danell *et al.*, 1999) and woody legumes are underrepresented on island floras (Parker, 2001) and relatively slow to invade from their tropical origins (Crews, 1999). *Frankia* (Burleigh & Dawson, 1994) and presumably *Rhizobium* are dispersed by birds, earthworms, water and possibly wind and may have to reach a certain density threshold in order to promote expansion of host plant populations (Parker, 2001). It appears that N limitation in temperate and boreal regions may have both biogeographic and ecological explanations.

The presence of legumes or actinorhizal plants (ones potentially infected with *Frankia*) does not guarantee an available source of N. *Frankia* can survive without a vascular plant host and appears to be more capable of forming nodules in older, more fertile soils (Burleigh & Dawson, 1994). About 20% of all species of legume have been tested for nodulation. Of those species, many do not fix N (Sprent & Sprent, 1990), fix in low amounts that are insufficient even to meet their own needs for N (Holter, 1984), recycle their N efficiently, but without contributing N to the soil or to other plants, or compete with coexisting species for other limiting nutrients or water, thus offsetting any potential benefit of N fixation (Walker & Vitousek, 1991). In most cases, then, the largest additions of N from an N-fixer may come only when the plant dies (examples are *Dryas* shrubs on a glacial moraine (Kohls *et al.*, 1994) or *Lupinus* forbs on a volcano (del Moral, 1993b)).

Despite the belief that most N contributions in primary succession are from vascular N-fixers, Walker (1993) found that vascular N-fixers were present in 77% of 141 primary seres but were a dominant part of the vegetation in only 28%. Dominance (having a rank of third or higher in cover or importance value) was most common for shrub-forming N-fixers such as *Myrica* trees and *Alnus* or *Dryas* shrubs, all of which are infected by the bacterium *Frankia*. Both woody and herbaceous legumes (infected with *Rhizobium*) were less dominant. Dominance might be related to mesic environments such as boreal glacial moraines that permit the establishment of large-leaved shrubs such as *Alnus* (Matthews, 1992; Walker, 1995, 1999b). *Myrica* dominated mesic microsites in coastal dunes in Virginia (U.S.A.) (Young, 1992). It is also possible that N-fixers are most successful in low-N environments (Holter, 1979; Vitousek & Walker, 1987), but Walker (1993) found no relationship between initial soil N status and dominance by vascular N-fixers. Grubb (1986) suggested that very low N levels could inhibit the establishment of N-fixers that need N from their environment to begin the energetically expensive process of N fixation. Similarly, a lack of P, Mo or Fe might limit N fixation (Silvester, 1989; Vitousek & Howarth, 1991) or colonization by N-fixers (Holter, 1979). Fertilization with P demonstrated that N fixation was P-limited on young volcanic substrates in Hawaii (Vitousek, 1999) and on an Alaska (U.S.A.) floodplain (Uliassi & Ruess, 2002). Salt intolerance limited both

N fixation (Sande & Young, 1992) and seed germination (Young *et al.*, 1992) of *Myrica* on the Virginia (U.S.A.) coastal dunes. The various restrictions on the establishment of vascular N-fixers limit their effects in primary succession.

Vitousek & Field (1999) addressed the conundrum that N limits growth in most ecosystems despite the common occurrence of N-fixers and the apparent advantage that N-fixers should have in N-limited systems. In fact, the cost of N acquisition and growth for N-fixers may be higher than for non-fixers under many circumstances (Gutschick, 1981). Shading, P availability or grazing may differentially limit growth of N-fixers relative to non-fixers (Vitousek & Field, 1999). Rates of N fixation can be reduced in older plants or in soils from later stages of succession, possibly by root herbivores such as nematodes (Oremus & Otten, 1981). Losses of dissolved organic N from leaching and of trace gases produced by nitrification further limit N availability and growth. When N-fixers are successful in primary succession, they clearly can have a significant impact on ecosystem productivity and succession (Walker & Chapin, 1987; Walker & Vitousek, 1991). When the N-fixer is alien to the system, dramatic shifts in soil nutrient levels and increased competition for light and water can favor other alien, N-demanding species and alter successional trajectories (Vitousek & Walker, 1989; Peloquin & Hiebert, 1999).

Another interesting pattern suggested by Walker's 1993 survey was a variation in abundance of N-fixers by habitat. Vascular N-fixers were most abundant on glacial moraines and mudflows and least abundant on volcanoes and rock outcrops (Table 4.3). Nitrogen-fixers were of intermediate abundance on mined lands, landslides, floodplains and dunes. Explanations of these patterns remain speculative. The extreme conditions of volcanoes and rock outcrops suggest that surface stability or development of organic matter and water-holding capacity are more important than N additions or that the very low levels of available N hinder establishment of N-fixers. Rates of N fixation are highest in warm and mesic habitats such as coral reefs, tropical forests and desert streams and lowest in cold habitats such as tundra and Antarctic streams (Grimm & Petrone, 1997).

Different life forms of N-fixers are likely to colonize different substrates. Grubb (1987) suggested that woody plants colonize rock cracks and herbs colonize gravels, whereas grasses colonize silts (see section 4.3.1 on soil texture). This generalization does not appear to hold for N-fixers in primary succession, as many woody N-fixers colonize silty floodplains (Walker, 1993, 1999b), and an herbaceous legume preferred fine silt to gravel on a Japanese floodplain (Nakatsubo, 1995). Walker (1993)

Table 4.3. *Number of studies of primary succession (from a total of 141) where vascular nitrogen-fixers were absent, present, abundant or dominant*

Percentages in the abundant and dominant categories were calculated within successional types. Types of succession: MO, glacial moraine; MU, mudflow; MI, mined lands; LA, landslide: FL, floodplain; DU, dune; VO, volcano; RO, rock outcrop. Abundance categories: absent, not mentioned in species list; present, mentioned and < 5% cover; abundant, in several stages and > 5% cover; dominant, one of three species with the highest cover or importance value. When several nitrogen-fixers were present, only the highest abundance category was used.

	Type of succession (number of studies)								
Abundance category	MO	MU	MI	LA	FL	DU	VO	RO	Total
Absent	0	0	2	1	6	5	14	5	33
Present	1	1	3	3	9	9	7	6	39
Abundant	5	2	3	6	6	2	5	1	30
Dominant	7	3	6	1	8	9	4	1	39
Total	13	6	14	11	29	25	30	13	141
Percent abundant and dominant	92	83	64	63	48	44	30	15	49

From Walker (1993).

Table 4.4. *Number of studies of primary succession where nitrogen-fixing species from three life forms were found in eight types of sere*

All abundance categories (from Table 4.3) were combined. When several nitrogen-fixers were present in one sere, each was included. Abbreviations for types of succession are found in Table 4.3.

	Type of succession (number of seres)								
Life form	MO	MU	MI	LA	FL	DU	VO	RO	Total
Herbaceous legumes	8	5	9	8	15	18	9	4	76
Woody legumes	0	2	10	1	6	3	8	3	34
Other woody species	19	7	1	6	21	7	8	2	75

From Walker (1993).

found as many species of herbaceous legume on floodplains (silt, gravel) as on dunes (gravel); woody legumes more or less evenly dispersed among mined lands (gravel, rock cracks), volcanoes (rock cracks) and floodplains; and most species of non-leguminous N-fixer on floodplains and glacial moraines (silt, gravel; Table 4.4). Without more detailed evidence of local site factors, it appears that surface texture is not a robust predictor of the life form of N-fixers in primary succession.

4.5.2 Organic matter

Carbon fixation by plants and decomposition of the fixed C by microbes drive the accumulation of soil organic matter (SOM), without which soil will not develop. The principal input is, of course, plant litter (which includes leaves, stems and roots). Inputs of plant litter are positively correlated with productivity but are also impacted by rates of leaf and root turnover. Root exudates are also important sources of C. Litter accumulates whenever inputs exceed outputs. An extreme example that occurs in waterlogged and cold conditions results in deep layers of peat. Animal parts (hair, feces, horn, carcasses) are usually not the primary source of litter. During outbreaks of insects, animal contributions (mainly from insect feces) can provide 18–36% of all litterfall and 60–70% of N and P entering the litter annually (Hollinger, 1986).

The amount and quality of soil organic matter (SOM) affect plant growth primarily through their impacts on water and nutrient availability. Plant litter, in turn, directs soil development. Differences in litter quality can alter soil formation even on the same initial substrate. Dunes in The Netherlands with natural *Populus* forests had very different soil profiles (higher decomposition, no podzol formation) than dunes with planted *Pinus* forests after only 80 yr (Wardenaar & Sevink, 1992). In addition, spatial variation in soil profiles was greater in the *Populus* than the *Pinus* forests due to variation in aspect, slope, wind-driven deposition of leaf litter and the distribution of understory plants. In a nearby chronosequence of *Pinus* stands, development of SOM paralleled primary productivity, with initial increases in both, followed by a decline in productivity and little change in soil profiles after about 100 yr (Emmer & Sevink, 1994). Soil pH also declined over time. In dunes where blowouts had occurred, the loss of initial SOM led to slower accumulation rates than in the original forest stands. Emmer and Sevink concluded that decay dynamics, rhizosphere processes and atmospheric deposition were more important than chemistry in soil formation. In a third dune system in The Netherlands, litter and late successional vegetation were removed at different times in order to restore rare early successional species (Berendse *et al.*, 1998). In the ensuing chronosequence, low decomposition rates relative to plant turnover and few losses from leaching or denitrification led to rapid accumulation of SOM. On uplifted marine

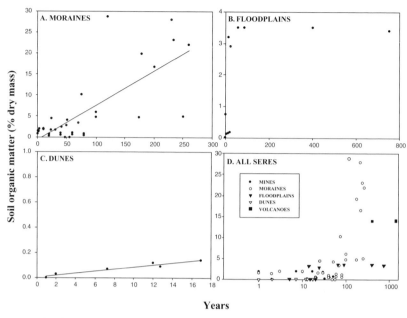

Fig. 4.12. Changes in soil organic matter (% dry mass) in surface mineral soils (10–30 cm deep) from 14 studies in five types of primary sere. (Data presented as % organic carbon were multiplied by 1.7 to convert to % SOM.) Lines represent significant ($p < 0.05$), first-order linear regressions. (A) Glacial moraines ($n = 7$ studies; $r^2 = 0.57$; $p < 0.0001$); (B) floodplains ($n = 3$; $r^2 = 0.25$; $p = 0.114$); (C) dunes ($n = 1$; $r^2 = 0.90$; $p = 0.004$); (D) all seres combined (including two studies of mine tailings and one of a volcano). The following references were used. Moraines: Persson (1964); Viereck (1966); Jacobsen & Birks (1980); Mellor (1986); Messer (1988); Ohtonen et al. (1999, two habitats). Floodplains: Van Cleve & Viereck (1972); Fonda (1974); Luken & Fonda (1983). Dunes: Wallen (1980). Tailings: Leisman (1957); Wali (1999). Volcanoes: Kitayama (1996a). Note the variable temporal scales on the x axes and the logarithmic x axis in (D).

terraces in Finland, SOM levels were also linked to changes in vegetation and environmental stress (Aikio et al., 2000).

In a survey of 14 primary seres, we found that SOM values in surface mineral soils were initially much less than 5% and did not increase substantially until about 60 yr following the disturbance (Fig. 4.12). Significant increases over time were detected for glacial moraines and in the one dune study, but not for other types of sere. On one mudflow (Sollins et al., 1983) and one floodplain (Fonda, 1974), SOM values remained below 10% even after 500 yr. On one volcano (Kitayama, 1996a), values did not change between 400 and 1400 yr, but did increase over the rest of the Hawaiian

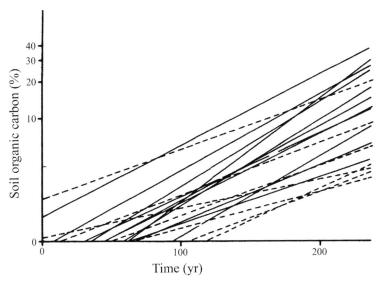

Fig. 4.13. Soil organic matter accumulation patterns from 18 Norwegian glaciers. Statistically insignificant chronofunctions ($p \leq 20\%$) are indicated with broken lines. From Messer (1988).

chronosequence (estimated to extend to 4.1 million yr). Highest values occurred on two moraines in northwestern North America (Jacobsen & Birks, 1980; Ohtonen *et al.*, 1999) and one in Norway (Mellor, 1986). The North American moraines are in humid, coastal climates with high primary productivity and potentially favorable conditions for decomposition. However, the Norwegian moraine is in a drier, less productive alpine setting, supporting Messer's (1988) suggestion that SOM accumulation on moraines is not related to climatic parameters. Messer (1988) found significant increases in SOM in 11 of 18 Norwegian moraines and large variation in SOM values (Fig. 4.13).

Decomposition rates determine the recycling of nutrients and thereby influence SOM accumulation, soil development and the rate of succession. Numerous processes influence litter decomposition rates. These include moisture, temperature, C : N ratio, lignin content, litter size and degree of burial. Interactions among the decomposers (see section 4.4 on soil biota) are also critical. For example, microarthropods influence decomposition by the type of decomposer fungi they consume (Klironomos *et al.*, 1992). Rapid decomposition occurs in aerated, moist and fertile soils. Nutrient release during decomposition stimulates additional decomposition, providing a degree of positive feedback to decomposing microbes (Vitousek & Howarth, 1991). However, increased decomposition may result in increased leaching of nutrients and slower soil development as for dune slacks in The Netherlands (Sival & Grootjans, 1996). Conditions that are poor for primary production are also generally poor for decomposition, leading to the accumulation of undecomposed organic matter on the surface of the ground and slower rates of soil development. For example, N availability limited root decay during early coastal dune succession in Virginia (U.S.A.) (Conn & Day, 1996). The balance between primary productivity and decomposition regulates SOM accumulation (Messer, 1988).

Litter can be classified by its mass and degree of incorporation into the soil. Partly decomposed portions of plant and animal litter are sometimes termed the light fraction (< 1.65 g cm^{-3}). This fraction increased for the first 600 yr of a chronosequence on a mudflow in Oregon, U.S.A., but then stabilized (Sollins et al., 1984). In the same study, the heavy fraction (organic matter that is adsorbed onto mineral surfaces) continued to increase from 600 to 1200 yr, with a projected continued increase, only offset slightly by periodic fires.

Floodplains provide a variety of habitats that demonstrate the influences of soil moisture and burial on decomposition rates and nutrient cycles. Litter in active channels can be transported or buried by flooding (Malanson, 1993). Litter can also be retained in slow-moving meanders or swamplands. Nutrient release (Mg, K > Ca, N) from leaf litter on a South Carolina (U.S.A.) floodplain was higher than in adjacent uplands, in part owing to higher water availability on the floodplain, but also to more readily decomposable leaves (Shure et al., 1986).

Litter quality is obviously important to decomposers and is frequently measured by C : N ratios and lignin content. High levels of litter N can inhibit microbes that decompose lignin (Carreiro et al., 2000). Variations in litter quality were more important than variations in microsite on Puerto Rican landslides (R. W. Myster & D. A. Schaefer, pers. comm.). But local substrate qualities can be important. Substrate levels of N limited decomposition of roots (and experimental cotton strips) in early succession on dunes in Virginia (Conn & Day, 1996).

Plants that produce litter that is slow to decompose help retain nutrients from being lost during primary succession. One good example is *Dicranopteris*, a vine-like fern that colonizes many severely disturbed habitats throughout the tropics including lava flows in Hawaii (Russell & Vitousek, 1997) and landslides in Puerto Rico (Guariguata, 1990; L. Walker, 1994) and St. Lucia (Wardlaw, 1931). *Dicranopteris* litter not only decomposes slowly, but also immobilizes N, P and other nutrients (Maheswaran & Gunatilleke, 1988). The litter also hosts N-fixing bacteria, although Russell & Vitousek (1997) found that N fixation was concentrated in certain hot spots and that total N contributions by the fixers did not exceed the N from rainfall in Hawaii. Finally, the litter often remains suspended above the ground surface, preventing contact with decomposer organisms and further delaying decomposition. These conditions delay both nutrient loss and soil formation. Low soil nutrients and the shade produced by the ferns delays the invasion of trees to the landslides (L. Walker, 1994).

4.6 Spatial patterns

Spatial heterogeneity is apparent at all scales for both abiotic and biotic soil properties during primary succession. However, changes in the nature of that heterogeneity may occur as soils develop. The number (richness) and relative abundance (evenness) of soil biota give one measure of heterogeneity (Armesto et al., 1991). Geostatistics measure the variation in soil characteristics from point to point within a plot (Robertson, 1987; Jackson & Caldwell, 1993). Global- to microsite-scale patchiness is found for the environmental controls over soil development (climate, parent material, topography, erosion) and local to microsite variation is important for the other physical variables (texture, compaction, water content, CEC, pH and nutrients). Physical variability originates with uneven weathering of parent material (that is itself heterogeneous) and the irregular surfaces that result. Chemical variability comes in part from differential leaching of nutrients on the micelle surfaces. Biotic variables add their irregular responses to heterogeneous physical and chemical environments so that complex surfaces and microsites develop. Slight differences are amplified by the interaction of multiple processes, not unlike a meander in a river beginning from differential flow of water around a rock. Knowledge of how each critical process interacts with others at various spatial scales is essential to understand successional change. For example, to restore degraded dune slacks in The Netherlands (Grootjans et al., 1998), one must understand nutrient availability to roots at the microsite scale (5 cm), patterns of SOM accumulation within and between slacks (5 m) and local (50 m) and regional (500 m) hydrology (Fig. 4.14).

Hierarchies of scale permit us to combine spatial scales both smaller and larger than the scale of interest, essentially assuming a relative degree of homogeneity at those other scales (O'Neill et al., 1986). Yet within a given spatial frame of reference or landscape, there is still a wide variety of

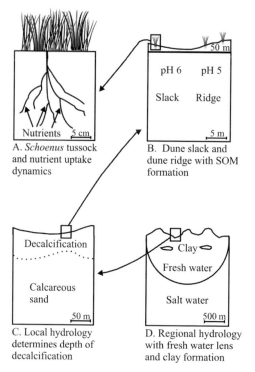

A. *Schoenus* tussock
and nutrient uptake
dynamics

B. Dune slack and
dune ridge with SOM
formation

C. Local hydrology
determines depth of
decalcification

D. Regional hydrology
with fresh water lens
and clay formation

Fig. 4.14. The importance of spatial scales for studies of SOM accumulation on dunes. Reprinted from Grootjans *et al.* (1998), with permission from Elsevier Science.

characteristics or patchiness of soil variables that must be addressed. On a small scale, aggregation of soil particles, tunneling of animals, variable inputs and leaching of nutrients all give rise to a lack of uniformity in soil characteristics. The spatial distribution of plants alters soil development and is determined in part by the spatial patterns of seed dispersal, differences in plant life histories, variation in light availability and disruption of soil by animals (Armesto *et al.*, 1991). In early primary succession, heterogeneity is generally associated with patches of higher fertility than the surrounding matrix. For example, soil development (and plant colonization; see section 5.3.2 on dispersal mechanisms) under plants can be more rapid than in adjacent open areas (Halvorson *et al.*, 1991). However, soil heterogeneity in later stages of succession is often linked to patches of disturbance or reduced fertility such as tree falls or landslides. The loss of vegetation, soil and organic matter on a landslide generally represents a low nutrient patch on the landscape level. Within a landslide, however, plant growth is often patchy as well. Myster & Fernández (1995) found that light and VAM as well as N and C defined patch structure on Puerto Rican landslides. Most studies of landslides confirm that more organic matter accumulates in the deposition zone at the base of the slide than on the slip face at the top (Adams & Sidle, 1987; Guariguata, 1990; see also section 4.2.4 on erosion), providing more patchiness via vertical zonation.

Microbial communities are usually more developed under plant canopies than in open areas on glacial moraines (Ohtonen et al., 1999), on volcanoes (Hirose & Tateno, 1984) or in deserts (Binet, 1981; Zak & Freckman, 1991) because of the dependence of the microbes on the production of plant litter as a carbon source. The patches of relatively high soil fertility under plant canopies presumably become more pronounced over time, although verification of the successional development of such patches is difficult. Bolling & Walker (2002) found that factors limiting desert shrub growth (N, P, SOM) on abandoned roads developed spatial heterogeneity (became concentrated under shrubs) faster than less limiting soil factors (texture, pH).

Disturbances can increase spatial heterogeneity (for example, when surfaces are roughened or new habitats are created). A vertical heterogeneity occurs when new floods interrupt floodplain soil development by depositing silt and multiple layers of buried organic horizons result (Fig. 4.15). Disturbances can also decrease spatial heterogeneity (for example, by smoothing previously rough surfaces or blurring distinctions between habitats). Spatial heterogeneity may increase during succession, as biotic diversity interacts with the varied physical environment (Wardenaar & Sevink, 1992; Alpert & Mooney, 1996). Spatial heterogeneity can also decline, if one or a few species become dominant and impact soil formation is a uniform way (Armesto et al., 1991). Yet links between above- and below-ground patterns of spatial heterogeneity depend on whether plant diversity promotes habitat heterogeneity (Hooper et al., 2000). Spatial heterogeneity of mycorrhizal infections was higher in severe disturbances and early successional soils than in later successional soils in secondary succession in fields in Ohio, U.S.A. (Boerner et al., 1996). Too few studies exist to make any robust generalizations about successional trends in spatial heterogeneity of soils.

4.7 Summary

Soil formation in primary succession results from the interactions of many processes including transport of material to and from the site, *in situ*

Fig. 4.15. Buried organic layers from alternating forest floor development and deposition of new silt layers in subsequent floods on the Tanana River floodplain, central Alaska, U.S.A.

changes in parent material and interactions between organisms and the physical substrate over time (cf. Fig. 4.11). Both macro- and microclimatic variables influence the rate of soil development, with faster rates under wet and warm conditions. Following a disturbance, recovery to pre-disturbance conditions (if it occurs at all) is often a disjointed process, with some variables responding more quickly than others. For example, on abandoned paved roads in Puerto Rico, Heyne (2000) found that litter mass, bulk density, soil moisture, soil organic matter and total soil nitrogen reached adjacent forest levels within 11 yr, but it took longer for nutrient pool sizes (30 yr) and soil pH (60 yr) to recover. Heyne also found differences in recovery rates for various vegetation parameters. Such patterns are common for other seres, with organic matter often recovering more quickly than other variables (Matthews, 1992), suggesting that ecosystem development is not a uniform process for all variables even within a given disturbance type and area. Comparisons of multiple disturbances of the same type (e.g. glacial moraines in Norway; Messer, 1988) also show wide variations in processes in soil development. Finally, comparisons of soil factors among different types of primary sere result in wide variations in rates and patterns of recovery (Figs. 4.5, 4.8, 4.12). With such spatial and temporal variation, generalizations must be tentative and are largely statements to be tested when further comparable data are assembled.

Some generalizations concerning soil formation in primary succession are possible. Organic matter and N accumulate, acidity and Al concentrations increase and P and cation concentrations decrease. Many of these changes are linked through positive feedback loops (e.g. higher nutrient availability, higher litter quality and higher decomposition rates) that provide some coordination among the variables of soil development and probably limit the rate of change of the interrelated factors. What is harder to predict is the trajectory of soil development, including any likely endpoint or equilibrium state. Will the site develop soils that resemble pre-disturbance soils, even though some of the state factors (e.g. microclimate, topography, adjacent organisms) may have changed? Because so many dynamic variables influence soil development, describing the changes one can measure rather than relying on general predictions of likely patterns is probably a more practical approach. One can then address trajectories that include the gradual loss of nutrients and the disruption of developing soil profiles from erosion, burial or other factors (Matthews, 1992). Time frames must also be specified. Much work on soil development in primary succession encompasses less than 200 yr, a phase that often includes an accumulation of nutrients and organic matter. The few studies that follow soil development over much longer time spans often find a period of nutrient depletion and immobilization (J. Walker et al., 1981, 2001; Kitayama, 1996a; Mueller-Dombois, 2000) coupled with declines in primary productivity and vegetation stature (Kitayama & Mueller-Dombois, 1995a,b; see sections 5.6.2 on

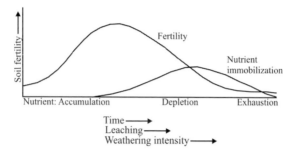

Fig. 4.16. Soil formation and degradation in tropical environments. From Fox *et al.*, *Allertonia* **6**(2): 132, figure 1 (1991), reprinted with permission of the National Tropical Botanical Garden.

under-saturated communities and 7.1.7 on retrogressive trajectories). Fox *et al.* (1991) suggested a model for such changes in tropical soils (Fig. 4.16), but the most thorough test of such patterns to date and of Walker and Syers' model for P (Fig. 4.9) showed only some congruence with the predicted patterns (Crews *et al.*, 1995). The longer the chronosequence, the more uncertainty there exists about not only soil formation but also about the validity of the chronosequence (Fastie, 1995).

5 · *Life histories of early colonists*

5.1 Introduction

Primary succession begins with the input of seeds, spores, animals and organic matter or by vegetative expansion from adjacent habitats. Distance alone plays a major role in determining what reaches a new site, so the landscape context of a new surface is crucial. Most species fail to disperse beyond some short distance and the total propagule density is low, so chance plays a large role in governing which species reach isolated surfaces. Thus the initial vegetation of isolated sites can be highly variable. Early species composition is governed by a complex suite of interacting forces, often leading to a chaotic mosaic of early species association (see sections 5.3.4 on predictability and 5.4.1 on stability).

In this chapter, we explore colonization and establishment phenomena and their consequences for primary succession. We focus first on pre-dispersal effects such as pollination and seed set, then on dispersal mechanisms. Finally, we explore the factors that affect establishment and the repeatability of the first species assemblages. Most primary seres are colonized by propagules dispersed by wind, water or animals. We ask, 'How do different degrees and types of isolation and substrate types affect colonization?' We then examine the factors that permit establishment and the consequences of differential longevity.

5.2 Pre-dispersal considerations

Primary succession lets us evaluate community assembly without the confounding influences of residual vegetation. We can search for patterns among sites of similar origin and between sites created by different processes. In this way, critical factors can be highlighted. As succession unfolds, controlling processes such as competition become increasingly similar to those of secondary succession.

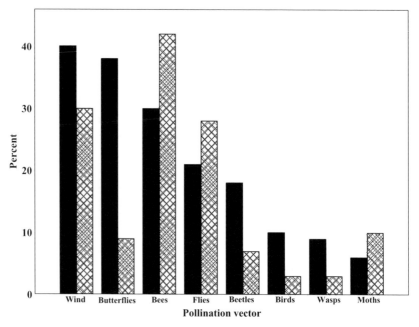

Fig. 5.1. Proportion of plant cover pollinated by different vectors in forests of the western (hatched bars) and eastern (solid bars) Cascades. From del Moral & Standley (1979). Totals exceed 100% because several taxa have more than one pollination mechanism.

5.2.1 Pollination and seed set

Pollination biology has rarely been studied in a successional context. Major pollination vectors that augment self-pollination include wind, insects, birds and bats. Climatic features and the successional stages of surrounding landscapes influence the pool of potential pollinators. The pollination mechanisms of most plant species can be inferred from flower morphology, and direct observations normally confirm these inferences. For example, del Moral & Standley (1979) compared the probable pollination vectors in open, dry conifer forests (East Cascades) to moist conifer forests with closed canopies (West Cascades). They demonstrated that dry coniferous forests had more beetle-, bird-, butterfly- and wind-pollinated species than did the moist coniferous forests, which were dominated by species pollinated by bees, flies and moths (Fig. 5.1). These differences often lead to a flora with different dispersal type spectra.

Off-site pollination
The successional status of the surrounding landscape affects colonization rates and types of pollinator. For most primary seres that start within landscapes experiencing early secondary succession, pollination rarely limits the availability of suitable colonizers because pollinators of such species are common. However, if vegetation adjacent to the new site is mature and contrasts sharply with the new primary surface, it may have few pollinators that can service typical colonizing species. Off-site pollen and pollinator limitations may slow early primary succession, although we know of no studies of this phenomenon.

On-site pollination
Recently formed sites normally lack a diverse fauna so the absence of pollinators could limit the expansion of non-autogamous flowering colonists. Perhaps for this reason, successful pioneers are usually dominated by wind- and self-pollinated (autogamous) species. Rydin & Borgegård (1991) showed that most pioneers of recently formed islands in Sweden were self-pollinated. On Mount St. Helens, 18 of 21 invading species were either wind-pollinated or potentially self-pollinated (del Moral & Wood, 1993a). The seed set of the widespread nitrogen-fixing pioneer shrub *Dryas*, found on glacial moraines throughout subarctic and arctic habitats, was unaffected by pollinator limitations (Wada, 1999). Such findings suggest that pollinator limitations rarely affect the rate or trajectory of early succession in temperate or arctic settings (cf. Prach *et al.*, 1997). This is partly due to the frequency of species that are self-compatible (Rydin & Borgegård, 1991) and to the usual general nature of pollinator requirements of pioneers (Larson & Barrett, 2000). However, pollinator limitations may be important for some animal-pollinated, obligate out-crossing tropical species or in some specialized habitats such as mangroves. For example, seed set of two tropical *Ficus* tree species found on Anak Krakatau (a recently formed volcanic island in Indonesia) was reduced significantly by pollinator shortages (Compton *et al.*, 1994). They did not suggest successional consequences, but did emphasize consequences for frugivorous birds and bats. The absence of pollinators must alter vegetation structure in mature tropical forests. Vertebrate and invertebrate pollinators on tropical islands are among the most threatened species (Cox & Elmquist, 2000). The loss of these pollinators strongly suggests that tropical primary succession in the future will not resemble that of the past. Partial self-compatibility of the widespread colonizing

tree *Metrosideros* on volcanic surfaces in Hawaii led to adequate levels of seed production (Carpenter, 1976). However, higher levels of seed production were obtained when endemic birds were permitted access to flowers. In contrast, the almost complete self-incompatibility of the giant perennial forb *Argyroxiphium* on the upper slopes of Hawaiian volcanoes has led to sharp declines in plant populations as the native insect pollinators have declined (Carr *et al.*, 1986; Powell, 1992). The absence of suitable pollinators in the recovery of isolated tropical sites should be considered when rehabilitation of tropical sites is being planned (Walker & Powell, 1999a) (see section 8.3.1 on dispersal).

On-site seed production
Stressful habitat factors can limit the seed production of invading species. A plant may become established, but be prevented from flowering or producing mature seeds owing to soil infertility or drought. This pattern was observed repeatedly on Mount St. Helens (del Moral, 1993b) where the forb *Epilobium* is common, but is rarely observed to mature on barren pumice (Fig. 5.2A). In contrast, this and other pioneer species thrive on residual soils (Fig. 5.2B). Restoration ecologists recognize that stress retards recovery, but their rationale has involved biomass production, not reproductive failure. The success of rehabilitation projects may be improved if more attention is paid to reproductive success, not just survival and vegetative expansion.

5.2.2 Seed banks

Seed banks are crucial to secondary succession, because they influence initial community composition. The numbers of species and seedlings produced from seed banks decay with time (Thompson, 1978; Roberts & Vankat, 1991) unless they are replenished by the same species. During primary succession, the opposite occurs: seed density and species richness of the seed bank increase as new species invade and the populations of each species increase and reach maturity (Houle & Phillips, 1988; Looney & Gibson, 1995; Duncan & Duncan, 2000), although the seed bank initially may not resemble the developing vegetation (Titus, 1991). Grandin & Rydin (1998) described the seed bank on islands that formed in Lake Hjälmaren in 1882. The seed banks in the most deeply buried soils were pioneer species (annuals with good dispersal), whereas surface seed banks at each vegetation analysis from 1886 to 1995 were similar to the vegetation at the time of analysis. The similarity between the vegetation

A

B

Fig. 5.2. Epilobium angustifolium plants. (A) Plants that established readily on deep pumice failed to flower for many years. (B) Plants growing on shallow pumice where roots could reach the old soil flowered and set seed profusely.

in 1995 and the seed bank declined with soil depth, and several seed bank species no longer occurred in the vegetation. Clearly, the soil retains a memory of previously occurring species, but where conditions change rapidly, recruitment must come from outside the system.

No seed bank may develop in dynamic seres where surfaces are eroded or deposition is rapid (Walker et al., 1986). Tekle & Bekele (2000) found little similarity between compositions of the seed bank and the vegetation on eroded Ethiopian hills. They suggested that this was because disturbance affects vegetation and seed banks differentially. In Hawaiian forests, a similar lack of congruence occurs. Most plant cover and seed rain was from native species but most of the seeds in the seed bank were from alien species (Drake, 1998), which suggested that future disturbances would favor establishment of these aliens. Parker et al. (1989) support the proposition that disturbance integrates many variables that can explain disparities between vegetation and seed bank composition. One quite unusual disparity occurs on cliff faces in Ontario, Canada. Here, seeds are transported from the undisturbed forest above the cliff by wind and gravity and accumulate on these exposed cliffs to produce higher species richness for the seed banks (Larson et al., 2000).

Severe disturbances remove the seed bank, eliminating any signal to the new sere. However, seed banks from pre-existing vegetation sometimes can be important in primary succession. Propagules can be transported with the disturbance (e.g. landslides, glaciers, lahars or dunes) or remain in place until they are exposed by erosion. Soils with viable seeds, or viable spores of moss, ferns and microbes, are often associated with melting ice and can have an impact on primary succession on Antarctic fell fields (Smith, 1993) or glacial moraines (Matthews, 1992). Seeds in the lower deposition zone of landslides can accelerate primary succession (Guariguata, 1990; Walker et al., 1996) and lahars may incorporate seeds that start primary succession (Nakashizuka et al., 1993). Shifting dunes are an unlikely place to find seed banks (Ehrenfeld, 1990), but buried seeds do persist in some dune systems (Zhang & Maun, 1994). As a sere develops, its seed bank changes and can influence later stages and trajectories.

Seed banks that remain in situ can be exposed when water levels fall. This occurred when Spirit Lake (Mount St. Helens) was lowered to expose sediments that rapidly developed wetland vegetation (Titus et al., 1999). Juncus and Scirpus graminoids, which presumably existed as seed banks, dominated succession. A Norwegian lake lowered in 1987 revealed barren sediments that contained a large seed and spore bank of species incapable of germinating before the draw-down (Odland, 1997). Similarly, the removal of sod in dune slack vegetation in The Netherlands revealed a dynamic seed bank in the soil layer containing both viable early successional species and accumulating late successional species (Bekker et al., 1999).

A special type of seed bank is the short-term persistence of seeds in the canopy (Noble & Slatyer, 1980). For example, some *Pinus* trees retain closed cones for several years. Many tree species retain mature seeds within protective fruits for long periods prior to dispersal. Should a major destructive event occur that either sterilizes the soil or deposits a new substrate without killing these arboreal seeds, they may be shed on to a receptive surface. Lahars and tephra deposits may kill trees while leaving them relatively intact with a viable seed crop in the canopy. A lahar on Mount Rainier (Washington, U.S.A.) smothered the roots of existing trees without uprooting them. Because the riparian *Alnus* trees retained mature fruit, viable seeds subsequently were shed upon a virgin surface, leading to rapid recolonization (Frenzen et al., 1988). Similar events occurred on Mount Ksudach (Kamchatka, Russia) where tephra 1–3 m deep killed *Betula* and *Alnus* trees, yet seeds persisted in the canopy. Because the substrate was deep and infertile, few of these seeds germinated successfully, but on the margins of this deposit many stems were new seedlings, not regenerating adults (Grishin et al., 1996).

5.2.3 Vegetative reproduction

Vegetative expansion along margins can accelerate primary succession in small sites such as rock faces, landslides and animal feces. This diffusion has strong spatial constraints. Diffusion can also be important where the disturbance affects a narrow corridor (e.g. a riparian zone). For example, vegetative reproduction by such herbs as *Phragmites*, *Iris* and several *Carex* sedges permitted rapid recolonization along the banks of the Rhône River (France; Henry et al., 1996). Yet vegetative expansion was limited by inundation and physical damage near a Canadian river channel and by silt burial and perhaps decreased soil aeration at further distances from the channel (Douglas, 1987).

After the initial establishment from seed, vegetative expansion permits a species to expand and thrive in a harsh habitat where further seedling establishment would be rare. On pumice and lahars of Mount St. Helens, *Penstemon* and *Luetkea* (Fig. 5.3) are two common species of subshrub that have increased disproportionately by clonal expansion after their initial

Fig. 5.3. Luetkea pectinata on pumice at Mount St. Helens. This species is typical of those that have expanded greatly by vegetative means.

establishment (del Moral, 1999a; del Moral & Jones, 2002). Vegetative expansion is very common on young volcanic landscapes, with small shrubs in the Ericaceae particularly common in temperate and boreal regions.

There is a gradation between primary and secondary succession and many disturbances can create a mosaic within which the two form a matrix. Residuals may survive immediately adjacent to barren sites, but seeds of residual species may not be adapted to establishment on primary surfaces (Fuller, 1999). However, vegetative expansion of residuals along the edges of lahars or landslides, or in erosion features (Fig. 5.4) initiates succession sooner than otherwise possible. In 1980, buried roots of the forb *Lupinus* sprouted on lahars on Mount St. Helens to form 'nascent foci' for subsequent expansion. In Puerto Rico, the fern *Dicranopteris* invades landslides (Walker *et al.*, 1996), while on Mauna Loa, Hawaii, *Dicranopteris* dominates the understory during primary succession because of its strong rhizomatous growth (Russell *et al.*, 1999) (Fig. 5.5). The balance between sexual and vegetative reproduction therefore depends on residual plant parts, life forms and physical conditions. On an Alaskan

Fig. 5.4. Erosion removed silt and permitted the recovery of plants, which subsequently expanded into the barren regions (Pine Creek, Mount St. Helens, 1980).

floodplain, Krasny *et al.*(1988) found vegetative reproduction dominated in unstable (erosive or frequently flooded) and dry sites. Seedlings were only important on mesic and stable surfaces.

5.3 Dispersal

Dispersal is an essential adaptation because the worst place for most seeds to fall is beneath a parent (Rey & Alcántara, 2000). Most plant species have dispersal abilities that permit only local expansion (diffusion) because short-distance dispersal confers sufficient advantage for success. Few species are adapted for routine long-distance dispersal. After a catastrophe, however, dispersal from a distant point must occur before ecosystem development can proceed. Therefore, early seres are 'donor controlled' (Wood & del Moral, 1987). As the surface is colonized, it comes under local control: the first colonists produce the great preponderance of new seedlings in the immediate vicinity. In general, poor dispersal ability means that relatively few species can spearhead an invasion. However, because these few mobile species have difficulty dominating barren primary successional sites, the early trajectory still may be determined by species that are less able to disperse but that arrive before their establishment

Fig. 5.5. Dicranopteris ferns invade (A) landslides in Puerto Rico and (B) forest understories in Hawaii.

can be inhibited. In this section, we explore landscape factors that affect dispersal into newly formed habitats.

5.3.1 Dispersal parameters

Distance alone can affect species that will colonize. Therefore dispersal affects all aspects of primary succession. We summarize model approaches to dispersal and investigate the effects of distance on the kinds of species reaching a site. We then summarize the experimental work documenting seed shadows and ask how models help to explain colonization.

Dispersal models
Empirical studies suggest that most species have limited dispersal distances (Malanson & Cairns, 1997). These studies imply that primary succession should be slow and constrained by dispersal limitations. Although this may be true for dispersal into stressful habitats in a short time frame, pollen records indicate that tree species migrated much faster after the last glacial retreat than experimental data suggest. For example, Fastie (1995) observed that *Picea* trees migrated 400 m yr^{-1}. This rate is consistent with the rate needed to explain post-glacial northward migration rates of

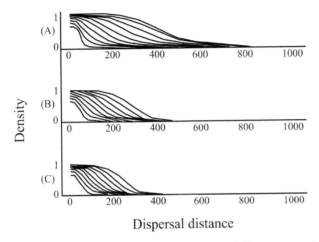

Density

Dispersal distance

Fig. 5.6. Models of population spread under different assumptions. The average dispersal distance is the same for each case, but in the top panel the maximum distance is greater; this gives rise to a population with an accelerating rate of spread. From Clark *et al.* (1998). Density and distance units are arbitrary.

temperate forests based on the pollen record. Reid's paradox (Clark *et al.*, 1998) describes the conflict between usually observed migration rates and dispersal rates calculated in experimental studies. Clark *et al.* (1999) advanced a solution to this paradox, although we note that the presence of relict sites is not precluded by this solution. They proposed a mixed dispersal model that fits short distance dispersal patterns well, but which also has a 'fat' tail, meaning that a measurable fraction of the seeds falls at long distances (Fig. 5.6) and are capable of establishing populations that are disjunct from the main population. Species are characterized by their dispersal efficiency. The '2Dt' model of Clark *et al.* (1999) combines local with long-distance dispersal models and assumes that the Gaussian distance factor varies randomly. It predicted seed rain more accurately across a range of species from several biomes than did the classical model. The model predicts that many species can migrate long distances in the absence of barriers. It also predicts that migration is characterized by long jumps followed by expansion from the locus of establishment. These 'nucleation' processes are common in primary succession (Clarkson & Clarkson, 1983). Bullock & Clarke (2000) suggested another way around Reid's paradox. They measured seed dispersal in several low heath species and found that seeds were trapped up to 80 m from the shrubs. These authors suggested that better sampling protocols and better models would indicate that most species can disperse over greater distances than often assumed.

Augspurger (1986) modeled the effects of seed morphology on dispersal ability based on aerodynamic qualities and mass. For 34 tree species, she determined mean dispersal distances of 22 to 194 m, assuming release from the canopy in a modest wind. Although the tails of typical curves with means in this range would be substantial, this study did not account for the presence of adjacent plants that would further reduce dispersal distances. Indirectly, this study suggests that invasion of primary sites in the moist tropics will be dominated by species in the immediately adjacent vegetation and therefore will be highly variable. Parrotta & Knowles (1999) confirmed this expectation on mined surfaces in Brazil. These conclusions based on trees apply more strongly to low shrubs and herbs because these growth forms normally disperse seeds from lower heights within a dense canopy.

Empirical studies
Whereas models help to explain enigmas about dispersal, empirical studies of seed dispersal permit direct comparisons among species and estimates of seed rain. Most wind-dispersed species disperse only for short distances. Distance–dispersal curves rapidly approach zero. The tails of such curves are imprecisely known, but tails determine the rate of migration and the chance of an occasional single jump. Seed shadows vary spatially and temporally and are difficult to predict. The large variability found in seed trap studies (Rabinowitz & Rapp, 1980; Wood & del Moral, 2000) implies that seed arrival early in succession is stochastic. Fort & Richards (1998) found that sandy flats (playas) received a substantial seed rain (over 50 seeds $m^{-2} d^{-1}$) at distances over 700 m from the source. However, this unusual seed rain involved high winds, smooth surfaces and few barriers. Platt & Weis (1977) found that the mean dispersal distance could be predicted according to seed appendages. Species without plumes were restricted to less than 3 m; those with large plumes averaged about 20 m. All these dispersal curves had fat tails, suggesting that jump dispersal over long distances is probable.

Studies by Nathan *et al.* (2001) developed a detailed wind-dispersal model and tested it empirically using *Pinus halepensis*. They found that the model accurately predicted dispersal patterns determined by seed traps. Wind velocity was more important than such biological factors as the height of seed release, terminal fall velocity and number of seeds released. These authors suggested that synchronization of seed release with strong winds is the most effective biological mechanism to increase dispersal distance.

Table 5.1. *Summary of types of seed dispersal*

Dispersal mechanism	Variants	Examples	Relative distance
Passive mechanisms			
Gravity (*barochory*)		*Malus, Juglans*	Very short
Wind (*anemochory*)	Minuscule	Orchidaceae, Ferns	Very long
	Parachute	Asteraceae, Salicaceae	Long
	Parasail	Aceraceae, Pinaceae, Betulaceae	Moderate
	Tumbler	*Polygonum, Salsola*	Short to moderate
Water (sea = *thalassochory*)		*Cocos, Rhizophora, Cakile*	Very long
Active mechanisms			
Ballistic (*ballochory*)	Dehiscent	Brassicaceae	Short
	Explosive	*Eschscholtzia*, Fabaceae, *Arceuthobium, Impatiens*	Short
Ants (*myrmecochory*)		Fabaceae	Short
Vertebrates (*zoochory*)			
Birds	Internal (*endochory*)	*Juniperus, Solanum*	Moderate to long
	External (*epichory*)	*Quercus, Lemna, Juncus*	Moderate
Mammals	Internal (*endochory*)	Lauraceae, *Cistanthe*	Short to moderate
	External (*epichory*)	*Bidens, Ambrosia, Plumbago, Castanea*	Moderate to long
Humans (*anthrochory*)		Weeds	Moderate to extremely long

5.3.2 Dispersal mechanisms and their consequences

The nature of dispersal provides the first hint about how a sere will unfold. Here, we will discuss both passive and active dispersal mechanisms. We summarize these dispersal mechanisms in Table 5.1. Relative distances in the table are typical of local dispersal, but any species may achieve extremely long dispersal distances by a variety of contingent factors (e.g. hurricanes and airplanes).

Passive dispersal
Passive dispersal agents are those powered by external, non-biological forces such as wind, water and gravity. Willson *et al.* (1990) summarized the distribution of dispersal adaptations in 35 studies of temperate zone

plant communities to produce dispersal spectra (percentages of a flora) for each category. They categorized very small seeds, scatter-hoarded seeds and censer plants (wind-shaken seeds from pods) as having no particular mechanism because they could not categorize the mechanism based on morphology alone, but most of these would be dispersed passively. They found unaided and wind-dispersed species to be the most common, whereas ballistic mechanisms and transport by animals were rare. They found that frugivore dispersal varied from 10% of the flora in most studies to 60% in New Zealand. Primary succession near glaciers had almost no frugivore-dispersed species, perhaps because moraines rarely attract birds. Animal dispersal increased during succession in this study as dispersal mechanisms became more specialized. In contrast, early primary succession is dominated by wind-dispersed species (cf. Nakamura, 1984; Chapin, 1993).

The spores of lower plants are effectively wind-dispersed because they are minuscule and buoyant, so mosses or ferns often start primary succession in mesic habitats (Griggs, 1933; Brock, 1973; Lewis Smith, 1993; L. Walker, 1994). In some seres, such as on glacial moraines (Worley, 1973; Matthews, 1992) and in disturbed arid lands, cryptogamic crusts form and stabilize the exposed surfaces (see section 4.3.1 on soil texture). Although the stabilizing effect of crusts is important, crusts are not usually required for succession. Vascular plants routinely colonize barren substrates without indirect facilitation (Winterringer & Vestal, 1956; Veblen & Ashton, 1978).

Wings or other buoyancy mechanisms facilitate long-distance dispersal, particularly during intense storms (e.g. hurricanes). Often overlooked are species whose fruits (or the entire plant) tumble across the ground, sometimes for long distances or across formidable barriers. *Polygonum* forbs produce many small fruits and then the entire shoot dries and breaks from the rootstock. In this mode, seeds can travel several kilometers in a strong wind. The widespread annual forb *Salsola* has a similar tumble dispersal method and is an effective colonist of newly exposed surfaces such as mined lands or road embankments in arid climates (Vanier & Walker, 1999). Some taxa have prodigious dispersal abilities. The Orchidaceae possess tiny seeds that form a major component of the aeolian seed pool, along with ferns, some mosses and a few animals. Empirical (Kalliola *et al.*, 1991; Kadmon & Pulliam, 1995) and theoretical (Clark *et al.*, 1999) studies agree that wind dispersal can provide longer jumps than animal-mediated dispersal (but see below).

Rivers and ocean currents effectively move propagules, but usually only to the shore. Seeds and flotsam arrive at the beach, but usually do not advance inland. Andersen (1993) reported that water-dispersed species dominated the Danish coast, but were restricted to the strand. Only 23 species reached the new volcanic island of Surtsey (1963), 30 km south of Iceland, by 1992. Six were sea-borne (Fridriksson & Magnusson, 1992) and the remainder wind- and bird-dispersed. On Rakata (Indonesia), sea-dispersed species dominated immigration during the first 40 yr. There-after there were few water-dispersed species that had not yet arrived and there were few open habitats that could readily be colonized (Whittaker & Jones, 1994a).

Freshwater dispersal typically dominates newly formed islands. In a Swedish lake formed in 1882, 42% (47 species) were transported by water (Rydin & Borgegård, 1991). For many riparian species, there is no alternative to water transport. Johansson *et al.* (1996) found that species with the best floating mechanisms were most frequent along rivers (Sweden). They also found that terrestrial barriers could have profound effects on early communities. Floods and streams normally disperse riparian species to initiate succession, maintain gene flow and replace species losses. When passive water dispersal was disrupted by dams (Sweden), mean richness in each impoundment was lower and the between-impoundment variation higher than on a free-flowing river (Jansson *et al.*, 2000). Water also can move seeds over ice and snow, to cross otherwise un-bridgeable barriers and deposit seeds in unlikely places (Ryvarden, 1971, 1975).

Some taxa disperse actively. Dehiscent species produce seeds within fruits that split. Examples include mustards and some geraniums. The latter produce achenes that can bury themselves. More dramatic are species that explode with little provocation to disperse their seeds up to 2 m from the parent (e.g. *Cytisus* shrubs and *Eschscholtzia*, *Euphorbia* and *Lupinus* forbs).

Active dispersal by animals
Although primary succession is usually initiated by wind-dispersed species, birds and small mammals often transport large seeds (e.g. *Pinus* and *Fagus* trees) that colonize bare substrates (Grubb, 1987). Animal dispersers are especially important in the tropics. Birds transport seeds moderate distances and have even transported seeds between hemispheres. Many birds are tireless frugivores, resulting in endozoochory. Inedible seeds pass through the gut to emerge primed to germinate in a nutritious

spot. Wetland species are transported in mud clinging to wading birds (epizoochory). This directs dispersal from one wetland to the next. Other plants produce large seeds that are transported and cached by species of woodpeckers and related birds. *Quercus*, *Nothofagus* and *Castanea* trees benefit from seed burial by birds. Wilkinson (1997) suggested that many plants that are considered wind-dispersed might be dispersed over longer distances by birds. Travel by exploring naïve birds, carrying seeds in odd directions, may be more important than migrations and could explain how species with limited dispersal ability migrate rapidly after glacial retreats (Clark, 1998).

Mammals transport seeds and fruits both internally and externally. Primates (Chapman & Onderdonk, 1998) and bats frequently disperse tropical fruits after eating them. Fruit bats can transport seeds up to a few hundred kilometers (Shilton *et al.*, 1999) and the absence of bats can slow or alter succession (Whittaker *et al.*, 1997). Ungulates transport seeds and defecate them far from the source in fertile clumps (Wood & del Moral, 2000). The loss of many large herbivores from the Americas at the end of the Pleistocene has surely altered vegetation patterns and successional pathways. In addition, the reintroduction of livestock has resulted in significant changes in plant dispersal patterns. Weedy plants are now routinely distributed along trails in pristine areas via the dung of horses (Campbell & Gibson, 2001). Animals also transport seeds externally with barbs, spines or sticky glands, as dog owners and cattle herders know. Among other oddities is the transport of seeds across long oceanic distances by tortoises (Fenner, 1985) and the dispersal of cactus by cattle.

Ants move seeds that have elaiosomes (attached lipid bodies) for short distances. More than 70 plant families include elaiosome-bearing (myrmecochorous) species, with many in Australia and South Africa. Hughes & Westoby (1992a,b) demonstrated that these seeds were normally transported less than 4 m and that seed predation by ants was high. Ants play only a small role early in most primary seres.

Humans are the most adept – if often unconscious – seed dispersers and merit special mention. We have reshaped the biological landscape by the global transport of biota. Plants that humans introduce either intentionally or inadvertently are often wind-dispersed, although there are many examples of animal-dispersed introductions in both temperate (e.g. *Rubus* shrubs and *Crataegus* trees) and tropical (e.g. *Lantana* shrubs) habitats. However, species that successfully invade new habitats usually do not have exceptional dispersal abilities. Newly formed surfaces will receive immigrants from the immediate surroundings regardless of where the colonists originated. Therefore, previous dispersal of plants to the surrounding landscapes by humans will strongly affect early primary succession because the colonists will be drawn largely from the immediate surroundings.

Humans also create dispersal corridors of many types, most of which facilitate species well adapted to disturbed or open habitats. Transportation corridors such as railway and highway edges offer abundant opportunities for ruderal species (Tikka *et al.*, 2001). Power transmission corridors often traverse undisturbed habitats, permitting species otherwise barred from undisturbed landscapes to invade natural plant associations.

Dispersal of animals
Animals are important vectors for plants, but of course they also must disperse into new habitats if complete ecosystems are to develop. Wind is a powerful force for dispersal that molds early ecosystem development (Howarth, 1987; Edwards, 1988; Ashmole *et al.*, 1992; Edwards & Sugg, 1993; Antor, 1994; Sugg & Edwards, 1998). Jet streams and storms transport aerial plankton that may fortuitously find a barren site (Greenslade, 1999); such waif dispersal also occurs in water (Cheng & Birch, 1987). Both the adult and reproductive stages of animals (eggs or larvae) are routinely transported. Animals often reach a site before seeds germinate and establish pre-vegetation communities (Howarth, 1979; New & Thornton, 1988; Thornton *et al.*, 1988). Spiders invaded pumice barrens on Mount St. Helens in massive numbers, although only six of 125 observed spider species established viable populations (Crawford *et al.*, 1995). Initially, without plants or prey, spiders ate one another. Along with pollen, dust and arthropods, spiders contributed significantly to early soil development. Spiller *et al.* (1998) found that spiders recolonized small islands in the Bahamas immediately after destruction by a storm surge. They also suggested that the lack of lizards was due to their poor dispersal ability. L. R. Walker (pers. obs.) found spiders landing on still warm lava in Hawaii, an example of precocious 'neogeoaeolian' invasion.

Extremely isolated islands such as New Zealand, Hawaii and the Galápagos lacked terrestrial mammals before human colonization. Less isolated new islands such as Anak Krakatau still lack monkeys, but they could be expected to reach there eventually by rafting. Many islands contain terrestrial animals that established by swimming narrow water barriers, but most populations are relicts that have persisted on land bridge islands. For example, foxes on Santa Cruz Island (California), a subspecies of the mainland population, are believed to be relicts cut off from the mainland by the rising post-Pleistocene seas.

Although flightless land vertebrates have difficulty in crossing larger water barriers, they sometimes manage impressive feats. Deer and bears, for example, can swim several kilometers to reach islands. Normal movements or seasonal migrations may introduce large animals to barren sites, but most flightless terrestrial animals require assistance to colonize isolated barren sites. Rafting on mats, trees or other flotsam provides an avenue for animals to reach developing sites (Fig. 5.7). Rafting is a lottery whose winners may include rodents, lizards and monkeys. Small mammals will explore primary surfaces, but establishment cannot occur until plants develop or there are relict sites that offer refuge.

Many animals disperse actively as part of their normal movements and migrations or to seek new habitats. The migrations of birds (e.g. golden plover, ruby-throated hummingbird and many swallow, sparrow and finch species), bats (e.g. *Lasiurus*, a species of which occurs in Hawaii) and some insects (e.g. the monarch butterfly) cover remarkably long distances. These and many other insects, birds and bats colonize new habitats and may be important vectors for plants colonizing new surfaces on islands. Insects may reach barren sites before plant development, hastening soil development by introducing carbon and nitrogen. Birds are attracted to oases or refugia and contribute to early ecosystem development by importing nutrients and seeds. The endangered Hawaiian goose (nene) is an important disperser of the native *Vaccinium* shrub on new volcanic surfaces (Mueller-Dombois, 2000). In the tropics, the absence of a bat species can limit pollination or seed dispersal in figs, thus altering the succession (Thornton, 1996). Bats can cross narrow water barriers, but they are normally absent until the vegetation matures. Accidental bird and bat dispersal can have cascading effects on succession by introducing unlikely plants early in the colonization process. Early introduction may lead to unusual trajectories by the priority effect (see section 7.1 on types of trajectory). Other vertebrates may cross short distances to reach refugia. These animals must navigate unsuitable habitats to reach even marginal ones. Deer mice (*Peromyscus*) crossed wide barren areas to recolonize relicts and wetlands on Mount St. Helens (C. Crisafulli, pers. comm.). Elk traverse these barrens, pausing at springs and thickets, but spend most of their time in nearby lush vegetation.

A

B

Fig. 5.7. Rafted vegetation may aid in dispersal of plants and animals on (A) the Tanana River in Alaska; (B) a landslide in Puerto Rico.

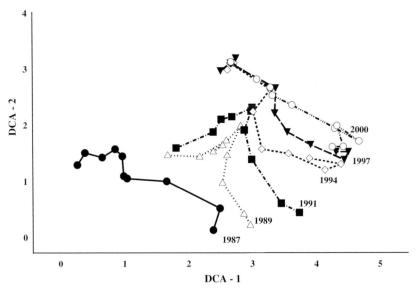

Fig. 5.8. Vegetation dynamics between 1987 and 2000 after the draw-down of a Norwegian lake. Each vector is the detrended correspondence analysis (DCA) score of the transect in successive years between 1987 and 2000, as shown. The position of the transects moves to the right on DCA-1 through the years and moves generally down DCA-2 with transect elevation. Modified from Odland & del Moral (2002).

Diffusion

Gradual dispersal along a front from an edge is a common mode for vegetative expansion onto stressful environments. On a small scale, diffusion occurs by centripetal invasion on rock outcrops (Ware, 1991; Meirelles *et al.*, 1999). On a larger scale, invasions along lake margins are also centripetal invasions (van Noordwijk-Puijk *et al.*, 1979). Phase transition, for example lake eutrophication, is a good example of pure diffusion. Populations expand slowly into lakes by vegetative means. However, when the environment changes suddenly, vegetative expansion is swift. Odland & del Moral (2002) determined that *Equisetum* horsetails invaded a former lakebed at a rate of 1.8 m yr^{-1} (Fig. 5.8).

Most examples of diffusion come from broad-scale expansions across landscapes. At large scales, gaps are irrelevant, but diffusion on a small scale is a result solely of distance. Any transect from intact vegetation onto newly formed sites is characterized by declining richness, density and size – distance by itself alters the seed rain. Evidence for this effect is common (Robertson & Augspurger, 1999).

Animal species with well-documented patterns of diffusion across a landscape include the muskrat (*Ondatra zibethicus*) in Europe, starlings (*Sturnus vulgaris*) in North America, fire ants (*Solenopsis geminata*) in the southern U.S.A., and house finches (*Carpodcus mexicanus*) and the opossum (*Didelphis virginiana*) in the western U.S.A. *Bromus* (a grass) and *Senecio* (a forb) are plants with well-described expansion patterns. Both spread by local diffusion after large jumps that were assisted by human activities. Late successional species may have a difficult time crossing fragmented landscapes because they depend primarily on local diffusion. Hiroki & Ichino (1993) showed that *Castanopsis* trees only advanced from mature trees along the margins of a lava flow 40 yr old. This slow diffusion was in sharp contrast to the jump dispersal of *Machilus* trees, which established only under the crown of the wind-dispersed pioneer shrub *Alnus*.

Tagawa (1964) showed that wind dispersal resulted in differential effects. A lava surface 48 yr old (Sakurajima, Japan) had high diversity downwind of an old surface, but low diversity upwind from the same surface. There was also a strong asymmetric pattern of density that corresponded to wind direction and force. Lahars on Mount St. Helens differed only in their proximity to intact vegetation. After eight growing seasons, mean cover was significantly lower on the isolated lahar. Dispersal limited density of two conifer species that dominated the adjacent woodland. *Abies* tree cover was initially 0.7%, whereas on the isolated lahar it was less than 0.08%; by 1998, cover had increased to 14% on the adjacent lahar but only 0.6% on the isolated one. *Pinus* cover increased from 0.2% to 9% on the adjacent lahar, but only from 0.01 to 0.4% on the isolated lahar. Plant species dispersion strongly depends on proximity to the woodland. On the adjacent lahar, *Abies* density declined sharply with distance, but *Pinus* was scattered and demonstrated no density gradients (Fig. 5.9). On the isolated lahar, there were no gradients because distances to sources were effectively the same for all points in the sample.

Seed availability can limit diffusion in isolated habitats that contrast sharply with their matrix. Small isolated rock outcrops may lack suitable species to invade as soils develop so that seres remain arrested until suitable species eventually colonize.

Other stresses may limit diffusion. Houle (1996) showed that seed germination sites for *Elymus* grasses were limited and it expanded on dunes only by clonal growth (cf. Poulson & McClung, 1999). Maun (1994) explored mechanisms of plant establishment on dunes and emphasized the importance of physiological tolerance mechanisms and, in contrast to Houle (1996), the importance of nurse plants. Conditions can be so

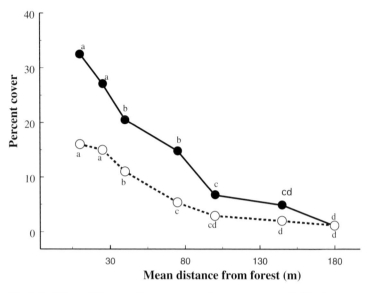

Fig. 5.9. Effect of distance from an intact forest on cover of *Abies lasiocarpa* (filled) and *Pinus contorta* (open) in 1998 on a lahar created in 1980.

severe that diffusion even by rhizomes is restricted. Ishikawa *et al.* (1995) showed that rhizomatous species were confined to less stressful shoreline environments and could not invade sites occupied by more stress tolerant species even though the growth rate was maximal at their leading edge (Ishikawa & Kachi, 1998). The parent plant can transfer resources and support diffusion into a hostile habitat. *Polygonum* herbs on scoria on Mount Fuji expanded vegetatively to provide a major stabilization force (Adachi *et al.*, 1996a). This study demonstrated that N accumulated by older parts of a clone was transported to the margins to support expansion.

Diffusion from established plants by vegetative means or from intact vegetation by propagule dispersal is effective, but spatially limited. It is insufficient to explain migration rates. It is also insufficient to explain the rates of early succession.

Jump dispersal

Many primary surfaces are neither infertile nor very stressful. They may be isolated from sources of colonization or they may be next to intact vegetation, but inhospitable only to the species found there. Rapid diffusion is precluded, but long-distance seed dispersal can lead to local colonization points. Because establishment conditions are not restrictive, expansion

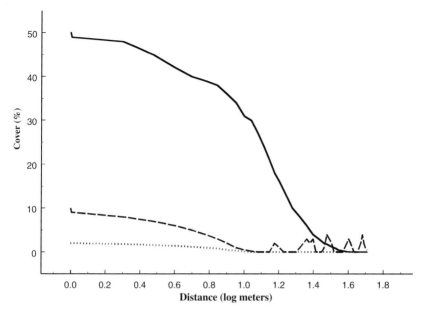

Fig. 5.10. Dispersal as a function of stress: theory. Solid line, normal dispersal by diffusion from a proximate source; dashed line, limited diffusion, with infrequent establishment by jump dispersal into a moderately stressful habitat; dotted line, rare establishment under stressful conditions.

from these initial colonizers is rapid. It seems likely that colonists that combine effective seed dispersal with strong vegetative expansion will dominate early stages of many seres (Bazzaz, 1979).

Diffusion occurs when proximal species can invade the primary surface. Alternatively, if there are no suitable species nearby and conditions are not very stressful, initial cover will be low and distributions will be characterized by spikes at random distances (Fig. 5.10). These spikes result from the infrequent establishment of lucky colonists followed by local diffusion. If suitable colonists are distant and the site is stressful, diffusion will contribute little and there will be only small cover changes with distance. Stressful conditions will limit the rate of secondary spread from the few colonists. In contrast, if the surroundings remain diverse, it is likely that many species will be able to invade and that densities will build more quickly than in systems of low biodiversity (cf. Grime, 1998).

Combined dispersal

Invasion from a distant source is the rule in primary succession simply because no propagules survive and adjacent species are unsuited to colonization. Only a few propagules reach an isolated site and most of these are wind-dispersed species. This reality suggests that an effective colonization mechanism is to establish, then expand by local seed dispersal. Of course, this requires maturation. Many species are diplochorous, with two principal forms of dispersal (Ohkawara & Higashi, 1994). They may hitchhike across barriers with vertebrates, then diffuse from founder individuals. Although the winged fruits of the forb *Cistanthe* are adapted for buoyancy, their mass and prostrate plant morphology suggest only local wind dispersal. However, elk eat these fruits and transport the seeds endogenously over several kilometers before defecating. Primates also form an important dispersal vector (Andersen, 1999). As tropical tree species usually have poor dispersal, directed dispersal by primates can transport seeds among forest fragments or to sites recovering from degradation (cf. Whittaker *et al.*, 1997).

Combined dispersal was true for 16 of 18 species tested on pumice at Mount St. Helens (del Moral & Jones, 2002). A simulation study demonstrated that emerging spatial patterns were consistent with rare colonization from distant sources, followed by diffusion from nascent foci. Plants expand from founders as ripples in a pond expand from a thrown rock (see, for example, Game *et al.*, 1982). This pattern is common when the invaded habitat is isolated and the barren habitat is inhospitable.

5.3.3 Barriers

In order to understand fully the assembly of communities on newly barren surfaces, characteristics of the landscape and the surrounding biota must be known. Landscapes form a mosaic of challenges for dispersal. Schmitt & Whittaker (1998), after Pickett *et al.* (1987), presented a hierarchy of causes and mechanisms of succession. Their analysis showed that the size and severity of disturbances, the propagule pool, dispersal abilities, climatic factors and disturbance timing all influence dispersal. White & Jentsch (2001) reviewed how disturbance regimes control ecosystem change, including primary succession. They proposed that in addition to disturbance size and severity (or magnitude), the duration and abruptness of the impact dominate the future of the system. On any landscape, disturbances are common. They include small-scale, low-impact disturbances that have no successional consequences, and moderate impacts that create gaps best understood through patch dynamics approaches that may include secondary successional systems. Only intense, high-magnitude disturbances remove

Table 5.2. *Relative percent cover in each of five dispersal categories on four grids at Mount St. Helens showing how isolation affected dispersal to a site*

Isolated sites were initially dominated only by species with good to excellent dispersal.

Dispersal category	Adjacent lahar	Isolated lahar	Isolated pumice	Very isolated pumice
Poor	6.8	6.8	5.1	8.8
Modest	58.5	27.2	19.5	20.7
Moderate	7.4	18.0	14.8	6.7
Good	16.0	32.1	23.9	13.8
Excellent	11.5	16.6	33.7	50.0

From del Moral (1998).

all biomass to initiate primary succession. These disturbances can so profoundly alter a site that the recovering trajectory never approaches conditions found in undisturbed sites. The floristic context, sometimes ignored, is significant because it restricts the species that can invade. Fastie (1995) noted that seed sources on a glacial moraine included refugia of undamaged trees and successional stands that represented post-glacial invaders.

Barriers to colonization accentuate isolation caused by mere distance (Table 5.2). They create differential filters, so knowledge of potential and actual barriers assists in predicting the first colonists. Physical barriers include water, mountains and inhospitable habitats. Newly formed, isolated wetlands on Mount St. Helens were dominated by wind-dispersed species and were highly variable. This suggested a stochastic effect mediated by the upland barriers surrounding each wetland (Titus *et al.*, 1999; del Moral, 1999b). Less obvious barriers that affect dispersal include the direction of wind and currents.

Oceans bar most species from colonizing new habitats. Island floras can be dominated by wind and bird dispersal or by water dispersal, depending on distances and currents. Currents around Krakatau often convey unlikely passengers such as monkeys (Thornton, 1996). Because Indonesian currents are seasonal, another variable is the direction of currents or winds when seeds are released. Sea-water toxicity also filters the pool of colonists.

Subsequent dispersal of water-dispersed species into suitable upland is often thwarted by the absence of mammals. Andersen (1993) studied shore communities along the Danish coast and found different dispersal

Table 5.3. *Fraction of the vascular flora dispersed by birds in a variety of successional habitats*

Habitat	Percent bird dispersed	Examples	Reference
Hawaiian Islands	75%	*Bidens*	Carlquist, 1974
Surtsey	64%	*Cardaminopsis*	Fridriksson, 1987
Krakatau	34%	*Ficus, Solanum, Pandanus*	Whittaker & Jones, 1994a; Thornton, 1996
Peruvian Amazonia	16%	*Inga, Phytolacca*	Kalliola *et al.*, 1991
Lake Hjälmaren (Sweden)	12%	*Rubus, Solanum*	Rydin & Borgegård, 1991
Wetlands, Mount St. Helens	0%	None	Tu *et al.*, 1998

types along a gradient from the sea. Water dispersal dominated the shore, with wind dispersal becoming more common inland. Where vegetation was developed, animal dispersal, especially by ants and external transport, became more common. Wind-dispersed species can cross moderate ocean barriers and ocean currents can help to distribute species among island groups.

New oceanic volcanoes demonstrate the effects of isolation on seed dispersal. Birds routinely reach less isolated islands, dispersing most of the plant species that reach them early in succession. Table 5.3 shows examples of the percentage of the bird-dispersed flora in selected islands. The Hawaiian Islands are extremely isolated so that wind and sea routes are problematic. Even some normally wind-dispersed species may actually have arrived as passengers on a muddy bird (Carlquist, 1994). In contrast, Anak Krakatau is bracketed by Java and Sumatra, large islands that provide ready sources of seeds.

A new caldera with a fresh-water lake formed on Long Island (Vitiaz Strait, 55 km north of New Guinea) in about 1645. In 1968 a new island called Motmot formed in this lake. Motmot is young, with very unstable, porous and infertile soil in a low rainfall region (Harrison *et al.*, 2001). It is isolated by sea, land and fresh water. As a result, its flora of 45 species is about 12% that of Long Island and very different from comparable young, nearby island volcanoes (e.g. Anak Krakatau) that have access to sea-borne biota (Thornton *et al.*, 2001).

Lake Hjälmaren in Sweden surrounds many small islands that have undergone more than 100 yr of primary succession (Rydin &

Borgegård, 1988b). Bird-dispersed species were rare in early successional sites, but increased slightly on islands that developed woody plant dominance. Harsh establishment conditions may be a more stringent barrier than a lack of effective dispersal. For example, Surtsey, a young, harsh and isolated volcanic island, has a limited seed input. Ocean currents provide one vector, but the vegetation appears confined to the strand. Much of the island was barren in 2001, and upland vegetation was confined to bird colonies and a few other ameliorated habitats. River bars along the Peruvian Amazon are not isolated and birds form only a small component of the migrants (Kalliola *et al.*, 1991). Here the vegetation is dominated by wind- and water-dispersed species. The proportion of wind-dispersed species remained constant, confirming that wind is the more effective agent for crossing moderate barriers. Young wetlands forming on pumice and pyroclastic deposits of Mount St. Helens are isolated from intact and relict vegetation and had no bird-dispersed species in 1995 (Tu *et al.*, 1998). This wetland vegetation lacked bird-dispersed species through 2000. In an anthropogenic disturbance (logged islands), Kadmon & Pulliam (1995) found that the number of bird-dispersed woody plant species decreased by 50% as isolation increased from 100 to 500 m.

Studies of terrestrial invasions across barriers all suggest that maximum colonization rates occur when diffusion and jump dispersal combine. Most species have multiple dispersal mechanisms, although one may predominate. This allows plants to hitchhike with animals or effective seed buoyancy mechanisms to be combined with local spread by vegetative expansion, wind, gravity or water. Combined dispersal often produces heterogeneous vegetation initially dominated by species with efficient dispersal, but soon shifting to those that expand vegetatively. Clark's (1998) model explained rapid migration of temperate zone woody plants after glaciation by combining nucleation resulting from jump dispersal with diffusion. The expansion of invading species on primary substrates agrees with this model (cf. Clarkson & Clarkson, 1983). Barriers filter the available flora by restricting those species with poor dispersal abilities from crossing.

Barriers cause the initial vegetation to be an unrepresentative sample of the flora. Compared with mature vegetation, there will be too many species with excellent dispersal mechanisms (usually by wind) and too few with large seeds or specialized dispersal mechanisms. Dispersal by birds onto barren sites is uncommon because such directed dispersal requires some attraction (resources or perches). However, when early successional trees or shrubs do attract birds, the seed rain (e.g. on landslides in Puerto

Rico; A. Shiels, pers. comm.) becomes dominated by bird-dispersed species (Walker & Neris, 1993). Many common functional types will be excluded simply because they cannot reach a site. The absence of certain plant species can result in missing animals. Disharmony on oceanic islands is well known, but less appreciated is that disharmony is apparent in most early seres and can persist. Recovering ecosystems can sustain more species and more rapid development than is usually apparent. This suggests opportunities for rehabilitation that will be explored in Chapter 8.

5.3.4 Predictability

Prediction is the goal of any science. Early studies of succession assumed directionality that was combined with determinism (see section 3.3 on holism). Although primary seres sometimes develop in predictable ways, both natural and human-created vegetation often develop in surprising ways. Here we investigate the factors that may lead a sere to develop contrary to the conventional wisdom.

Chance and prediction

The distance a bird might disperse a seed is less predictable than distances achieved by passive transport by wind. If we consider that most typical early pioneer species will be drawn from a suite of species with good wind dispersal, then on a large scale, the composition of early successional stages is predictable. By contrast, dispersal to isolated sites via birds is less predictable; the pool of potential colonists is large, the number of seeds being introduced is small and the vectors are few (Ward & Thornton, 2000).

In addition to normal dispersal methods, lottery events are so rare as to be unpredictable. Amphitropical distributions of taxa in the southwestern U.S.A. and in Chile and Argentina imply a few extremely rare long-distance dispersal events (Morrell *et al.*, 2000). Migrating birds sometimes are blown off course (e.g. geese in Hawaii); should they carry a seed, a new introduction might occur. On a smaller scale, no one who has studied *Lupinus* on Mount St. Helens (e.g. Wood & Morris, 1990; Halvorson *et al.*, 1992; Bishop & Schemske, 1998) can offer a better explanation for its appearance on barren pyroclastic material within one year of the eruption than this: capsules were blown onto snow and then transported with the snow melt. Seeds started a few scattered populations that expanded into 'patches.'

Both stochastic and deterministic processes operate during primary succession. Environmental factors gradually assert control over the distribution and growth of individual species, forging tighter links between the vegetation and light, moisture and nutrient availability. However, even as the environment begins to exert control on the vegetation, the composition of the subsequent wave of colonizers may be *less* predictable. This suite of later colonists will include those with a low probability of invasion, so the identity of successful species in any given case can be uncertain. McCune & Allen (1985) found that chance led to a series of different dominants in similar valleys after severe fires in Montana. Forces that affect the degree of determinism appear to be operating in most seres, leading to a variety of trajectories.

However, predictability may also be affected by the degree to which a relationship develops between species composition and the environment. Initially, the environment of a site is poorly related to species composition. Lack of competition permits species to occupy a broad range of environments. Bowers *et al.* (1997) found that vegetation on young debris flows in the Grand Canyon was highly variable, but that each could be explained in terms of seed dispersal and local climate factors. del Moral (1999b) found that the predictability of wetland vegetation after 18 yr increased dramatically compared with the same sites sampled six years earlier (Titus *et al.*, 1999). Similarly, del Moral (1999a) found that initial composition of small isolated communities on pumice was highly variable. Later studies revealed that there was an increasing, although still tenuous, relationship between composition and the environment.

The biological legacy, nearby species that survived the disturbance, can reduce stochastic effects because the seed rain will be denser, species invading will be more consistent and more species will be represented. The presence of legacies (Franklin & MacMahon, 2000) may accelerate early primary succession. However, while residual plants, animals and diaspores are important to secondary succession, their effects in barren primary landscapes are constrained. Fuller (1999), working on Mount St. Helens, found that the effects of refugia within which forest understory species survived were limited to a distance of less than 50 m from the refugia. Surviving species did not invade barren pumice, but the margins of refugia had soil that provided an ideal habitat for invading species. Whereas pioneers on the barrens produced few seeds, those that established on refugia were robust and reproduced copiously. Their progeny were well placed to colonize the surroundings. The refugia not only contributed a greater seed rain of barren zone species, but also attracted birds and rodents and exported organic matter to the adjacent plots. In Hawaii, similar processes occur when fresh lava flows are recolonized from kipukas (islands of forest that survive). However, only a small subset of the forest species can colonize the lava.

The importance of safe-sites, regeneration niches (Grubb, 1977) and amelioration is related to the degree of stress (e.g. infertility, drought, salinity or toxicity). As a result, trajectories on initially stressful sites are less predictable than those on more benign sites situated in similar landscapes, even though the initial colonists may be predictable. Turner *et al.* (1998) developed a qualitative model of the predictability of succession. Predictability is reduced as disturbance size, intensity and frequency increase. All of these factors act against a shifting background of landscape factors, a changing biota and other stochastic elements. Savage *et al.* (2000) modeled forest dynamics as a function of disturbance. They concluded that in regimes with moderately frequent disturbance, many alternative communities developed, reflecting closely balanced competitive properties.

Habitat size
The vegetation of large habitats will be less predictable than small sites because the interior composition of large sites will be constrained by dispersal and subject to stochastic processes. A small suite of species with good dispersal and rapid reproduction can invade large newly created habitats rapidly and expand quickly, but which members of the suite are successful often has a large stochastic component. In terrestrial habitats, wind-dispersed species predominate because there is little to attract animals. The seed rain and initial population densities will be low. Therefore, although the predictability of the first wave of colonists will be good, spatial heterogeneity and dispersal gradients will be high. As colonization progresses, heterogeneity may decline as the initial colonists expand from their initial loci and colonization continues. However, subsequent colonists may be unexpected and a generally rare species may become dominant in a given sere. This potential paradox of decreasing long-term predictability even as the vegetation becomes more homogeneous needs to be examined carefully by permanent plot methods.

Habitat stress
Stress restricts successful colonization strategies and may lead to improved predictability, at least with respect to functional types. Díaz *et al.* (1998) found that there are consistent relationships between the species pool and environmental factors for most traits in all 13 Argentine habitats investigated. García-Mora *et al.* (1999) found that unstable dunes were colonized primarily by species with spreading root systems. We conclude that

stressful habitats (e.g. salt pans, sand dunes and mined lands) will be invaded by a small subset of the species pool and initially will be more predictable than less extreme habitats. In general, predictability will be high if the pool is small and if the stress is extreme. However, subsequent development from the small pool of species may be less predictable.

Habitat isolation

Early dispersal may be more deterministic in some isolated sites because only a few species can become established; however, it is locally stochastic. Later, it may become more stochastic because many factors influence which seral species may arrive. The availability of dispersal agents in tropical habitats is a crucial bottleneck (Ward & Thornton, 2000). The presence or absence of birds for dispersal will profoundly alter the initial and ultimate composition of the vegetation. Islands with different degrees of isolation may differ profoundly simply because bird dispersal agents may, or may not, reach them (Table 5.3). Priority effects may alter the nature of subsequent colonization. In temperate regions, habitat fragmentation leads to colonization by different sets of species, alternative trajectories and ultimately to multiple stable states (cf. Jacquemyn *et al.*, 2001).

5.3.5 Dispersal conclusions

Primary succession normally requires propagule dispersal because seed bank survival is rare. Dispersal is the first major filter that governs primary succession. Suitable species may be precluded from participation because they are initially incapable of reaching the site. Distance alone is an important isolating factor because typical dispersal distances are limited. Jump dispersal occurs for most species, but is stochastic for less able dispersers because so few propagules reach a site. Predicting early and mid-course succession requires knowledge of the locations of the available species pool. Physical barriers differ in their ability to filter species, so knowledge of the landscape is also required to predict the immigrant pool. Aquatic barriers are more difficult than inhospitable terrain because there are no opportunities for refugia or stepping-stones for terrestrial species.

Ultimately, the rate of species change declines because no additional species can reach the site and the dominants can successfully reproduce. Islands retain their unbalanced biotic composition, but it is unclear to what extent disharmony is reflected in mature mainland primary seres. Migration by diffusion occurs if barriers are minimal and new surfaces are not significant barriers to establishment. Jump dispersal appears to be a common feature of invasion whether or not diffusion occurs. Isolated or extreme sites are characterized by limited jump dispersal. However, dispersal from these initial colonists then follows a diffusion mode. Most primary succession sites have a limited ability to sustain biomass because they are stressful (*sensu* Grime, 1979). Many more species can be pioneers than actually are. These species are often stress-tolerant, but owing to their poor dispersal, they rarely reach sites early in succession. In contrast, species that do form the first wave of immigrants are usually poorly adapted to the site and readily displaced by subsequent invaders. The degree to which the initial species composition of a site may be predictable is a function of the scale of observation, the degree of isolation and the diversity of the available flora.

5.4 Establishment

Propagules that reach a barren site have passed one filter, but they are immediately confronted with others. While dispersal characteristics and isolation will dictate the number and types of species that arrive, site characteristics determine which of these can establish. Many species are limited by both dispersal and microsite availability (Eriksson & Ehrlen, 1992), but other factors such as seed predation and competition from established plants often regulate subsequent seedling density. A breeding population may develop immediately, but is more likely to be delayed by harsh environments. We will explore early life history traits to assess their effects on establishment.

5.4.1 Germination

Amelioration

Walton (1990) suggested that dispersal occurs without regard for the suitability of the habitat. Despite a large seed rain, immediate establishment may be impossible until after the development of the new substrate. However, the act of dispersal itself is a major and early form of site amelioration, leading to higher probability of success for subsequent immigrants. Ashmole & Ashmole (1987) showed that the biological fallout on very young lava flows in the Canary Islands supported rich arthropod communities and Edwards & Sugg (1993) demonstrated that similar fallouts are a major form of physical amelioration. Aeolian fallout leading to significant populations of invertebrate predators, parasites or detritivores have been shown in communities as diverse as glacial moraines (Hodkinson *et al.*, 2002), volcanoes (Wurmli, 1974), dunes

(Goralczyk, 1998), landfills (Judd & Mason, 1995) and coal spoils (Mrzljak & Wiegleb, 2000). Dissolved nutrients, whose origins may be very distant from the site (Chadwick *et al.*, 1999), are also added through rain. All such inputs significantly affect soil development. Many other forms of organic matter, including leaves, pollen, spores and feces, can reach a site and contribute to its amelioration.

Safe-sites

Seeds of *Salix* shrubs waft over barren sites on Mount St. Helens (Wood & del Moral, 2000), but establish only near springs and seeps. Seeds of *Anaphalis* forbs fall into a wetland and perish. Each invading species will respond differently to the same environment and germinate under different conditions. There has been little research conducted on the fraction of the incoming seed rain that contains species able to germinate under the conditions of a barren surface. On most substrates, conditions remain inhospitable for many years, so that successful germination can only occur on a small fraction of the habitat. Germination success can be enhanced by manipulating the site, thus demonstrating that the lack of microsites restricts germination (Greipsson & Davy, 1997; Titus & del Moral, 1998a). However, even in the presence of suitable microsites, propagules will not germinate unless all other physical requirements are met (Smyth, 1997).

As physical amelioration proceeds, the initial homogeneity of surfaces such as desert playas, smooth lavas (pahoehoe), rocks, pumice and mud flats breaks down. Particularly well-favored microsites eventually develop that can sustain establishment. The forces that create initial heterogeneity are physical. Although erosion is often a destabilizing and destructive force, it can also expose residual soil. On Mount St. Helens, erosion created small rills in featureless barren pumice that became foci for early seedling establishment (del Moral, 1983). Even though most plants had been killed, this old material was more readily invaded, and the erosion released survivors before they died. Freeze–thaw cycles fracture rocks to create microsites where seeds, moisture and dust may accumulate. Wet–dry cycles perform a similar function on fine-textured surfaces, creating cracks between smooth surfaces. These cracks trap seeds and provide protection from drying that permits seedlings to establish. *Umbilicus* forbs on Mount Etna pioneer cracks in weathered lavas very early in the sere and can persist for centuries (Fig. 5.11).

Sites such as talus, scoria and sand dunes are intrinsically unstable by the nature of their formation (see section 4.2.4 on erosion). Here, succession is routinely reset by erosion. Colonizing plants are buried, killed by

Fig. 5.11. Umbilicus rupestris grew well in crevices in lava on Mount Etna (Sicily). Note that lava surfaces sustain lichens in exposed microsites, mosses in more protected sites and various herbs in areas that have accumulated soil since the deposition of this lava in 1636.

exposure or torn from the surface by wind. However, wind also can stabilize a surface. It removes fine-textured particles and leaves behind a stable surface that traps seeds (Tsuyuzaki *et al.*, 1997). This 'desert pavement' conserves moisture and enhances seedling survival. Water erosion sometimes removes unstable surface materials to reveal a stable substrate ready for colonization (del Moral & Bliss, 1993). Species with vigorous rhizomatous growth such as the grasses *Leymus* and *Ammophila* often stabilize dunes and other loose materials to initiate succession. These various amelioration effects act in concert, gradually or rapidly preparing a site for successful colonization. It is rare that large sites develop homogeneously or simultaneously. In nearly every case, some microsites are more favored, and these are destined to harbor the first colonists.

John Harper coined the term 'safe-site' and described its significance in a comprehensive series of papers culminating in his classic book (Harper, 1977). Safe-sites often contrast sharply with the background surface and can harbor seedlings. They gradually become favorable to more species and can support greater productivity. As the sere develops, they comprise an increasing fraction of the landscape. These physical sites provide relief from drought, moderate temperature, reduce wind, collect nutrients, trap seeds and otherwise satisfy the regeneration requirements for at least some species that reach the site (cf. Grubb, 1977). A few propagules eventually lodge where they might germinate, usually in such physical features as small rills, fracturing rocks or eroding unstable material. Eventually, the microsite offers a chance of success to a germinating seed. Gradually, seedlings establish.

Stability
Microsites differ in their stability. Stable sites offer a long window of establishment opportunity. Bishop & Chapin (1989a) showed that rehabilitation of gravel pads was relatively easy because established plants responded to fertilization and were not subject to erosion. Such sites filled in quickly as the planted species set seed. In contrast, Frenot *et al.* (1998) found that glacial outwash plains supported low densities of plants that only expanded slowly. The sites lost fine material through wind erosion, so scant germination could occur on the eroded surfaces. Houle (1996) found that dune stability was required for germination and that only germination determined establishment patterns.

Physical features can be used to identify safe-sites in barren habitats. Jumpponen *et al.* (1999) characterized safe-sites in front of retreating glaciers. They could predict the unique combination of physical traits that constituted the safe-site for each species. Stocklin & Baumler (1996) showed that safe-sites became less common along a chronosequence associated with a retreating glacier as the surface developed. Lavas develop different microsite heterogeneity patterns. A'a (Figs. 2.4B, 5.12) is fractured and contains many apparent safe-sites, whereas pahoehoe (Fig. 2.4A) is smooth. On Hawaii, Aplet *et al.* (1998) showed that under moist conditions, succession was more rapid on a'a, but pahoehoe supported more

Fig. 5.12. Broken a'a lava field on Mount St. Helens (400 yr old). Rock surfaces are dominated by lichens and mosses; cracks support the invasion of conifers and other woody species.

rapid development in dry conditions. Thus the quality of a safe-site is conditional and safe-sites differ in their quality, general suitability and size (Kroh *et al.*, 2000).

Safe-sites are not immutable. Continuing physical amelioration increases their abundance and quality, but they may also erode, desiccate, become wetter, be buried (del Moral 1993b; Tsuyuzaki & Titus, 1996; Tsuyuzaki *et al.*, 1997) or be claimed by an adult plant (Walker & Powell, 1999a). Safe-sites often disappear under dominance by pioneer species and they may not be replaced by safe-sites suitable to other species. Stocklin & Baumler (1996) demonstrated that what constituted a safe-site differed on a glacier moraine transect and that older habitats were drier and contained few safe-sites for pioneers. Del Moral (1993a) described microsite relationships in suites of pioneers on pumice at Mount St. Helens. As the terrain matured and initial species spread vegetatively, sites suitable for germination became scarce (del Moral & Jones, 2002). Walker (1989) demonstrated that N and organic litter accumulation on developing Alaskan floodplains restricted the number of species that could establish and this led to succession. Early floodplain colonists typically depend on seed dispersal during an ephemeral period of receding water that provides a moist yet well-drained seed-bed (Nechaev, 1967; Walker *et al.*, 1986). Immediate germination and rapid growth assure that seedlings will not wash away or become desiccated. When safe-sites become scarce, succession slows.

The nature of safe-sites in early succession is usually obvious and based on physical attributes. Many species may be capable of establishment in a particular safe-site. Thus, early primary succession is a lottery in which arrival is the prime determinant of the early course of events. In mature vegetation, safe-sites and regeneration niches may depend on subtle biological interactions as well as physical conditions and therefore are difficult to recognize and quantify. Similarly, replacement of individuals in a safe-site may be driven only by chance, such that turnover is common, but not directional. The broad tolerance of pioneers and their lack of safe-site specificity are characteristics of the carousel model (see section 3.9 on models). This view of early stochastic development in primary succession implies that primary succession is not predictable on a species level, although it may be on a functional level. The early trajectory will be conditioned by what might be termed stochastic priority effects.

Dormancy mechanisms often prevent inappropriate germination and permit germination only during a favorable season. Environmental cues permit seeds to germinate when they are more likely to establish successfully. Environmental factors that can break dormancy include light, high and low temperatures, moisture and some combinations of these factors. In cool, temperate habitats, many species require stratification (long periods under cold, moist and dark conditions). In dry habitats, saturating rains frequently break dormancy, or seeds may be abraded as they are blown along grainy surfaces. Many ruderal species remain dormant in soil until some soil disturbance exposes them to a brief flash of light. This triggers a sequence of events that breaks dormancy and initiates germination. Baskin & Baskin (1998) provide an excellent compendium of the dormancy and germination ecology of seeds. Seed germination depends on several internal and external factors.

Niederfringiger-Schlag & Erschbamer (2000) found that seedling establishment on moraines in the Austrian Alps was enhanced where there was existing vegetation and that safe-sites including microsites modified by adult plants or stones each enhanced success. Established plants on older moraines inhibited establishment. Late successional species could germinate on young moraines, suggesting that dispersal limits and not germination failures slowed succession.

5.4.2 Growth

Once a seed has reached a safe-site, it must survive and grow in order to succeed. On most barren primary surfaces, it is initially only the most favorable sites that permit seedlings to mature, and the nature, distribution and abundance of safe-sites is therefore important not only to germination but to establishment. For example, Kochy & Rydin (1997) demonstrated that species diversity on new islands was related to the degree of habitat heterogeneity, not to any area or distance effects, thus emphasizing the importance of safe-sites for growth as well as germination. Safe-sites are also crucial in exposed habitats such as cliffs (Houle & Phillips, 1989) where wind and erosion preclude seed capture. Hilton & Boyd (1996) showed how slow growth constrained the development of species on dunes.

The extent to which chance effects persist is unclear, but it is likely that they are never fully erased. For such effects to be eliminated would require that there be a very close connection between each species and the physical environment, strong competitive interactions such that the superior species predictably predominates, and relatively homogeneous habitats to provide few refuges for inferior species. These conditions are uncommon.

Abiotic conditions

Most primary successional surfaces are stressful (see Chapters 4 and 8). Plants of infertile surfaces such as sand dunes (Olff *et al.*, 1993), floodplains (Walker *et al.*, 1986), glacial moraines (Chapin *et al.*, 1994), tephra (Hirose & Tateno, 1984), pyroclastic materials and mined lands (Game *et al.*, 1982; Marrs & Bradshaw, 1993) are all characterized by slow growth. Nutrient deficiency retards seedling development. Slow growth usually translates into a failure to develop a sufficient root system and seedlings succumb to drought (cf. Morris & Wood, 1989).

Drought is a feature common to many primary seres. Obvious examples include coastal and inland sand dunes, rock outcrops, scoria and lava. Less obviously, drought effects can occur on river gravel bars, near bogs and on wetland margins. Algae colonizing gaps in vegetation in rocky intertidal zones can even experience desiccation in the larger gaps (Kim & DeWreede, 1996). García-Fayos *et al.* (2000) demonstrated differential seed germination on eroded 'badlands' in Alicante and Murcia, Spain, where the summers are dry and winter rains are intermittent and scant. Seven of eight species tested suffered reduced germination at 0.34 MPa, and none germinated at 0.99 MPa. Bog expansion can be limited by a lack of sufficient water (Grosvernier *et al.*, 1997). Excessive water rarely affects primary succession, although *Sphagnum* moss may expand as water tables rise (van Breemen, 1995). 'Hydrarch' succession may be impeded by seasonal drought or alterations of the water table. A drop of 15 cm in the water table led to an invasion of raised *Sphagnum* bogs by *Pinus* trees and *Calluna* shrubs in Germany (Frankl & Schmeidl, 2000). Wetlands such as those in Azraq, Jordan, face an accelerating danger of desiccation due to draw-down of ground water for irrigation and drinking (Sampat, 2000).

Chronically extreme temperatures can limit establishment and biomass accumulation. Low temperatures retard the accumulation of organic matter on moraines, and hence slow succession (Frenot *et al.*, 1998). Thawing permafrost in Manitoba led to rapid succession from *Sphagnum* bogs to hummocks invaded by *Picea* trees (Camill, 1999). Low temperatures, unusually short seasons or persistent snows retard upward migration of timberline species, but high temperatures limit growth more frequently than low temperatures. Seedling death on tephra is usually due to heat, not drought (D. Chapin, 1995). Eggler (1963) noted that soils on Parícutin Volcano, Mexico, were hot during much of the growing season, thus intensifying drought and slowing recovery. Similarly, mined lands suffer temperature extremes during the growing season that can interact with drought or nutrient stresses.

Acidity often limits plant growth in both natural and human-created systems. Extreme pH levels immobilize P and other minerals (see section 4.3.4 on pH). Volcanic tephra is often acidic (pH 4–4.5; Wagner & Walker, 1986). Many industrial wastes are either very acid (pH 3.5 or less for coal mine tailings) or very alkaline (pH 9 or higher for Leblanc waste; Ash *et al.*, 1994). A goal of rehabilitation (Chapter 8) is often to accelerate amelioration of extreme pH levels, for example by adding lime or slag to very acid wastes or compost, manure or sewage sludge where pH is high.

Some natural sites suffer from excessive salinity. Invasion or growth in inland playas is limited to extreme halophytes, and growth on coastal dunes may be limited by salt. As lakes retreat within isolated basins, soil salinity increases. West & Young (2000) describe the transition from typical upland *Artemisia* shrubs through *Atriplex* shrubs to *Sporobolus* grasses and finally *Salicornia* forbs in the western U.S.A. Desert wetland systems may contain normal hydrophytes (e.g. *Typha* and *Scirpus* graminoids) but their margins are often saline (e.g. dominated by *Distichlis* or *Puccinellia* grasses) and terminate in a barren hypersaline playa (Hamilton & Auble, 1993). Studies of the salinity of coastal dunes have produced conflicting results. Dune plants are often moderately salt tolerant, with the most tolerant plants found nearest the shore (Barbour *et al.*, 1985). Salt spray and saline soils may or may not (Van der Valk, 1974) play an important role in controlling coastal dune succession.

Anthropogenic substrates often contain heavy metals that retard succession. Mining wastes or soil near smelters often contain copper, arsenic, lead, zinc, chromium or other heavy metals at levels too extreme for plant growth. Shrubs and trees are rarely adapted to these materials. However, Bradshaw (1952; see Bradshaw *et al.*, 1978) developed lead-tolerant strains of herbs to reclaim lead mine tailings. In the absence of remediation, derelict sites remain dominated by a few species and undergo succession very slowly. Humans may accelerate primary succession through many amelioration methods (cf. Salonen *et al.*, 1992), but the practical problem is often which forms of amelioration are appropriate (see Chapter 8).

Pre-reproductive growth

Germination does not assure success, and seedling death is common. Chapin & Bliss (1989) demonstrated that seedling survival of two tolerant perennials growing on tephra varied substantially, between 10 and 40% for two cohorts of *Eriogonum* and *Polygonum* herbs over three years. That these species are among the most tolerant in this habitat suggests that survival of most species is much lower. Annuals, to be successful, must reproduce immediately. Most do not. Perennials risk mortality for several years prior to reproduction, but survivors can accumulate resources and eventually be assured of successful reproduction. Even if the individual establishes, it remains at risk for many years. Conifer seedlings occur on the Pumice Plains of Mount St. Helens, but have not achieved reproductive success. Successful seedling establishment near mature plants is constrained by any of several factors. Plants may flower, but fail to set seed due to a lack of pollinators. Fruits may be produced, but not dispersed. Fruits may be dispersed, but not into an appropriate habitat. The individual persists in an effectively vegetative state, gradually spreading out over the terrain, but never producing seeds.

Growth to maturity

The growth phase continues with the development of a reproductively mature plant. This phase continues in ways that vary with the life history of the species in question. For annual plants, the growth phase merges into a reproductive phase. For herbaceous perennials, it may be several years before the individual accumulates sufficient reserves to flower. For any species, this phase depends on local stresses outlined above. For shrubs, this phase also can last from several to many years. For most trees, there may be a long growth phase. Tree species that reproduce quickly may be at an advantage over those that mature slowly. For example, *Pinus contorta* produced their first cones on high-elevation lahars on Mount St. Helens within 18 yr (del Moral, 1998); no other conifer there has yet reproduced on new substrates. The consequences are not yet evident, but small *Pinus* seedlings have become more abundant than *Abies* seedlings on lahars more than 100 m from the intact forest.

Growth forms
Growth form spectra change through succession and differ among succession types. Under severe nutrient stress, early succession is dominated by clonal species with good wind dispersal (Prach & Pyšek, 1999). A few examples follow which will contrast growth forms in very early succession with growth forms from later stages in several primary seres.

Turnover in early succession may be a result of simple death of short-lived pioneers, autogenic changes in the environment, competition from later-arriving species, grazing or other factors. Turnover results in growth form changes, which implies that under most circumstances, biotic interactions drive change (see Chapter 6).

On glacial moraines, species that dominate older sites are more persistent and deep-rooted. On the sub-Antarctic Kerguelen Island, a few shallow-rooted, short-lived herbs occupy the youngest terrain (Frenot *et al.*, 1998). A more robust perennial grass with deeper roots replaces them. Finally, a prostrate herb with deep, widely spreading roots achieves dominance. Only seven species occurred during the estimated 200 yr sequence, and only the annual did not persist. Rebounding coastal Arctic plains also show a shift in dominant growth forms. Small grasses and sedges are replaced along a toposequence by a robust perennial grass with low herbs and finally by mat-forming willows, rhizomatous grasses and prostrate perennial herbs (Bliss & Gold, 1994). In severe environments species accumulate, there is a preference for persistent, spreading species, and successional states are marked more by dominance shifts than by complete turnover.

Growth forms also change as dunes develop. In a simple coastal dune system on the Gulf of Cadiz, García-Mora *et al.* (1999) found that unstable dunes were dominated by perennials with deep, spreading root systems and by species capable of withstanding frequent burial. Winter

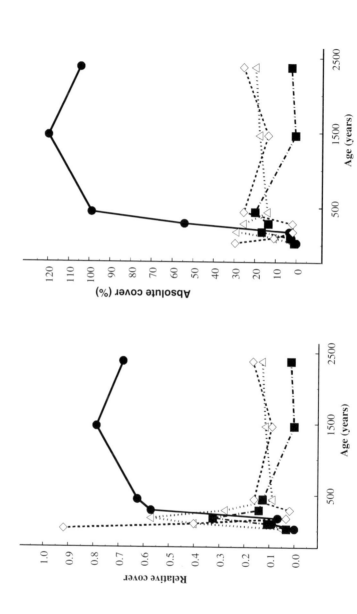

Fig. 5.13. Changes in relative and absolute cover of various growth forms along a chronosequence on Lake Michigan (U.S.A.). Filled circles, phanerophytes; open triangles, chamaephytes; filled squares, hemicryptophytes; open diamonds, geophytes. Modified from Lichter (1998). See text for more explanation.

annuals similar to inland vegetation dominated stabilized dunes, a reversal of typical progressions from annuals to persistent perennials.

Lichter (1998, 2000) studied a chronosequence of Lake Michigan (U.S.A.) sand dunes in which he sampled dunes ranging in age from 25 to 2375 yr. We categorized Lichter's species on selected dunes into Raunkiaer life forms and determined the cover of each along the chronosequence (Fig. 5.13) expressed in relative terms. Phanerophytes (woody plants taller than 50 cm) became dominant within 300 yr (*Picea*, *Thuja*, *Abies*), and continued to increase in dominance for another 1000 yr (*Tsuga*, *Quercus*, *Acer*). Geophytes (herbs that die back annually to buried perennating organs) such as *Ammophila*, then *Calamovilfa*, both sturdy rhizomatous grasses, stabilized these dunes and dominated the second stage along with chamaephtyes (woody or perennial plants with buds near the ground). A mixture of hemicryptophytes (perennial herbs that die back to the surface) dominated the mid-stage understory, then declined as tall woody species began to dominate and then virtually disappeared. Chamaephtyes (*Arctostaphylos* and *Juniperus*, followed by the ericaceous shrubs *Vaccinium* spp., *Gaultheria* and *Gaylussacia*) dominated middle stages, and then declined in relative terms, though they remained common throughout the chronosequence. This chronosequence demonstrates a typical shift in growth from dominance from geophytes to chamaephtyes and hemicryptophytes to phanerophytes. In this case, therophytes (annuals) were lacking.

Another study of dunes looked only at the first 30 yr (Olff *et al.*, 1993). Immature Dutch dunes demonstrated different life-form patterns according to the geomorphic position. The plains changed little, except that geophytes became increasingly common and hemicryptophytes invaded after 25 years. On the dry slope, phanerophytes and geophytes increased at the expense of therophytes (annuals). On the dune, hemicryptophytes (binding grasses) dominated early stages, but therophytes were virtually absent. Chamaephytes were common in the middle phase, but declined as phanerophytes invaded. The shrubby phanaerophyte *Hippophaë* appeared to exclude grasses and forbs by shading. In this study, geophytes became dominant where soil developed, but they were uncommon on dunes.

Lavas, like dunes, can require millennia for full development, requiring a chronosequence approach to study growth form changes. Poli & Grillo (1975) summarized development on lavas on Mount Etna that were 550 yr old and compared them to younger flows. We combined that work with earlier work by Poli (1965) to assemble a life-form sequence. Young lavas were dominated by therophytes (e.g. *Briza*, *Filago*, *Plantago*) with few phanerophytes or geophytes and a few hemicryptophytes (e.g. *Isatis*, *Rumex*) and chamaephytes (e.g. *Centranthus*, *Senecio*). Stable lavas developed vegetation dominated by hemicryptophytes (e.g. *Daucus*, *Lactuca*, *Crepis*) and chamaephytes (*Micromeria*). As lavas matured, small phanerophytes, notably *Genista* and *Spartium*, began to invade and geophytes (e.g. *Asphodeline*, *Leopoldia*) became common along with many species from exposed sites. Mature lavas at mid-elevations were dominated by phanerophytes (eventually by *Quercus*), with sparse understories of chamaephytes and hemicryptophytes, and few therophytes or geophytes.

Landslides may be productive and are more likely to demonstrate species and life-form turnover due to competition. Introduced herbs initially dominated communities under *Kunzea* (*Leptospermum*) *ericoides* (teatree) in New Zealand. As shade intensified, ferns and robust native herbs became dominant. After 50 yr, the rhizomatous graminoid *Microlaena* dominated the understory (Smale *et al.*, 1997). Tropical landslides are more complex. Stable mineral soils are typically colonized by herbs, then overwhelmed by climbing ferns such as *Dicranopteris* (in Puerto Rico), then colonizing trees such as *Cecropia* and *Casearia* and finally by fully mature tree species (Walker *et al.*, 1996). Here there was a distinct turnover of species and growth forms based on competitive replacement.

During early primary succession, there is normally a shift from smaller, more ephemeral species to persistent species with strong vegetative growth, based primarily upon differential success at filling empty space. Subsequent community assembly is characterized by the accumulation of taller species and competitive replacements, based on changing environmental interactions.

Biomass accumulation

Plant growth during early primary succession is limited by local factors such as infertility and drought. Isolation alone may also limit biomass development in the same way that it does species richness. For example, if one were to plant suitable species above the advancing timberline of Mount Fuji (e.g. *Arabis*, *Calamagrostis* or *Polygonum* herbs), biomass accumulation would be accelerated. Distance, not climate, has retarded the upward migration of plants, as evidenced by the gradient of alpine vegetation from tree line to almost no plant cover along an elevational gradient of just 80 m on Mount Fuji (Masuzawa, 1985). Dlugosch & del Moral (1999) demonstrated that successional development of even-aged lahar sites was correlated to elevation. Distance from colonizing sources limited invasion rates while the length of the growing season controlled biomass development.

Biomass will increase during much of a sere, although accumulations may stabilize long before species composition has stabilized. Biomass increased in each of three dune habitats studied by Olff *et al.* (1993) over 28 yr in permanent plots, but then stabilized. Biomass on the dunes increased from 200 to 1400 g m^{-2}. The shrub *Hippophaë* became dominant in dry sites. There was a complex interplay between species dynamics and biomass. Although it is normal for biomass to increase to a stable endpoint, biomass can decrease if, for example, a sod-forming grass replaces a dominant shrub.

Most species reach reproductive maturity through the accumulation of biomass. If conditions remain relatively open, these species are well positioned to dominate the site. Dominance should continue until they no longer produce successful offspring and they succumb to age or competition from later invaders.

Functional groups

Functional groups (or types) consist of species that respond in a particular way to a particular perturbation (Gitay & Noble, 1997). Díaz *et al.* (1998) compared functional traits to environmental factors across a broad region and found a strong filtering effect due to the environment. However, we are unaware of such studies conducted within one region to demonstrate either disturbance effects or more subtle environmental effects. Shao *et al.* (1996) did demonstrate that plant functional groups were correlated with patterns along a coastal barrier island. Three functional groups included (a) foredune grassland vegetation, which is adapted to dry, less stable conditions, (b) woody thickets and forests, adapted to older, stable sites, and (c) marshes dominated by hydrophytes adapted to hydric conditions. Much work remains to evaluate plant functional type changes within successional seres. In particular, this question bears on the degree of predictability in seres and on rehabilitation (Chapter 8). We will consider functional groups in more detail in Chapter 6.

5.5 Persistence and longevity

5.5.1 Persistence

The ability of an individual to persist and expand as soils mature and environmental conditions change has important consequences for primary succession. Diversity of a site can decrease when a few pioneer species persist and expand, while most pioneers die without successful replacement. Later, conditions may be so altered that even initially persistent

Fig. 5.14. A large, solitary *Arctostaphylos nevadensis* spreading on a 20 yr old lahar, Mount St. Helens, Washington (U.S.A.).

species are replaced by larger or more tolerant species. Therefore, the traits of successful invading species change through time. The ideal pioneer disperses ably, grows rapidly and therefore can produce vigorous vegetative growth. Such a species tolerates stress well and can persist indefinitely. Prach & Pyšek (1999) compared pioneers of newly formed mine spoils in the Czech Republic and found the successional dominants to be tall, wind-pollinated, capable of strong lateral spread and responsive to nutrients and moisture. This monopolist syndrome is comparable to Grime's (1977) 'competitor' strategy, not the 'ruderal' strategy common in secondary succession. Annuals were rare in these primary seres, although annual grasses are common in many other seres where conditions are less severe.

Plants will better withstand repeated disturbance if they possess strong vegetative growth. As was described above, vegetative reproduction permits local expansion from intact vegetation (e.g. edges and relics). Clonal growth also permits individuals to occupy space well before they may be able to produce viable seeds. Long-lived species such as *Arctostaphylos* shrubs (Fig. 5.14) occur sporadically on lahars on the eastern flanks of Mount St. Helens. They flower and bear fruit, but close inspections from 1995 to 2001 have never found young plants. Clones that are disturbed by browsing or erosion usually survive and recoup the damage. Future success of seedlings is also uncertain because dense moss carpets now cover most surfaces. Eventually, even these clones senesce and are invaded by woody species and robust herbs.

Sexual reproduction permits rapid colonization of the immediate surroundings if suitable conditions occur. Because the first colonists may ameliorate the local environment, they may be the center of nucleation – that is, they can facilitate the invasion of other species. To persist, particularly when secondary invaders include species that can grow over the original colonists, these colonists must produce seeds to expand the local population and to inhibit establishment of other species in favorable microsites (see Chapter 6).

5.5.2 Longevity

The ability of a species to persist is limited by its genetic potential and by the nature of the environment. Most primary seres are marked by transitions to longer-lived species. For example, Rydin & Borgegård (1991) noted a strong shift in persistence among species invading newly formed islands. Sixty percent of pioneer species were annuals, but no annuals occurred in later stages. The abundance of biennial species also declined with site age. Changes in life-history dominance imply that replacement by the same species is uncertain and persistence is crucial. A long-lived species that occupies an expanding area will produce many seeds. Paradoxically, the longer the species persists, the less likely it is that its own seedlings will replace it. The local area may be depleted of resources, there may be pathogens that preclude establishment, herbivore densities may increase or site amelioration may change conditions such that the species can no longer establish from seeds.

Longevity is also affected by environmental stresses. Species may live for prolonged periods under chronic nutrient or drought stress, simply occupying the site. This is analogous to stunted saplings, common in dark forests, that persist until some canopy disturbance affords them an opportunity to grow. Once conditions improve, persistent, but not reproductive, plants will have an advantage over those that disperse from a distance. *Metrosideros* forests in Hawaii persist for thousands of years. Individuals change leaf morphology and, under pressure from increased environmental and competitive stress, differ along the chronosequence. This species persists even though there is substantial species turnover in the understory (Kitayama & Mueller-Dombois, 1995a).

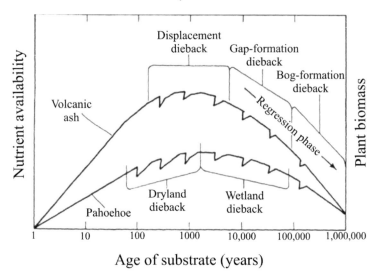

Fig. 5.15. Forest dieback during forest succession in Hawaii. From Mueller–Dombois (1986). With permission, from the *Annual Review of Ecology and Systematics* Volume 17 © 1986 by Annual Reviews. www.AnnualReviews.org

Chronic disturbance may cause the death of individuals and open a system for species not yet established because these conditions are normally less stressful than the initial conditions. Therefore, disturbance is an essential part of any sere. Herbivory can disrupt the system and may, depending on circumstances, arrest, retard or accelerate succession (see Chapter 7).

Cohorts sometimes die simultaneously. *Lupinus lepidus* forbs provide one example. Stem borers and leaf miners periodically devastate this species, which has a short life span. When dense concentrations develop, Fagan & Bishop (2000) demonstrated that herbivory hastened cohort demise and slowed the rate of population expansion. Because this species is pivotal in contributing nitrogen to infertile habitats, herbivory slowed succession. *Metrosideros* is the dominant native tree species on lavas of all ages on the island of Hawaii (Mueller-Dombois, 1987a). It experiences simultaneous cohort death perhaps due to a combination of causes such as extreme weather conditions that cause prolonged flooding or site-specific nutrient limitation (Fig. 5.15) (Balakrishnan & Mueller-Dombois, 1983; Mueller-Dombois, 1985, 2000; Gerrish *et al.*, 1988). Biotic agents (e.g. the fungal pathogen *Phytophthora* and the beetle *Plagithmysus*) may also hasten dieback in some instances. Death of the dominant canopy tree opens up the community for subordinate species such as a climbing fern,

Dicranopteris or for regeneration of different morphotypes of *Metrosideros* tree (Mueller-Dombois, 1986). Similar mass deaths due to intrinsic causes have been documented for other tree species including the European conifers, *Juniperus*, *Pinus*, *Abies*, *Quercus* and *Betula* in North America, *Abies* in Japan, and *Nothofagus* in New Zealand (Mueller-Dombois, 1987b).

5.6 Successional consequences of dispersal and establishment

5.6.1 Under-saturated early successional communities

Early seres are under-saturated compared with older ones of similar vegetation structure because stressful environments may preclude the growth of many species. Equilibrium conditions are unlikely to develop because the putative equilibrium number of species is increasing as the habitat develops more rapidly than new species can colonize. Many species that could establish simply do not reach the site.

Open habitats on Mount St. Helens were used to test the hypothesis that equilibrium species numbers are related more to dispersal than to habitat values. Four species–area curves for an isolated pumice barren (Abraham Plain) and two for a moderately isolated lahar are shown in Fig. 5.16. Data were obtained in 1999, 20 growing seasons after the sites formed. The curves flatten quickly and after fifty 10 m × 10 m plots were sampled. Thereafter, only a few rare species are added. There are more species on the lahars, which are closer to intact vegetation and are composed of old, reworked substrates. The isolated pumice has low mean cover (< 6%) and there appears to be room for invasion by additional species. However, all suitable species from the immediate area already occur. These data suggest that even 20 yr after their initiation, moderately stressful, moderately isolated sites are not in biogeographic equilibrium. While there are few new species to invade and increase the number of species, there is ample opportunity for existing species to expand so that the curves will be steeper and level off sooner.

5.6.2 Under-saturated late successional communities

Late primary successional sites should also be under-saturated because normal dispersal will not permit a full complement of species to reach a site before it is structurally mature. There will be fewer species present

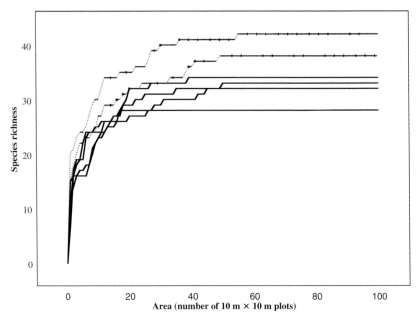

Fig. 5.16. Species – area curves in two sites on Mount St. Helens in 1999: Abraham Plains (solid lines); lahars (dotted lines). Values are cumulative number of species in one hundred 10 m × 10 m plots for each line.

than in similar mature vegetation in the immediate area that has not experienced significant disturbance for several generations of the dominant species. Data for this hypothesis are limited, but Clarkson (1990) reported that volcanic communities are significantly depauperate in species compared to older landscapes. Tussock lands and herb fields on Mount Taranaki lacked many species common in the same vegetation elsewhere. Isolated primary successional sites may not receive all species capable of growth in the developed conditions even after many decades, although species with at least moderate dispersal will arrive within a few years (Peterken & Game, 1984). This view is similar to that of Lawlor (1986), who suggested that very isolated habitats would have a very shallow species–area curve because immigration rates are so low, while extinction rates change little with isolation. In addition, the experience of restoration ecologists suggests that even with active introduction of species, mature restored habitats have fewer species than natural ones. Very old seres often begin to lose species, probably owing to site infertility (Walker *et al.*, 2000). Such ancient landscapes as those on dunes in eastern Australia (Walker *et al.*, 1981) may be analogous to anthropogenic sites that are so degraded that they cannot recover from major disturbance (see Chapter 8).

5.6.3 Novel species assemblages

Because landscape and dispersal effects can be pronounced, particular plant communities developing on new substrates may be new combinations. Plants from different habitats are thrown together and may persist for extensive periods before competitive factors reduce heterogeneity. Added to this is that many species did not occur in a region during previous major disturbances. Clarkson (1990) pointed out that early in New Zealand seres, the limited competition on recent volcanic substrates permits strong overlap among species that rarely co-occur elsewhere in mature vegetation. Tagawa (1992) noted that shrub vegetation dominated by *Ardisia* in southwest Japan formed a stable and nearly monotypic community on the smallest of three islands, but occurred with several species on the larger islands. He suggests that dispersal limited other species, permitting this novel community to persist and to resist invasion.

Early in vegetation assembly on Mount St. Helens, del Moral (1993a) noted that novel species groups had formed. These included the close and unusual association of *Lupinus lepidus* and *L. latifolius* on lahars (Braatne & Bliss, 1999). The European forb *Hypochaeris* continues to persist in association with native forb colonists including *Anaphalis*, *Epilobium* and *Hieracium*. This community has only recently formed, confined to abandoned dirt logging roads. Several wetland communities, dominated by *Salix* shrubs, with *Equisetum* horsetails or with *Juncus* rushes, are also novel (del Moral, 1999b).

5.6.4 Priority effects

If initial colonization has a stochastic component, do subsequent events override chance to produce more predictable vegetation during succession? Another way of stating this is to ask whether assembly order affects the course of succession. Assembly rules are general principles that determine how species combine to form communities (Belyea & Lancaster, 1999). If rules are strong and based on species composition, then convergence during succession should occur. If rules are strong, but based on growth form or other traits, then convergence is not inevitable, yet the sere will be predictable along functional lines. If meaningful rules do not exist, then any findings of pattern result only because there are only a few possible ways for a community to assemble (given, for example, climate or dispersal limitations). Weiher & Keddy (1995) noted that, although the experimental wetland communities they created were different from random and appeared to follow assembly rules, there was still a strong stochastic component. Species rank orders were inconsistent and many species occurred sporadically. In the context of early primary succession, so strongly affected by stochastic factors, any rules may be hidden.

Samuels & J. Drake (1997; see also D. Drake, 1990) suggested that priority effects could retard the development of communities that were predictably structured. They further asserted that assembly rules could only describe species interactions, not environmental controls. Implicit in this view is that chance leads to alternative sets of colonizers that can persist despite any putative assembly rules (see sections 3.8 and 7.1.1 on species assembly).

Tagawa (1992) noted that the chance priority arrival of *Neonauclea* trees strongly conditioned the subsequent development of forests on the three Krakatau islands. Rakata (the youngest) differs strongly from Panjang and Sertung. *Neonauclea* formed a forest on Rakata, but not on the others. These forests have resisted invasion by tree species such as *Timonius*, *Dysoxylum* and *Ficus* and therefore appear to keep assembly rules from being followed. In contrast to exclusionary processes suggested by Tagawa, Eriksson & Eriksson (1998) determined experimentally that species sown after another species had a greater probability of establishment success, suggesting that facilitation could accelerate changes and that there may be sequence rules. However, these rules are dependent on species interactions, not environmental factors. It is clear that in isolated habitats of many types, developing seres will accumulate species depending on landscape and stochastic factors and that the initial colonists may dictate subsequent development for decades. Rules involve biological interactions, that are themselves contingent on so many variables, that consistent prediction based on these rules remains unlikely over the course of primary succession.

5.6.5 Disharmonic communities

Disharmony in a community results when the vegetation is a biased accumulation of species from the surrounding landscape. The bias is the result of isolation, which selects differentially for distinct dispersal modes. However, it may also result from extreme stresses that permit only a few kinds of species to establish.

Surtsey, a small, isolated Icelandic island that formed in 1963, remained so barren in 1990 that Fridriksson (1992) could map each individual of all species except the forb *Honkenya*. The flora was an unrepresentative sample of the nearest large islands and of Iceland. Of the 24 species present in 1990, 72% were from the nearby Westman Islands and 28% from Iceland or beyond. Most (64%) were introduced by sea birds, with 27% carried by ocean currents. The remaining species were wind-dispersed. This study emphasized that geographic location, combining distance across a sea barrier with unfavorable winds, is a main factor that determines the invasion of new habitats.

Several research groups studied the Krakatau Islands extensively where Whittaker *et al.* (1997, 1999) described disharmony. They found that 46 families (represented by 91 species) found on nearby islands were not represented. The lack of suitable dispersal mechanisms was considered the most important factor, but breeding systems might also be important. For example, there may be a lack of specific pollinators (see section 5.2.1 on pollination) or the species may be obligate out-crossers.

Whittaker & Jones (1994a, b) proposed a model of recolonization on Rakata that helps to explain how dispersal leads to disharmony. The first phase was dominated by rapid colonization along the shore by sea-borne dispersal units and colonization of the interior by wind-dispersed ferns, grasses and Composites. During this early phase, animal dispersal to the interior was limited because there was little to attract birds. Sea transport continued to dominate the second phase, but wind dispersal became less important and animal transport became more important. Trees and shrubs developed into woodlands and fern dominance was reduced. Frugivores were increasingly attracted to the woodlands and accelerated the pace of colonization. In the last phase, there was little further invasion along the strand, but some species began to expand inland. These species are diplo-chorous (sea-borne), but potentially bird- or bat-dispersed; they may have formed an enticement to birds and bats that facilitated the introduction of purely endochorous species. There is now little available shore habitat and most species in the species pool have arrived. Forests have developed nearly continuous canopy and arboreal ferns and orchids have arrived. Woody species introduced by frugivores formed the largest group during this most recent phase. There were also limitations to invasion in this situation that are instructive. Species unlikely to invade the developing vegetation included additional sea-borne species because there was no available habitat, wind-dispersed species with large-winged seeds that were unlikely to cross the water barrier and large-seeded species that

Table 5.4. *Rates of colonization for primary dispersal mechanisms on Rakata Island since 1883*

Phase I, 1883 to 1897; Phase II, 1898–1919; Phase III, 1920–1989. Secondary spread is not considered, but includes at least 24 sea-colonists and 15 human-introduced species. The total was calculated from the number of years in each phase and includes all species that were found in the phase, even if they later disappeared.

	Colonization rate, species per year			
Phase	Sea	Wind, spores	Wind, other	Birds and bats
I (16)	1.64	1.0	1.0	0.14
II (21)	1.18	0.18	0.86	1.32
III (69)	0.29	0.83	0.83	1.26

Data derived from Whittaker & Jones (1994a).

are bat-dispersed or dispersed by land mammals. Table 5.4 summarizes the patterns of invasion in terms of annual rates of species colonization, derived from several sources and summarized by Whittaker & Jones (1994a).

Disharmony has evolutionary consequences. Kitayama (1996b) described the vegetation on Mount Haleakala, a basaltic volcano 800,000 yr old in the Hawaiian group. He found that species richness and turnover in wet forests was much lower than in less isolated comparison sites. Species turnover along landscape gradients was very low on Haleakala, a reflection of the generally broad amplitude of most species and their lack of specialization relative to the comparison flora. Adaptive radiation has been limited on this very isolated volcano.

5.6.6 Biogeographical effects

MacArthur & Wilson (1967) proposed the equilibrium theory of island biogeography (ETIB) that has been applied to primary succession in several ways. Rydin & Borgegård (1988a) studied islands in Lake Hjälmaren, Sweden, that were formed by lowering the lake 1.3 m in 1882. They found that a predictable species log-area curve was the best description of species richness. Interestingly, the distance effect predicted by ETIB disappeared over time. Although area was always the primary determinant, distance effects were important only for the first 20 yr. Thereafter, only area and habitat complexity were important. This suggests that all species able to grow on these islands had reached them within two decades and that thereafter distance was unimportant. The largest islands, over 0.3 ha, appeared to have equilibrated by 1985, with extinctions equal to immigration. Thirty previously recorded species did not occur in 1985, apparently due to competitive interactions. Small islands remained unstable, precluding equilibrium.

In stunning contrast to these Swedish islands is Motmot, a caldera island formed in 1968. It lacks littoral species because it is surrounded by fresh water and has not received many *Ficus* species owing to the lack of dispersal agents (Thornton *et al.*, 2001). The flora remains well below putative saturation levels and it is disharmonic, as would be expected for a young, extremely isolated habitat. As yet, the surrounding Long Island (350 yr old) also remains substantially under-saturated. Owing to its harsh physical environment, it remains in a quasi-stable equilibrium of species richness, well below that of other islands more favorably located in the region. The combined example of Long Island and Motmot cautions us not to expect plant communities undergoing primary succession to rapidly approach equilibrium numbers predicted by ETIB.

5.6.7 Establishment conclusions

All vegetation on this planet results from primary succession, although the initial effects may have long since been erased. Barren sites resulting from massive disturbance or upheavals will be colonized, but how they are colonized is determined by landscape factors and life-history properties of the proximate biota. This phase is strongly stochastic, leading to initially variable species associations. For plants, each life-history stage may have a fatal weakness that prevents success. Proximity to the newly available site is the most important factor, from which much else follows, but proximity alone does not dictate primary succession. A species in the donor pool must produce seeds at an opportune time, be dispersed by effective abiotic (wind, water or gravity) forces or have suitable animal vectors available to it. Distance and barriers to the newly created site restrict the types of initial colonist. The barren landscape must trap seeds and offer opportunities for seedling germination and establishment. The degree of habitat stress governs the initial rate of biomass accumulation and ultimately of species turnover. Once an immature plant is established, it must withstand physical stresses (e.g. drought, low nutrients, high temperatures, instability) and persist. Usually, a viable population is built by successful on-site reproduction. However, the pioneering species may sometimes only serve to ameliorate the site and fail to reproduce successfully. Successful early colonists will expand vegetatively and by seed to rapidly colonize the site.

Communities assembling early in succession are governed largely by chance. This can lead to novel and to disharmonious communities. Disharmony has ecological consequences when a few species continue to dominate a sere, and evolutionary consequences when such taxa undergo adaptive radiation. Most primary seres remain under-saturated for a very long time because colonization cannot keep up with local extinction and because many competent species fail to reach the site.

Arrested succession occurs when better competitor species fail to arrive. Rarely, the population of a potential dominant may not develop due to a lack of pollinators. There also may be priority effects, so that the sere is strongly regulated by the initial colonists. As a system develops, the relationship between species and their environment becomes more predictable in part because competition intensifies to eliminate less adapted species. However, which set of species may occur in later stages of a sere remains less predictable.

The first colonists do not normally persist, although where the pool of immigrants is restricted and conditions are stressful, they may. More typically, ameliorating conditions enhance invasion probabilities of less stress-tolerant species. Species turnover results from the combination of competitive exclusion along a temporal gradient and the elimination of safe-sites for colonist species. After the initial rush of establishment in open habitats, primary succession is driven largely part by biotic interactions. We will explore these in Chapter 6.

6 · *Species interactions*

6.1 Introduction

Species interactions are pivotal to succession because species change is not merely the independent response of each species to physical changes in its environment. Species' impacts on their environment affect other species. Even in the barren landscapes of early primary succession, species interactions are important. Exceptions include some dunes, floodplains or volcanoes where persistent abiotic disturbances reduce the importance of interspecific interactions (Partomihardjo *et al.*, 1992; Houle, 1997; but see Shumway, 2000) or isolated areas where dispersal is the principal limiting factor (Fridriksson, 1987; Lukešová & Komárek, 1987; see section 5.3 on dispersal). Sometimes overall patterns of succession are best explained by changing life histories and life forms of dominant species without a need to invoke species interactions (Walker *et al.*, 1986; McCook & Chapman, 1993; Chapin *et al.*, 1994). However, interactions generally alter both the rate and trajectory of primary succession.

Plants selectively concentrate soil nutrients, transport water from the soil to the atmosphere and add organic matter when they decay. Plants also remove resources and impact their neighbors that need the same, often limiting resources. Animals alter soils by burrowing, feeding, defecating and dying. Animals use plants as food, nest sites and protection and affect each other as they search for these resources. All these interactions have a profound impact on succession. Studies of interactions in primary succession are particularly productive because of the opportunity they provide to observe how community interactions develop without the complications of legacies from earlier communities.

Succession is driven by a continuum of factors ranging from those that clearly originate from outside the system of interest (allogenic) to those that clearly originate from within (autogenic; Tansley, 1935). These agents of species change are variably called causes or mechanisms, depending on the organizational level of concern (Pickett *et al.*, 1987).

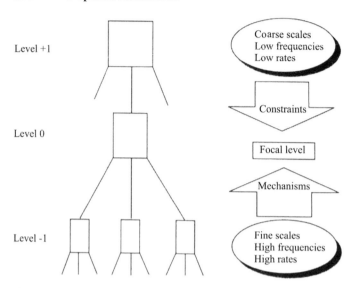

Fig. 6.1. Hierarchies of causes and mechanisms. Reprinted from Pickett *et al.* (1999), with permission from Elsevier Science.

For example, a volcano may 'cause' primary succession, but the mechanism (one hierarchical step down) is species change. Similarly, species change may cause succession, but competition is the mechanism, or competition may cause succession, but nutrient depletion is the mechanism (Fig. 6.1) Clearly, the spatial scale of concern needs to be defined in order to determine what is within the system of interest (Tilman & Kareiva, 1997; Pickett *et al.*, 1999). Similarly, the hierarchical level of interest (community, population, individual) needs to be designated. Van Andel *et al.* (1993) suggested that from the perspective of change in plant species, allogenic factors could be described as causes and autogenic factors as mechanisms. Although this approach only defines the hierarchical level and leaves the spatial scale to be defined by the investigator, we adopt it for the purposes of this chapter. Species interactions within a community will be considered autogenic drivers of succession. First we look at how plants and microbes interact to promote soil development. Then we focus on the interactions of established plants and animals. We will examine the successional implications of positive (facilitative) and negative (competitive, allelopathic) interactions among plants, interactions between plants and animals (herbivory) and interactions among animals (heterotrophic succession).

6.2 Plant–soil and animal–soil interactions

In Chapter 4 we focused on how soils develop in primary succession. Plant, microbial and animal processes were viewed from a decidedly belowground perspective. In this section, we return to the more conventional aboveground perspective and discuss how plants and animals change the environment in which they live. This concept was termed reaction by Clements (1928), who considered reaction the key to all succession because he believed that it explained the progression of stages. He distinguished it from the response and adaptation of an organism to its environment and noted that organisms directly affected the physical environment, thereby indirectly affecting each other. We will discuss three types of reaction: the impacts of organisms on soil formation, stabilization and microclimate (Matthews, 1992).

6.2.1 Plant impacts on soils

Plants alter their environments in many ways that have successional consequences (van der Putten, 1997). We described some of the effects of plants on soils in section 4.4. Here we elaborate on the broader successional implications of plant reaction.

The decay of plant litter and addition of carbon is the biggest contribution of plants to soil formation in the earliest stages of primary succession and to soil modification in later stages. Plant litter contributes to humus formation, water retention, nutrient addition and acidification (see section 4.5.2 on organic matter). However, even these simple generalizations must be qualified. In very wet conditions, plant litter and the development of organic soils may actually dry out soils through a build-up of the soil surface and improved drainage (Clements, 1928). Plant growth can also displace soil water through retention in plant tissues and transpiration. Further, plant root exudates add nutrients but plants absorb and immobilize nutrients during growth. Acidification of soils results from accumulation of organic acids in plant litter (see section 4.3.4 on pH). Each of these effects of plant litter has indirect successional implications because of the variable tolerance of other plants to the altered conditions.

Litter quality has an important influence on succession through its feedback to nutrient availability (Moro et al., 1997). In infertile ecosystems, plants grow slowly, utilize nutrients efficiently and produce only small amounts of nutrient-poor litter that decays slowly and may immobilize nutrients, providing a negative feedback loop to further slow soil development (Chapin et al., 1986). The opposite occurs in productive ecosystems with more rapid decomposition and higher rates of N mineralization (Hobbie, 1992) and soil development. Early stages of primary succession are usually N-limited. Nitrogen additions (from litter of an invasive N-fixer) increased litter decomposition and N mineralization in early primary succession on volcanic tephra in Hawaii (Vitousek & Walker, 1989). Soils develop more rapidly in such high-nutrient sites, but species turnover may not be more rapid, depending on plant–plant interactions. Fast-growing species that monopolize resources at high-nutrient sites can inhibit succession (Walker & Vitousek, 1991; see section 6.3.2 on inhibition). Sometimes litter inputs also promote species change. Leaf litter of *Populus* trees on the Tanana River in Alaska had negative effects on N-fixation, decomposition and N mineralization in soils from an earlier stage dominated by the shrub *Alnus*, thus promoting the change from *Alnus* to *Populus* (Schimel *et al.*, 1996, 1998).

Direct effects of plant litter on succession include impacts on the germination and growth of other species. Large leaves or other plant parts can crush or cover young seedlings and reduce water infiltration. Litter that is suspended takes longer to decompose (see section 4.5.2 on organic matter). Depth of litter can impact colonization and growth of seedlings or vegetative sprouts through physical means (Guzmán-Grajales & Walker, 1991; Berendse *et al.*, 1998) or through chemical leachates (del Moral & Muller, 1970).

Plants have other long-term effects on soil development than carbon additions through litter (van der Putten, 1997). These include nitrogen fixation, mineral weathering by exudates, maintenance of soil organisms, influences on atmospheric deposition (e.g. of S, N) and impacts on the physical properties of soils. Positive feedbacks for plants may occur as a result of species with significant ecosystem impacts (sometimes called bioengineers; Jones *et al.*, 1994; van Breemen & Finzi, 1998; see section 8.3.1 on dispersal). Plant fitness and plant effects on soil properties can be tightly or loosely coupled, or indirect (Fig. 6.2) (Binkley & Giardina, 1998). Tightly coupled processes might involve a positive feedback loop where properties of leaf litter of a particular species, for example, maintain soil fertility at levels that give that plant a competitive advantage over other species (e.g. conifers that create acidic, low-nutrient soils from their litter). A loosely coupled interaction might involve an N-fixer gaining a temporary but not a long-term benefit from increased N levels in the soil. Uncoupling of plant fitness and soil properties could occur in systems with frequent disturbances or simple opportunism of soil biota to resources regardless of their source (Binkley & Giardina, 1998). We suspect

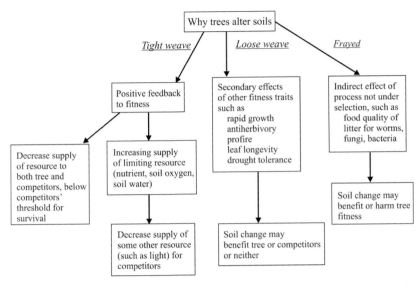

Fig. 6.2. Hypothesis diagram for the types of connections that weave trees and soils together. From Binkley & Giardina (1998), with kind permission from Kluwer Academic Publishers.

that plant–soil interactions in primary succession begin uncoupled and gradually develop closer interactions until greater coupling occurs in the later stages of succession.

Spatial variability in plant effects is most obvious in early stages of primary succession in harsh environments (Alpert & Mooney, 1996). The establishment of a colonist initially ameliorates the environment in the immediate vicinity by shading and moderating potentially extreme ambient temperatures. Subsequent impacts such as entrapment of wind-blown materials, root effects and nutrient accumulations accentuate the differences between areas with and without plants (Eggler, 1963; Wright & Mueller-Dombois, 1988). Islands of relatively high fertility can develop where the colonists are long-lived (Fig. 6.3) (Garner & Steinberger, 1989). These islands become centers of colonization from which subsequent colonists expand (see Chapter 5) in a process Yarranton & Morrison (1974) called nucleation. They suggested that clump sizes of colonists grow until they coalesce, then decline in size as clumps of the second wave of colonists grow and coalesce in turn. The tendency of organisms to clump (Tilman & Kareiva, 1997) is just one of many spatial dynamics that affect colonization and species interactions (Silvertown & Antonovics, 2001).

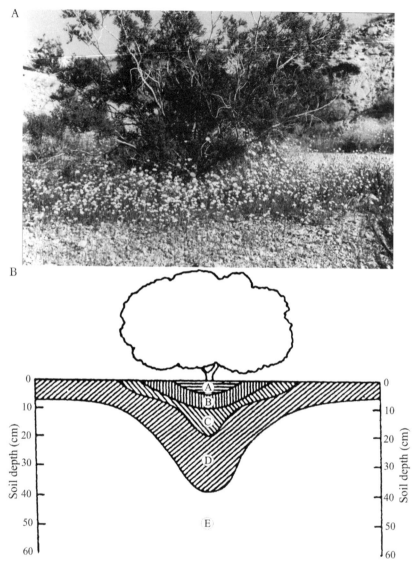

Fig. 6.3. Desert plants can act as nurse plants, favoring growth of understory species, and can also concentrate nutrients in fertile islands of soil. (A) A *Larrea* shrub in the Mojave Desert (U.S.A.) with annual plants growing underneath it but not in adjacent open areas between shrubs. (B) A diagram of a fertile island under a desert shrub, showing decreasing levels of total soil nitrogen with distance from the roots. Zone A, more than 0.06% N; B, 0.05–0.06%; C, 0.04–0.05%; D, 0.03–0.04%; E, 0.02–0.03%. Modified from Klemmedson & Barth (1975) and Garner & Steinberger (1989).

Table 6.1. *Types of interaction in primary succession*

Interaction type	Description	Successional implications and mechanisms	Sample references (see text for more)
One-way			
+	Obligatory facilitation	Succession only if facilitator present	Turner, 1983
(+)	Facultative facilitation	Accelerates succession	Jacquez & Patten, 1996; McAuliffe, 1986; Harris et al., 1984
	Direct	Herbivore protection	
	Indirect	Site amelioration	Poulson, 1999; Bruno, 2000; del Moral & Wood, 1993a; Jumpponen et al., 1998b
	Immediate		
	Delayed	Delays succession	
−	Competitive inhibition		Chapin et al., 1994; Nakamura et al., 1997; Poulson, 1999
	Direct	Removes light, water, nutrients	
	Indirect	Changes disturbance regime	
+/0	Commensalism	Nurse plants	Yeaton, 1978; Eccles et al., 1999; Holzapfel & Mahall, 1999
Two-way			
+/+	Double facilitation (or facultative mutualism)	Mutualisms, nurse plants that benefit	Moro et al., 1997; Eccles et al., 1999
+/−	Contramensalism	Facilitator later out-competed	Gaynor & Wallace, 1998
−/+	Competitive inhibition (or competitive displacement)	Inhibition benefits inhibitor	Young et al., 1995
−/−	Non-hierarchical	Delays succession	
	Hierarchical	Effects depend on life histories	Benedetti-Cecchi & Cinelli, 1996
		Increasing competitiveness during succession	Horn, 1976
Three-way	Indirect facilitation or competition	Promotes species diversity	Breitburg, 1985; Levine, 1999
	Differential responses	Promotes species diversity	Bellingham et al., 2001
	Cyclical	Promotes species diversity	McAuliffe, 1988; Valiente-Banuet & Ezcurra, 1991; Eccles et al., 1999

Spatial gradients of soil fertility within the clumps (with highest values midway toward the center; Moro *et al.*, 1997) can interact with gradients of light (with highest values near the edge) to increase spatial heterogeneity and further impact species replacements and succession. Eccles *et al.* (1999) reviewed causes of clumping in desert vegetation in a South African desert. If clumps are due to dispersal limitations, and competitive interactions occur within clumps, then over successional time the clumps should dissipate and spatial patterns of plants should become random and then regular (Phillips & MacMahon, 1981; Prentice & Werger, 1985). If clumps result from cyclic replacements, then asymmetrical competition should exist first between the nurse plant and the understory, then between the emergent understory and the original nurse plant (Valiente-Banuet & Ezcurra, 1991; see section 6.3.1 on nurse plants). The relationship between a nurse plant and its understory can consist of a unilateral positive impact of the nurse plant and no effect of the understory on the nurse plant (commensalism) or a facultative mutualism (Table 6.1; see section 6.3.1 on facilitation). Temporal patterns in soil fertility in deserts are difficult to determine and often involve millennia (McAuliffe, 1988). Bolling & Walker (2002) found no temporal changes in patterns of soil fertility in the Mojave Desert on abandoned roads over an 88 yr chronosequence, but control plots had more clearly defined fertile islands (Fig. 6.4) suggesting the very slow development of the fertile islands in that desert.

The first plant colonists in primary succession stabilize surfaces by retention of wind-blown or water-borne particles aboveground and by the anchoring effect of roots (Matthews, 1992). Microbial surface crusts on glacial moraines (Worley, 1973; Wynn–Williams, 1993), dunes (Forster & Nicolson, 1981; Grootjans *et al.*, 1998) or deserts (Belnap, 1995) reduce erosion and can provide a stable surface for retention and growth of mosses and vascular plants or hinder their establishment (Chapin *et al.*, 1994; Grootjans *et al.*, 1998). Roots of larger vascular plants stabilize substrates at larger scales. *Alnus* shrubs at Glacier Bay reduced gully formation (Cooper, 1931) and later-successional trees resist all but landscape-level surface disturbances such as riverbank erosion or permafrost rise and resultant wind throws (Noble *et al.*, 1984; Viereck *et al.*, 1993). Sediment deposition in floodplains is highest where tree trunks are the densest (Nanson & Beach, 1977) and increasing elevation of floodplain surfaces promotes succession by permitting invasion of less flood-tolerant species (Menges & Waller, 1983; Johnson *et al.*, 1985; Walker *et al.*, 1986). Small, mat-forming herbs such as *Raoulia* on rocky New Zealand floodplains

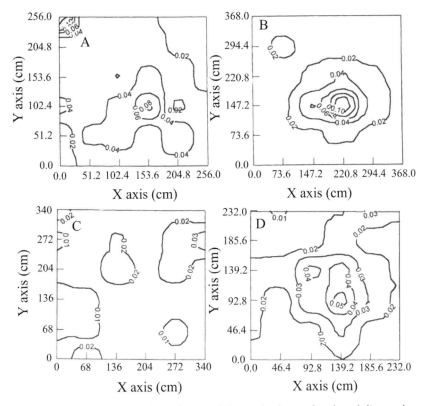

Fig. 6.4. Patterns of total soil N (%) around *Larrea* shrubs on abandoned dirt roads in the Mojave Desert. The shrub base was in the plot center and the axis labels represent sampling locations ($n = 36$ points per shrub). (A) Road 21 yr old; (B) undisturbed desert control for (A); (C) road 88 yr old; (D) undisturbed desert control for (C). From Bolling & Walker (2002).

trap silt, which provides a moisture-retaining seed bed for larger plants (Gibb, 1994).

Plants modify their microclimate by altering humidity and light (Smith *et al.*, 1987; Franco & Nobel, 1989; Garner & Steinberger, 1989; Schlesinger *et al.*, 1996). Reduction of wind speed, increased transpiration, condensation of fog, entrapment of wind-blown rain or snow and development of a litter layer and soil all modify humidity. Soil temperature extremes can be moderated by vegetation (e.g. by a stand of *Betula* trees on a glacial moraine in Iceland; Lindroth, 1965). Light reduction is dependent on leaf and branch morphology and plant height and density, which can result in gradually decreasing light availability from the edge to the center of a canopy (Moro *et al.*, 1997). Microclimate changes can alter the rate and trajectory of primary succession by impacting the balance of competition and facilitation and the success of species at each life history stage.

6.2.2 Animal disturbances

Animals are essential to soil development (see section 4.4.4 on animals). They play a key role in the compaction and/or aeration of soil, distribution of litter, nutrient addition and decomposition. Their role in herbivory is discussed in section 6.4.2. Large animals compact soils, particularly on trails and around water holes and salt licks. These activities can break down and mix litter particles into the soil surface, thereby increasing decomposition and potentially nutrient loss through leaching. Large animals can increase soil temperatures and decrease soil moisture, or they can reduce nutrient loss through compaction and reduction of erosion (Willig & McGinley, 1999). Compaction reduces opportunities for seedling establishment. Animals also aerate soils by digging and burrowing. Small invertebrates such as ants, termites, earthworms and nematodes are, of course, extremely important in soil mixing. Desert litter 'decomposes' faster than expected because termites move it quickly underground (Whitford *et al.*, 1981). Animals add nutrients through their feces and carcasses. These nutrient hotspots can become foci for colonization, especially when seeds or spores are transported in the feces (see section 5.3.2 on dispersal mechanisms), as illustrated by iguanas on new volcanic surfaces in the Galápagos Islands (Hendrix, 1981). Heterotrophic succession on carcasses is discussed in section 6.5.

6.3 Interactions among plants

All plants need the same resources of light, nutrients, water and space, yet these requirements differ among species and change during the life history of each species. Most plants modify their immediate environment in some way that can impact establishment and growth of both other species and other individuals of the same species. Differential species responses to these environmental changes can drive succession. For example, Wright & Mueller-Dombois (1988) found that early colonists on volcanic cinder in Hawaii were adapted to higher light intensities and had deeper roots than the subsequent group of colonists, in conformance with Tilman's (1987) model of differential tolerance. Habitat amelioration under shrubs improved establishment only of the latter group. In the following section we describe various ways in which species interact and how those interactions affect succession processes. When species interactions appear to have little influence on succession, life history factors such as arrival times, growth rates and longevities drive successional change (Walker et al., 1986; Farrell, 1991; see Chapter 5).

6.3.1 Facilitation

Facilitation is broadly defined as any positive influence of one species on another. However, the term is most appropriately used in a successional context (Van Andel et al., 1993) to imply an influence that promotes species compositional change to the next stage (Connell & Slatyer, 1977). Facilitation is the successional consequence of Clements' concept of reaction. Obligatory facilitation occurs when the establishment and subsequent habitat modification by one species is essential for the establishment of a later species. This habitat modification may involve soil enrichment or structural changes. Certain red algae, for example, were necessary to provide structures for the establishment of surf grass (*Phyllospadix*) on the rocky coasts of Oregon, U.S.A. (Turner, 1983). When the presence of one species is not essential for later species, facultative facilitation is much more likely. In this scenario, all species can establish in early successional environments, but facilitation increases the likelihood of establishment. Implicit in most examples of facilitation is that the facilitator is incapable of replacing itself (Clements, 1928). This can occur because the facilitator is relatively short-lived, because the environment changes so that self-replacement is impossible, or because later successional species directly inhibit pioneer regeneration. These three mechanisms are not mutually exclusive. The third scenario has been called contramensalism (Table 6.1). For example, pioneer bryophytes on bare rocks facilitated colonization of the liverwort *Conocephalum*, which then smothered the pioneers (Gaynor & Wallace, 1998).

At the community level, facilitation is generally indirect. Resources may be augmented in ways that favor any nearby individuals or later arrivals through improvement of nutrient availability, substrate stability, microclimate or other variables. The effects of indirect facilitation can be delayed until long after the facilitator dies (del Moral & Wood, 1993b), making it difficult to exclude the presence of facilitation in short-term studies. Between two individuals, facilitation can have a relatively direct impact, as when a dense shrub physically shelters an annual from herbivory. Thus, facilitation does not necessarily involve modifications of the physical environment (cf. Jacquez & Patten, 1996). Another type of indirect facilitation is the improvement of species C that results when species A inhibits B and species B would otherwise inhibit C (Breitburg, 1985; Connell, 1990; Levine, 1999). Mutual or double facilitation (Table 6.1) is a form of mutualism (see section 6.4.1 on mutualisms) that can occur when a facilitator is also facilitated, as when herbs grow best under a shrub and return leaf litter that benefits N mineralization for the shrub (Moro et al., 1997).

Background

Clements and other ecologists in the first half of the twentieth century emphasized reaction as the critical driving force of successional change, particularly in the harsh environments of primary succession. The popular Clementsian paradigm of a directional succession to a climatically determined climax and successional equilibrium (see Chapter 3) lost some of its support during the middle of the twentieth century and alternative non-equilibrium views of contemporaries of Clements gained support (McIntosh, 1985; Glenn-Lewin et al., 1992). However, the role of facilitation was still considered essential, particularly in severe environments undergoing primary succession (Connell & Slatyer, 1977). Subsequent experimental approaches to succession failed to find evidence of obligatory facilitation in secondary succession. However, proving that one species must precede the next in succession is difficult because of the absolute nature of the statement and the complex interactions and often delayed implications of many factors. Variable site conditions and plant life histories can produce several alternative trajectories. If only one later successional species grows without the earlier successional facilitator, then the obligatory role of facilitation is nullified. Matthews (1992), after a comprehensive and global literature survey, concluded that there was no convincing evidence for obligatory facilitation on glacial moraines. Nor is there a broadly applicable facilitative role for N-fixers in all types of primary succession (Walker, 1993). On the other hand, evidence for facultative facilitation has accumulated in the last decade (Callaway, 1995), and it now appears that facilitation often influences succession in conjunction with inhibition (Connell & Slatyer, 1977; Walker & Chapin, 1987; Callaway & Walker, 1997). Although facilitation is presumed to be more prevalent in severe environments than in more favorable ones (Walker & Chapin, 1987; Bertness & Callaway, 1994; Callaway & Walker, 1997; see Fig. 3.7), too little comparative work has been done to adequately test this hypothesis (S. Wilson, 1999).

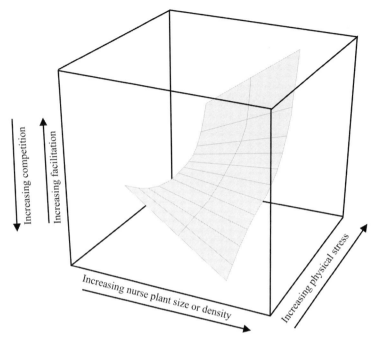

Fig. 6.5. Conceptual model of the effects of size or density of nurse plant(s) and physical stress gradients on the relative importance of facilitation and competition. Under high physical stress, increasing nurse plant size or density increases the relative importance of facilitation. Under more favorable physical conditions, increasing nurse plant size or density increases the relative importance of competition. Modified from Callaway & Walker (1997).

Environmental severity includes both stress (resource depletion) and disturbance (resource disruption), and species interactions may be differentially affected by these two factors (Brooker & Callaghan, 1998). Facilitation is also dependent on the amount of resources available. Although severe environments typically lack adequate levels of most resources for plant growth, some resources may be plentiful. For example, in a wet, oligotrophic system a species that increases soil water retention does not have the same facilitative effect as one that increases nitrogen (Gold & Bliss, 1995). A further influence on facilitation is the relative density and size of the interacting species. Dense thickets or large species are more likely to facilitate new plants in harsh environments and to inhibit their establishment in favorable environments (Fig. 6.5) (Callaway & Walker, 1997). Finally, all discussion of facilitation must specify a time frame, especially in relation to the life histories of the species involved.

Table 6.2. *Facilitation affects different life history stages of a plant*

Life history stage	Mechanism of facilitation	Sample references
Dispersal	Bird perches (birds defecate seeds)	Vitousek & Walker, 1989; Campbell *et al.*, 1990
	Entrapment of seeds/spores by vegetation	Day & Wright, 1989; Choi & Wali, 1995; Gagné & Houle, 2001
Germination	Scarification in animal guts	Khurana & Singh, 2001
	Site amelioration	Carpenter *et al.*, 1987; Morris & Wood, 1989; Blundon *et al.*, 1993; Gold & Bliss, 1995; Vazquez *et al.*, 1998
Establishment and growth	Site stabilization	Johnson *et al.*, 1985; Moreno-Casasola, 1986; Friedman *et al.*, 1996b
	Mycorrhizae	M. Allen *et al.*, 1999
	Decreased herbivory	Bryant & Chapin, 1986; Callaway, 1992
	Site amelioration	Birks, 1980b; Valiente-Banuet & Ezcurra, 1991; Adachi *et al.*, 1996b; Hodkinson *et al.*, 2002
Reproduction	Present alternative food for pollinators	Rathcke, 1983; Stone *et al.*, 1998
Survival	Decreased herbivory	Bryant & Chapin, 1986
	Decreased disturbance	Rosales *et al.*, 1997

Facilitation can affect specific life history stages of a plant (Table 6.2) (Tiffney & Barrera, 1979). Plant dispersal to a site is facilitated by perches where birds defecate seeds or by vegetation that traps seeds and spores. Germination can be facilitated by scarification of seeds as they pass through animal guts or by site conditions that increase soil water, nutrients or organic matter (see section 5.4.1 on germination). For example, algae retained water in dune slack systems and so facilitated germination of vascular species (Vazquez *et al.*, 1998) and fungi on Mount St. Helens concentrated nutrients that allowed other fungal species and mosses to colonize (Carpenter *et al.*, 1987). Establishment and growth can be facilitated by factors that stabilize a site, by the presence of mycorrhizae, by the reduction of herbivory or by improved soil water, nutrient and organic matter content. Stabilization of cobble beaches by *Spartina* grasses allowed other species to colonize (Bruno, 2000), providing the *Spartina* patches reached a critical size ('conditional facilitation'; Bruno & Kennedy, 2000). Bacterial biofilms growing on rock surfaces in streams may facilitate colonization by algae (Stevenson, 1983). *Fucus* growth on newly exposed intertidal rocks following ice-scour was facilitated by a thin growth of blue-green algae (but inhibited by the thicker growth of filamentous green algae; McCook & Chapman, 1993). Reproduction can be facilitated by the presence of alternative food sources for pollinators at the site. Survival of a plant can be facilitated by reduction in herbivory or disturbance. Combinations of effects occur across the life history stages of single species, making it essential that the net effects of positive and negative interactions be integrated. For example, *Alnus* shrubs in post-glacial succession inhibited germination of a long-lived tree such as *Picea* but the nitrogen added by *Alnus* facilitated growth of mature *Picea* (Chapin et al., 1994). In contrast, the N-fixing shrub *Myrica* facilitated recruitment, growth and reproduction of two herbs on dunes in Massachusetts, U.S.A. (Shumway, 2000).

Succession can be facilitated when species of different longevities are introduced. For example, on New Zealand floodplains, the invasive shrub *Buddleja* displaced a longer-lived native shrub (*Kunzea*), and this accelerated succession (Smale, 1990). Succession on abandoned mines was accelerated when *Kochia* herbs produced allelopathic substances that reduced herbivory but prevented its own reproduction. *Kochia* was soon replaced by the second wave of colonists (Iverson & Wali, 1982; Wali, 1999a), which benefited from the shelter provided by *Kochia*.

The timing of facilitative interactions during succession alters the effects of facilitation. Facilitation is most obvious during early succession when resources are limited, relative growth rates are high and life spans are short. Remnant plant populations, common in such primary seres as landslides, can facilitate succession by providing plant material and residual microbial or animal populations that speed recolonization of the new disturbance (Eriksson, 2000). New colonizers that are able to respond to such early facilitation but that are longer-lived can have an important and cascading effect on subsequent successional trajectories. Facilitation later in succession may have less substantial implications, merely shifting species composition within an established trajectory.

Types of direct facilitation between individuals include symbiotic relationships such as mycorrhizae, root connections and N-fixers. The successional effects of these relationships are discussed in Chapter 4 and section 6.4.1 on mutualisms. Another type of direct facilitation involves colonists that provide physical protection from herbivores for later arrivals. In the Sonoran Desert (U.S.A.) the establishment of the tree *Cercidium* was more likely where small *Ambrosia* shrubs protected it from rabbit herbivory (McAuliffe, 1986). Sometimes initial colonists provide safe-sites for regeneration of later invaders. Kelp regeneration off the California (U.S.A.) coast was facilitated by fast-growing algal stands following storm damage (Harris *et al.*, 1984) and early colonists of waterlogged dune slacks in The Netherlands facilitated other species by releasing oxygen through their roots (Grootjans *et al.*, 1998). On cliff faces in Spain, plants able to grow in cracks in the rock facilitated soil development and eventual colonization of species requiring a more stable supply of water and nutrients (Escudero, 1996).

Light

Shade can facilitate germination and survival of species that do not grow well in the open conditions of early primary succession (Moro *et al.*, 1997; see Nurse plants, below), often (Berkowitz *et al.*, 1995; Holmgren *et al.*, 1997) but not always (Alpert & Mooney, 1996) by ameliorating drought stress. Drought-resistant plants that are early colonists of both limestone and granite rock outcrops in the eastern U.S.A. are poor competitors for light (Baskin & Baskin, 1988; Ware, 1991). Frequent drought stress severely limits succession in these habitats (Anderson *et al.*, 1999). However, in mesic areas, growth is more often inhibited than facilitated by shade (Fig. 6.6) (Holmgren *et al.*, 1997; see section 6.3.2 on inhibition). The adaptation of plants to different intensities of light helps explain species replacements in succession (Tilman, 1985).

Nutrients

Indirect facilitation through habitat amelioration occurs in many ways, but, as noted above, it has rarely been shown to be obligatory. The most common examples are increases in nutrients, organic matter and water-holding capacity discussed in section 6.2.1, none of which is dependent on a particular species. Microbial colonizers of glacial moraines in Antarctica (Smith, 1993) and polar deserts in the Arctic (Gold & Bliss, 1995) may be precursors for bryophytes, lichens or vascular plants because they provide stability, organic matter and nutrients. Various mosses and vascular plants then invade the new soils (Viereck, 1966; Birks, 1980b), but the process is not deterministic or closely predictable. A shrub such

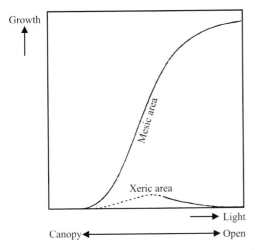

Fig. 6.6. Growth of establishing plants along a gradient from canopy shade to adjacent open areas. Note that in mesic areas, existing shrubs inhibit growth. In xeric areas, shrubs facilitate growth (negative slope indicating growth decreases with distance) under moderate to high light conditions, but potentially inhibit growth (positive slope) under shadier conditions. The dashed line suggests that deep shade rarely occurs in xeric habitats. From Holmgren *et al.* (1997).

as *Dryas* typically invades the moraine and serves as a collector of wind-blown silt.

Vascular plants with N-fixing symbionts (N-fixers) are frequently believed to facilitate primary succession because of the low initial N levels (Walker, 1993; Clarkson & Clarkson, 1995). The most comprehensive early evidence for facilitation in primary succession came from several sets of data from Glacier Bay, Alaska. Crocker & Major (1955) showed that *Alnus* shrubs increased soil nitrogen and SOM (soil organic matter) and decreased soil pH. They speculated that a drop in soil N after 100 yr resulted from the death of the *Alnus* stage and simultaneous uptake of soil N by the next successional dominant, *Picea*. They suggested that growth of *Picea* trees was normally dependent on N fixed by *Alnus*. Lawrence and colleagues (Lawrence, 1958, 1979; Schoenike, 1958; Lawrence *et al.*, 1967) showed that the addition of either leaves of *Alnus* or *Dryas* (the other N-fixer in the system) improved growth of *Populus* trees. These studies and that of Crocker & Major (1955) were widely cited as evidence of the importance of facilitation in primary succession. Bormann & Sidle (1990) confirmed the inputs of soil N by *Alnus*, and Chapin *et al.* (1994) demonstrated the positive influence of *Alnus* on *Picea* seedlings.

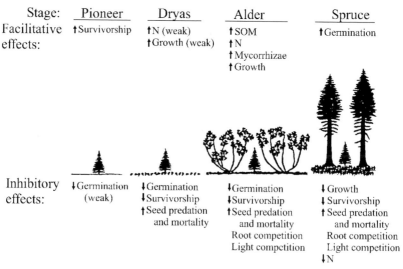

Stage:	Pioneer	Dryas	Alder	Spruce
Facilitative effects:	↑Survivorship	↑N (weak) ↑Growth (weak)	↑SOM ↑N ↑Mycorrhizae ↑Growth	↑Germination
Inhibitory effects:	↓Germination (weak)	↓Germination ↓Survivorship ↑Seed predation and mortality	↓Germination ↓Survivorship ↑Seed predation and mortality Root competition Light competition	↓Growth ↓Survivorship ↑Seed predation and mortality Root competition Light competition ↓N

Fig. 6.7. Successional impacts of post-glacial successional stages on establishment and growth of *Picea* seedlings at Glacier Bay, Alaska. From Chapin *et al.* (1994).

Chapin *et al.* (1994) also found that the *Alnus* stage inhibited germination and survival of *Picea* seedlings (Fig. 6.7). Although transplanted *Picea* seedlings grew better in *Alnus* thickets than in earlier successional stages, within the *Alnus* thicket, trenching around the seedlings (to reduce root competition) or tying the *Alnus* canopy back (to reduce light competition) further improved *Picea* seedling growth. Fastie (1995) questioned the validity of the chronosequence used by Crocker & Major (1955), suggesting instead that the *Picea* stands used by Crocker & Major did not begin with an *Alnus* stage. Sampling in the same stands as Crocker & Major, Fastie found much more rapid growth in younger *Picea* stands with an *Alnus* history than in older *Picea* stands where *Alnus* had not been an important component of the vegetation. Thus, the apparent decrease in soil nitrogen in the older *Picea* stands, used by many to explain facilitation of *Picea* by *Alnus*, was in fact due to different trajectories of succession. Walker (1993, 1995) suggested that soil N accumulation at Glacier Bay is unusually slow compared with other glacial moraines with vascular N-fixers. Low N levels coupled with a mesic environment favorable to the growth of large, leafy N-fixers such as *Alnus* may make facilitation more likely at Glacier Bay than at most other primary seres (Matthews, 1992; Walker, 1995). Because *Picea* does not require *Alnus* to colonize (Cooper, 1923) and may grow at least as well in *Salix* as in *Alnus* thickets (Cooper, 1931), obligatory facilitation can be ruled out.

The presence of *Alnus*, *Picea* and other species is also related to dispersal abilities of each species (Chapin, 1993). Therefore, for Glacier Bay, succession is driven by a combination of facilitative, competitive and life history factors (Chapin *et al.*, 1994).

There are many more examples of symbiotic (Walker, 1993) and non-symbiotic (Vitousek *et al.*, 1987; Walton, 1993) N-fixers that may facilitate primary succession. Perhaps the most convincing examples come from mined lands, where symbiotic N-fixers are common (Walker, 1993). Trees (Leisman, 1957; Schuster & Hutnick, 1987) appear to grow better on mines where there have been N-fixers. On dunes in The Netherlands, plant biomass increased following a stage dominated by the N-fixing shrub *Hippophaë* (Olff *et al.*, 1993), but *Hippophaë* did not substantially alter nutrient levels on dunes in Ireland (Binggeli *et al.*, 1992). Work on New Zealand dunes established the facilitative role of *Lupinus* herbs on *Pinus* trees (Gadgil, 1971), neither taxon being native. Habitat amelioration by N-fixers also occurs in primary succession on volcanoes (Vitousek *et al.*, 1987; Clarkson; 1990; del Moral & Wood, 1993a,b), floodplains (Van Cleve *et al.*, 1971) and other glacial moraines (Birks, 1980b; Blundon *et al.*, 1993). Experimental tests of facilitation of primary succession by N-fixers remain rare, and inhibitory effects may outweigh facilitative ones (Wood & Morris, 1990). Restoration efforts on mined lands in Ohio, U.S.A., suggested that although plantings of the N-fixing tree *Robinia* stabilized the tailings and added nitrogen, its presence was not beneficial for interplantings of desirable tree species such as *Quercus* or *Pinus* (Larson & Vimmerstedt, 1983). However, *Robinia* stands had the highest levels of volunteer species, suggesting delayed facilitation (Larson, 1984). Actual field measurements of N fixation are also uncommon in primary succession (Vitousek & Walker, 1989; Blundon & Dale, 1990; Uliassi & Ruess, 2002). One study of a glacial moraine in Canada (Kohls *et al.*, 1994) suggested that facilitation by the low-lying shrub *Dryas* was not on species establishment, but rather indirectly through N contributions to already established plants.

Nurse plants
Nurse plants (Fig. 6.3) facilitate establishment and growth of individuals within the influence of their canopy or root zone (Niering *et al.*, 1963; Franco & Nobel, 1988). Explanations for this positive effect include the influence of the increased SOM and nutrients under the nurse plant (Charley & West, 1975; Schlesinger *et al.*, 1996; Jumpponen *et al.*, 1998b; Chiba & Hirose, 1993; see section 6.2.1 on plant impacts on soils),

increased soil water (Valiente-Banuet & Ezcurra, 1991; Caldwell *et al.*, 1998), increased soil temperature (Jacquez & Patten, 1996), increased shade or wind protection (Iverson & Wali, 1982; Carlsson & Callaghan, 1991) and increased protection from herbivory (Niering *et al.*, 1963; McAuliffe, 1988). The facilitation can be mutual, if the nurse plant also benefits from the association (Pugnaire *et al.*, 1996; Moro *et al.*, 1997).

The effect of a nurse plant can be extensive where there are mat- or thicket-forming species. These thicket-formers can facilitate species change if they stabilize the substrate (see section 6.2.1) or ameliorate the environment by providing shade and nutrients. However, the positive effects of shade and organic matter are often offset by competition for light and nutrients from these dense and persistent plants (see section 6.3.2 on inhibition). Increased interception of fog also can improve growth of seedlings under nurse plants (Wright & Mueller-Dombois, 1988), especially when the nurse plant has deeper roots than the plants establishing underneath it. Fog drip also may contain allelopathic compounds that inhibit seedling establishment (del Moral & Muller, 1969). Hydraulic lift is the addition of water to surface soils from deep taproots of established plants, which may benefit shallow-rooted species under the nurse plant (Callaway *et al.*, 1991; Caldwell *et al.*, 1998). Sometimes it is the death of the nurse plant or a portion of its canopy that facilitates establishment of other species by providing protection and nutrients without competition for light or resources (Adachi *et al.*, 1996a,b; del Moral & Bliss, 1993). Facilitation can be accentuated when the nurse plant has dense branching (Mueller-Dombois & Whiteaker, 1990; Brittingham & Walker, 2000) or fixes nitrogen (Walker, 1993), but N-fixing abilities are certainly not a feature of most nurse plant relationships (Jumpponen *et al.*, 1998b). The benefits of increased branch or plant density are only likely in stressful environments. In less stressful habitats, dense plants are more likely to inhibit than facilitate (Fig. 6.5) (Callaway & Walker, 1997). For example, on fertile lava-derived soils on Mount Etna, the sparse canopy of *Genista* shrubs supported a relatively lush herb layer while the denser canopy of *Quercus* trees created a barren understory (R. del Moral, pers. obs.). Nurse plants in arid lands may not have a positive effect on understory plants during droughts, when interception of rainfall or nurse plant root competition for water inhibits rather than facilitates understory species (Fig. 6.6) (Tielbörger & Kadmon, 2000; L. Walker *et al.*, 2001). Because precipitation varies widely in arid lands, facilitative interactions are also likely to vary temporally, although successional implications of such variability are more difficult to determine (McAuliffe, 1988; Bolling & Walker, 2000).

Successional implications

Facilitation by plants is widespread during early primary succession. The balance between facilitation and inhibition is dynamic and varies seasonally, with the degree of site productivity, with variation in growth forms, with the rate of physical amelioration and many other factors. The principal successional implication of facilitation is that it accelerates the rate of species turnover. Species diversity may increase due to general amelioration of harsh site conditions (Stachowicz, 2001). Diversity may increase compared to an arrested succession (see section 6.3.2 on inhibition), or increase across the landscape (beta and gamma diversity) if the degree of facilitation is uneven and some patches undergo different rates of succession. Diversity also may increase if several species replace a single facilitator, but may decrease if the facilitated species eliminates all competitors and establishes a monospecific community. If a facilitator persists, species diversity may also increase. For example, the first colonists in tubeworm succession around benthic hot springs serve as a cue for later colonists to immigrate, but the early species are not out-competed when the later species arrive (Mullineaux *et al.*, 2000).

Species are spatially aggregated when facilitation is common. Nurse plants or clumps of early successional plants can serve as nuclei for later colonists (nucleation; Yarranton & Morrison, 1974). Positive feedback loops can develop between species and between species and the physical environment in primary succession that sometimes lead to cyclical pathways of succession. For example, on cliffs where early colonists promoted soil development that led to further colonization, eventual erosion from the increased weight of the soil will reset succession (Escudero, 1996). Floristically simple desert plant communities fluctuated between the first shrub that could colonize open spaces (*Ambrosia*), the second shrub colonist (*Larrea*), which appeared to be facilitated by *Ambrosia,* and open space that resulted from the death of either species (McAuliffe, 1988). Transition probabilities between the three states (dominance by either plant species or open space) were due to a variety of life history factors, facilitation and inhibition.

6.3.2 Inhibition

Inhibition is the negative effect of one species on another (Table 6.1). Within a successional context, the inhibitor slows or arrests successional change by preventing establishment of species in the next stage. This is often done by resource pre-emption (by fast-growing plants), but sometimes by direct antagonistic effects such as allelopathy (Tolliver *et al.*, 1995). Competitive inhibition can be distinguished from competitive displacement, where one species out-competes a co-occurring species that is also established at the same site. Competitive displacement can actually be a process that promotes rather than retards succession if the displacer is more vulnerable to replacement than the species it displaced. Such three-way or indirect interactions (Table 6.1) are common in successional dynamics (Breitburg, 1985; Walker & Chapin, 1987).

Background

Competition is a central paradigm for determining plant community structure (Clements, 1928; Aarssen & Epp, 1990), although how its importance varies with density (Berkowitz *et al.*, 1995) or productivity (Grime, 1979; Thompson, 1987; Tilman, 1987) is still unclear. The importance of competition to succession was postulated in an inhibition model best elucidated by Connell & Slatyer (1977) that incorporates life history attributes (arrival time and longevity) in addition to competition. They found more evidence to support this model than either the facilitation or tolerance models (see Chapter 3). Tilman (1985, 1988) suggested that succession is determined by the variable abilities of plants to compete under decreasing light : nutrient ratios. Walker & Chapin (1987) predicted that competition would be most important in the middle stages of primary succession when plant densities tend to be highest (Fig. 3.7). However, competition can also be important during the early stages of primary succession when extensive root systems are maximizing use of the limited nutrients available (Matthews, 1992; Marrs & Bradshaw, 1993). S. Wilson (1999) proposed that the early dominance of facilitative processes in primary succession was followed by an increase in first root and then shoot competition.

Inhibition, like facilitation, can affect any life history stage of a plant. Dispersal to a site can be inhibited by dense vegetation that stops either wind or animal dispersal. Surface litter layers or soil crusts can inhibit germination (Chapin *et al.*, 1994). Establishment and growth can be inhibited indirectly by resource preemption or depletion by other plants or directly by litter effects, allelopathy (del Moral *et al.*, 1978; Mahall & Callaway 1992) or promotion of herbivory. Reproduction can be inhibited by frugivory or perhaps competition for pollinators. Survival can be inhibited by anything that promotes herbivory or disturbance.

Inhibition can occur in any successional stage (Fig. 6.8). Early colonizing non-vascular plants can inhibit vascular plant colonization, as when

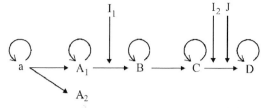

Fig. 6.8. Possible inhibitory impacts on the trajectory of succession.
a, Non-vascular plants; A_1 and A_2, concurrent colonists; I_1 and I_2, introduced
native species; B, C and D, subsequent colonizers; J, aggressive alien invader.
Competitive inhibition can occur early in succession (e.g. by non-vascular plants
inhibiting colonization of vascular plants or an invader inhibiting succession of
native species) or later in succession (e.g. by mid-successional species or invaders
inhibiting later successional species).

crusts form and reduce germination. Colonizing plants can inhibit similar
pioneer species that invade concurrently or subsequently or colonization
of mid-successional species, thereby promoting maintenance of the initial
colonists. Invasive species that are native or non-native can disrupt the
expected trajectories of succession by halting species change. Species at
any stage can inhibit establishment of the next stage and maintain the
current stage at least as long as the life span of the dominant species in
that stage. The notion of a self-replacing climax is actually an example
of inhibition as well, as the dominant species does not permit invasion
by other species (Van Andel *et al.*, 1993). Some species remain good
competitors (and colonizers) through many stages of succession and are
called 'stayers'. Examples include *Epilobium* herbs on emergent islands
in Sweden (Rydin & Borgegård, 1991) and *Metrosideros* trees on lava in
Hawaii (Aplet & Vitousek, 1994).

Examples of patterns that strongly suggest inhibition are common
in primary succession, but experimental evidence actually proving that
successional trajectories are due to inhibition is uncommon (Matthews,
1992). This stems in part from the typical focus on long-lived woody plant
species and in part from the complex interactions of inhibition with facilitation and life history traits (Aarssen & Epp, 1990; Callaway & Walker,
1997). A species can be inhibited but succession facilitated (or vice versa),
one trajectory of succession can be inhibited but another facilitated, or
one species can be promoted at the expense of another with a slightly
different set of requirements for establishment, growth or reproduction.
The broader successional context is also dependent on the nature of the
competition between species. Species vary in the extent to which they

need the same amount of a given resource and in their abilities to obtain that resource in the presence of other competing species (Aarssen & Epp, 1990). Resident species that monopolize current resources will effectively inhibit colonization by later successional species or non-native invasive species (Davis *et al.*, 2000).

Light

Competition for light is most likely to occur later in succession as plant densities increase shade (Clements, 1928; Walker & Chapin, 1987; Tilman, 1988) and successively larger organisms are introduced. Yet even dwarf shrubs and dense swards of herbs can resist the invasion of potentially competitively superior species. In seres where growth is limited by lack of nutrients, drought, extreme temperatures, salinity or toxicity, light rarely limits early processes (see section 7.4 on environmental feedbacks). Plants characteristic of rock outcrops survive only in high light. Habitat amelioration (particularly of drought) on rock outcrops generally results in replacement by species that are superior competitors for light (Sharitz & McCormick, 1973; Shure & Ragsdale, 1977; Ware, 1991). Less stressful habitats soon accumulate dense or tall vegetation that alters the surface light regime. The switch from an open habitat dominated by species of low stature to one dominated by taller woody species is a major functional transition. Shade may lead to declines in species diversity (Smale *et al.*, 1997). In wetlands dominated by *Salix* shrubs on Mount St. Helens, wetland herbs such as *Mimulus* and *Epilobium* were eliminated along with other pioneers of open habitats (del Moral, 1999b) but were replaced by only a few species. Even when dominant canopy species undergo self-thinning, low light intensities can be maintained by expanded lateral growth of surviving canopy trees or increasing dominance of secondary understory species that are more shade tolerant (e.g. tree ferns under *Metrosideros* trees in Hawaii; Drake & Mueller-Dombois, 1993). Many species, especially trees of commercial value, have long been compared by their relative shade tolerance, and one can explain many seres by an increasing shade tolerance of plants in each stage and inhibition of less shade tolerant species. For example, at Glacier Bay, *Dryas*, *Salix*, *Alnus* and *Picea* are increasingly shade tolerant and represent a common successional sequence (Lawrence, 1979). Plant size and growth form are important as well (Olff *et al.*, 1993). *Dryas* is a mat-forming plant that rarely exceeds 50 cm in height, *Salix* is an open-canopied shrub, *Alnus* is a closed-canopy shrub and *Picea* is a tall tree at maturity. Shade tolerance is therefore linked in this example to mature plant size. But shade inhibits as well as facilitates, depending on the optimal light intensity for a given species (cf. Fig. 6.6). *Picea* seedlings under *Alnus* survived the dense shade (as neither *Salix* nor *Alnus* seedlings could) but did better when released from it (Chapin *et al.*, 1994). Eventually, *Picea* overtops *Alnus* and excludes it.

Light requirements can vary at different successional stages and at different plant life history stages. Diatoms on ceramic tiles in the Mediterranean Sea near Greece were more susceptible to ultraviolet radiation in early than in late succession (Santas *et al.*, 1997). Some species need higher light (or higher red : far-red ratios) to germinate than they need to survive once established (Fenner, 1985). However, many species have similar light requirements, suggesting that shade tolerance alone is not enough to explain successional change. Nutrient requirements are also implicated.

Nutrients
Competition for nutrients is another logical explanation for inhibition by a superior competitor in succession. Increases in N availability, for example, have been proposed as an important mechanism driving change in primary succession (Crocker & Major, 1955; Vitousek & Walker, 1987; Olff *et al.*, 1993; see sections 4.3.5 and 4.5.1). However, organic matter accumulation was more important than N availability for species change on dunes in The Netherlands (van Mierlo *et al.*, 2000). Competitive inhibition occurs when a species monopolizes a resource and precludes the establishment of later successional species. Once later successional species do establish, however, they may end up dominating because they can survive, grow and reproduce at lower resource levels than the earlier species (Tilman, 1988). Connell & Slatyer (1977) described the exclusion of existing species by competition as part of the tolerance model, a process that can be considered competitive displacement (Walker & Chapin, 1987).

Thickets
Plants that form mats, thickets or simply dense stands often inhibit species change in primary succession by monopolizing resources (light, nutrients, water) and resisting replacement by other species. Thickets typically inhibit (Tolliver *et al.*, 1995; Nakamura *et al.*, 1997) but may facilitate germination of later successional species while inhibiting their subsequent growth (Allen & Allen, 1988; L. Walker, 1994). Invasive species can form thickets when their growth is unrestricted by herbivores or competitors

Table 6.3. *Examples of types of plant that form mats or thickets that inhibit primary succession*

Type of plant	Habitat	Sample references
Algae	Pools	Benedetti-Cecchi & Cinelli, 1996
	Rocks	McCook & Chapman, 1993
Cryptogamic crusts	Deserts	Kaltenecker *et al.*, 1999
	Dunes	Forster & Nicolson, 1981
Grasses	Moraine	Blizzard, 1931
Vines	Temperate and tropical zones	Webb, 1958; Melick & Ashton, 1991
Ferns	Landslides	L. Walker, 1994
	Lava	Wilmshurst & McGlone, 1996; Russell *et al.*, 1998
	Various	Gliessman, 1976; Fletcher & Kirkwood, 1979
Pine trees	Mudflows	Dickson & Crocker, 1953
Bamboo	Floodplains	Oliveira-Filho *et al.*, 1994; Nakamura *et al.*, 1997
Shrubs	Volcanoes	Tagawa *et al.*, 1985; Whittaker *et al.*, 1989; Walker & Vitousek, 1991; del Moral, 1998
	Moraines	Heusser, 1956; Chapin *et al.*, 1994
	Floodplains	Halwagy, 1963; Luken & Fonda, 1983; Walker & Chapin, 1986
	Dunes	Binggeli *et al.*, 1992; Young *et al.*, 1995
	Mine tailings	Prach, 1994a
	Landslides	Langenheim, 1956

and they grow taller or faster than native plants. Vegetative expansion from undisturbed habitats adjacent to a newly exposed surface can facilitate thicket formation (Ebersole, 1987). Examples of mats or thickets are shown in Table 6.3.

The persistence of thickets is dependent on the life span of the dominant species, its response to the local disturbance regime and its reproductive requirements. Climate can also be a factor. *Pteridium* fern thickets were more rapidly replaced by forest species in wet than in dry habitats following a volcanic eruption in New Zealand (Wilmshurst & McGlone, 1996). The demise of a thicket generally means that succession resumes, although not always along the same trajectory as when arrested by the thicket (del Moral & Bliss, 1993; Aplet *et al.*, 2002). If the thicket increases nutrient levels (Morris *et al.*, 1974; Clarkson & Clarkson, 1995; Russell *et al.*, 1998) or stabilizes a surface (Danin, 1991;

L. Walker, 1994), then succession ultimately may be facilitated. This is particularly true if seedlings of the next stage have prospered in the understory of the thicket from the nutrients, stability and perhaps protection from herbivory (Bryant & Chapin, 1986; Walker & Chapin, 1987) and are able to grow rapidly into the canopy following the demise of the thicket (Hosner & Minckler, 1963; Nanson & Beach, 1977). However, if it is the establishment and growth of the last species in the sere that is inhibited, the thicket merely delays succession. In Hawaii, the vine-like fern *Dicranopteris* (Russell *et al.*, 1998) inhibits the establishment and growth of *Metrosideros* trees, which represent the last stage of succession. The effectiveness of a thicket may depend on the thicket-forming species reaching a site and increasing in density before the second wave of species arrives. Otherwise, inhibition may not occur. This type of non-hierarchical competition or priority effect (see section 7.1.2 on diverging trajectories) was described for turf- and canopy-forming algae in shallow fresh waters in Italy (Benedetti-Cecchi & Cinelli, 1996). The turf-formers resisted the canopy species only if they arrived first.

Many seres are now initiated in biologically unique contexts. In many parts of the world, recently arrived alien species are available to colonize newly formed substrates. Ash from Mount St. Helens killed cryptogamic crusts in the Columbia River basin, allowing invasion by previously inhibited alien grasses (Harris *et al.*, 1987). In Hawaii, the invasive thicket-forming tree *Myrica* quickly invades newly formed lavas and inhibits the establishment and growth of native *Metrosideros* trees (Walker & Vitousek, 1991). Another invasive woody genus, *Tamarix*, forms thickets on flood-plains in arid lands throughout the world because it is more effective at tapping limited water supplies than its competitors (Busch & Smith, 1995). Its litter also makes the ground too saline for genera such as *Populus* and *Salix* . Because succession of woody dominants is limited to just a few stages under arid land riparian conditions, *Tamarix* effectively excludes both colonists and final-stage dominants. *Tamarix* thickets show no signs of dying and regenerate easily, so their impact may be more or less permanent (Walker & Smith, 1997). Their elimination requires active management (see Chapter 8). Alien species are often those with good dispersal powers and with more direct access to newly created sites so they can establish readily. They are less likely than natives to be grazed or attacked by insects and therefore often possess an additional competitive advantage. MacDonald *et al.* (1991) described how the remote volcanic island of La Réunion (Indian Ocean) has been devastated by alien species. There remain a few fragments of primary forests with potential colonists, but

recent lava flows were invaded by alien species, including trees (*Psidium,*
Ligustrum) and shrubs (*Rubus, Fuchsia, Lantana*). It appears that these aliens
will resist replacement by native species indefinitely. D'Antonio *et al.*
(1999) summarized invasions by aliens in disturbed habitats. They noted
that aliens were most likely to dominate anthropogenic sites, riparian
zones and some recovering wetlands, echoing observations by Crawley
(1997). The use of aliens in reclamation of disturbed lands often leads
to inhibition of colonizing natives (Densmore, 1992; see section 8.4.8
on alien plants) and should be avoided. However, on Israeli dunes, the
removal of thickets of woody species led to recovery of native understory
plants and small mammals (Kutiel *et al.*, 2000).

Successional implications
Inhibition can slow or arrest successional change for as long as the inhibitor can maintain its dominance. Sequential inhibition by a competitive hierarchy of species that are increasingly competitive may delay
succession (Horn, 1976). Non-hierarchical inhibition depends on the life
history characteristics (arrival time, growth rate, density) of the interacting
species. Species diversity may decline because monocultures dominate. If
the species is productive and has nutrient-rich litter, soil organic matter
will accumulate. If the dominant species grows more slowly and has recalcitrant litter, humus layers and soil organic matter will not increase, and
may even decrease. The stability of the site can also increase or decrease
depending on the interaction of the dominant species, site characteristics and the disturbance regime. The timing and manner of the death of
the inhibitor obviously impact successional trajectories. Self-thinning of
a monospecific canopy can allow understory species to invade, thereby
increasing diversity (Smale *et al.*, 1997). Self-inhibition (Iverson & Wali,
1982) can also promote successional change. New trajectories are possible
when the inhibitor outlives co-occurring species or alters resource levels that favor colonization by alternative species. Partial removal through
girdling of a dominant, thicket-forming and non-native nitrogen-fixing
tree (*Myrica*) in Hawaii led to its replacement by native tree ferns (Aplet
et al., 2002). In contrast, partial clear-cutting of the *Myrica* forest provided
more light and nutrients and resulted in invasion of another non-native
(*Rubus*). Succession can be dominated by a series of inhibitors that may
interact sequentially or spatially (each dominating a particular patch). The
importance of competition is also influenced by arrival times and relative
sizes and density of the interacting species. Inhibition is an important part
of many primary seres, but the dynamics must be evaluated in each case.

6.4 Interactions between plants and other organisms

Changes in species of plant during succession are accompanied by changes in species of microbe and animal as well. All of these groups interact by feedback loops, so changes in each group of organisms alter the other groups. Consequently, attempting to distinguish cause and effect may be futile. Plants provide carbon and nutrients for microbes. Microbes decompose the carbon and recycle the nutrients. Animals influence carbon production and redistribute it on the landscape. The role of microbes in soil development was discussed in section 4.4 on soil biota. Here we highlight mutualisms between microbes and plants that influence primary succession. We then discuss the facilitative and inhibitory influences of herbivores, detritivores and parasites on plant succession. Animals are often treated as passive responders to changes in plant succession (Bradshaw, 1983b), and to some degree this observation is correct. Plants provide the food and structure without which most animals cannot survive. However, we concur with those that see the role of animals in plant succession as essential and influential (Crawley, 1983; Majer, 1989a; Whelan, 1989). Plant succession is intimately tied to the animals that disperse, pollinate and eat the plants, redistribute nutrients and improve soil structure (Recher, 1989; see section 6.2.2 on animal disturbances).

6.4.1 Mutualisms

Plant species with obligate or facultative mutualists are at a disadvantage in colonizing primary surfaces because both partners must reach the site. Lichens need both the algal and the fungal partner. Although these partners often disperse together as soredia, lichens are not often among the earliest colonizers of primary seres (contrary to common belief; Cooper & Rudolph, 1953). However, lichens can become the dominant vegetation where poor retention of surface moisture limits competitors and atmospheric humidity is high, such as on boreal pumice deserts (Grishin et al., 1996), or in the tropics on lahars (Coxson, 1987), a'a lava (Kurina & Vitousek, 1999) or pahoehoe lava (Poli Marchese & Grillo, 2000). Lichen diversity decreased over time on landslides in a temperate rainforest, presumably because light decreased and mosses increased (Marks et al., 1989). The role of lichens in contributing to the weathering of rock surfaces is controversial and lichens may facilitate or inhibit establishment of other species (Woolhouse et al., 1985). Lichens may retard the growth of mosses and soil development on rock outcrops (McVaugh, 1943). Mosses (Delgadillo & Cárdenas, 1995), algae (Stevenson, 1983), liverworts

(Griggs, 1933) and fungi (Carpenter *et al.*, 1987) are common colonists of primary seres, although their ecological roles are often overstated.

Plants that are obligately mycorrhizal are similarly dependent on the arrival of their fungal symbiont, which can be constrained by dispersal considerations. Obligately mycorrhizal species were not among the earliest colonizers on landslides in Puerto Rico (Calderón, 1993) or on pumice on Mount St. Helens (Titus & del Moral, 1998b). The genus *Salix* has wide ecological amplitude on dunes in The Netherlands, perhaps due to infection by a seasonally (but not successionally) variable proportion of both ectomycorrhizae and vesicular–arbuscular mycorrhizae (van der Heijden & Vosatka, 1999). In Mexican dunes, some pioneer plants were mycorrhizal and others were not (Corkidi & Rincón, 1997), contrasting with the classical view (Janos, 1980) that only non-mycotrophic or facultatively mycotrophic species colonize early stages of succession. The N-fixing bacterium *Frankia* is very widespread and did not limit colonization of volcanoes in Hawaii (Turner & Vitousek 1987) or New Zealand (Walker *et al.*, 2003), so other environmental variables determine rates of colonization. We discuss the roles of mycorrhizae and N-fixers in soil development during primary succession in Chapter 4.

Pollinators can be important in determining which plants successfully reproduce in primary succession (see section 5.2.1 on pollination and seed set). Plants capable of self-fertilization, wind-pollinated plants, plants with generalist insect pollinators or plants whose pollinators are locally common are most likely to colonize a recent disturbance (Whelan, 1989). On rock outcrops in Georgia, U.S.A., two herb species in the genus *Talinum* coexisted but the one capable of self- or cross-pollination gradually replaced the other, which depended solely on cross-pollination (Murdy *et al.*, 1970). Legumes, commonly used in rehabilitation efforts for their N-fixing abilities (Urbanek, 1989), are dependent on insects for pollination. Species dependent on late-arriving animals for pollination (such as the Australian shrub *Dryandra*, which is pollinated by the honey possum, *Tarsipes*) are unlikely to be colonizers of primary seres (Majer, 1989b). The lack of animal pollinators in isolated habitats may produce trajectories dramatically different from those connected to mature vegetation (Whittaker, 1992).

6.4.2 Herbivores

Herbivores can affect the rate or trajectory of succession, species composition, primary productivity, the competitive balance between species, and their vulnerability to disease or drought (Edwards & Gillman, 1987;

Urbanek, 1989). Walker & Chapin (1987) proposed that vertebrate herbivory was more influential on species change in early succession and favorable environments, whereas invertebrate herbivory was more important in late succession and unfavorable environments (see Fig. 3.7). Herbivory may arrest, accelerate or deflect succession, depending on whether its impact falls more heavily on pioneers or seral species (see section 7.5; Davidson, 1993). In some cases the net effect of herbivory on succession is neutral, or other forces override herbivory (Benedetti-Cecchi & Cinelli, 1996). Seed mortality can be caused more by fungal pathogens than by herbivores (Myster, 1997). Detritivores also directly impact nutrient availability and indirectly influence successional trajectories and productivity (see section 4.4.4 on animals).

Herbivores' effects fall disproportionately on seedlings, leaves, roots and flowers, the least protected plant parts. Very early in primary succession, herbivory may be unimportant relative to physical stresses, but its cumulative effect may become substantial. It is likely that even moderate herbivory in early stages represents a large fraction of the overall community production, so that retardation of succession is the rule. Later, differential herbivory accelerates succession because more palatable species will be preferred, leaving longer-lived woody species to develop (Helm & Collins, 1997). Granivory can occur throughout a sere, and it may retard succession to the extent that late-seral, large-seeded species are preferentially consumed (Davidson, 1993). The successional consequences of herbivores eliminating early successional plant species depend on whether those plants were inhibiting or facilitating species change (Fig. 6.9) (Whelan, 1989; Farrell, 1991). Elimination of late successional species always delays succession. Below we outline significant impacts on the course of primary succession mediated by animals.

Vertebrates
Large vertebrates can retard succession in several ways. Clements (1928) described stable vegetation precluded from further successional development to the climatic climax by herbivores as a grazing disclimax. Examples include rabbits grazing quarry vegetation in England (Humphries, 1980), introduced rabbits and native wombats and wallabies grazing on railway rights-of-way in Australia (Newland, 1986), introduced nutria consuming delta vegetation in Louisiana, U.S.A. (Shaffer *et al.*, 1992) and deer browsing shrubs on landslides in Canada (Smith *et al.*, 1986) or on dunes in New Zealand (Smale *et al.*, 1995). Thickets that protect herbivores from predation also concentrate herbivore populations. Selective consumption

Model of succession

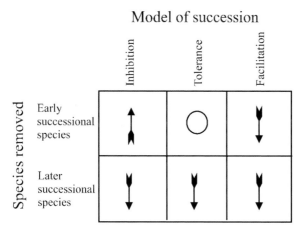

Fig. 6.9. Consequences of herbivory in succession. Arrows indicate that consumers increase (positive direction) or decrease (negative direction) the rate of succession. Circles indicate no effect. Models refer to Connell & Slatyer (1977). Modified from Farrell (1991).

of understory species of the next successional stage can effectively delay succession (Dahlskog, 1982).

Experimental studies have clearly demonstrated that vertebrate grazing can slow succession. Wallabies dominate a complex set of interactions with vegetation on dunes in southeast Queensland (Australia). Ramsey & Wilson (1997) conducted exclosure studies that suggested retardation of the conversion of *Spinifex* grasslands to woodlands in ways that were influenced by landscape factors and grass density. There was little impact on *Spinifex* except where it occurred at low density, but other species (especially N-fixers) responded to the absence of grazing in all *Spinifex* stands (Fig. 6.10). Succession on these dunes was related to nutrient accumulation, so grazing retarded succession through its effects on N-fixing forbs. *Spinifex* grazing also was intensified if the surrounding habitats provided good cover for wallabies, suggesting that landscape effects also may be important in determining the effects of animals on succession rates.

Direct effects can easily be gauged by experimental additions and removals, but indirect impacts of the early successional species (or the herbivores) on soil development may not be as detectable (Connell, 1983). For example, the pioneer species might be N-fixers that inhibit the establishment of later species but also increase soil N. Their removal from the system could impact later productivity, even as it facilitated species change. Similarly, the herbivores might be creating soil disturbances that

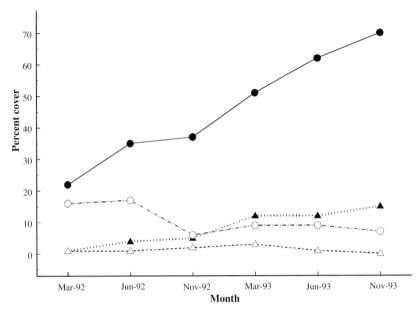

Fig. 6.10. Response of species other than *Spinifex* to protection from wallaby grazing in swards of moderate and dense *Spinifex*. Filled circles, exclosures in intermediate cover; filled triangles, exclosures in dense cover; open circles, grazed in intermediate cover; open triangles, grazed in dense cover. From Ramsey & Wilson (1997).

favor establishment of late successional seedlings or be dispersers of seeds of late successional species (Whelan, 1989). The presence or subsequent removal of the herbivore (when its food source is gone) could have complex implications for succession beyond simple facilitation or inhibition. Removal of herbivores later in succession is more likely to inhibit than facilitate succession, although intensity of the consumption is an important variable to consider (Fig. 6.9) (Farrell, 1991). Herbivory can therefore influence changes in trajectories and rates of primary succession.

Vertebrate herbivores can also facilitate primary succession by preferentially consuming or otherwise damaging early successional species. The preference of moose (*Alces*) for early successional *Salix* shrubs on Alaskan floodplains (Wolff & Zasada, 1979) increased the rate of transition to less palatable *Alnus* shrubs and *Picea* trees (Bryant & Chapin, 1986). Aboveground grazing on *Salix* also reduced its fine root biomass (Kielland *et al.*, 1997) and mycorrhizal infection (Rossow *et al.*, 1997). Moose browsing and defecation also altered soil organic content, which accelerated succession by enhancing soil nitrogen and carbon and further

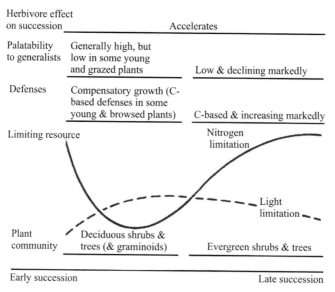

Fig. 6.11. Herbivory effects on succession on an Alaskan floodplain. In early to mid-succession, nitrogen limitation declined owing to the N-fixing abilities of *Alnus* and light limitation increased until the tall, narrow crowns of *Picea* dominated in later stages. Early successional shrubs were C-limited, which restricted their ability for compensatory growth or C-based chemical defenses. This made them more palatable than the heavily defended later successional species, so herbivory accelerated succession. From Davidson (1993).

reducing the competitive abilities of *Salix* (Kielland & Bryant, 1998). The increasing C : N ratio in plant tissues during this Alaskan floodplain succession led to increasing C-based plant defenses and decreasing palatability of plants, so herbivores that ate the more palatable early species accelerated succession (Fig. 6.11) (Davidson, 1993). Geese in coastal marshes also accelerate succession because they prefer pioneer herbs (Olff *et al.*, 1997), although they may preclude development in restoration programs by eating any available forage. Early successional species can be damaged, facilitating later successional species, as when early successional *Pinus* in the U.S.A. are attacked by pine beetles, which favors invasion by insect-resistant species of *Abies* and *Picea* (Amman, 1977). Rabbit grazing of *Schoenus* tussock grasses on coastal dunes in The Netherlands released N retained in the tussock biomass, facilitating N-responsive late-successional species (Grootjans *et al.*, 1998). Finally, herbivores can create gaps that favor invasion by late successional species (Lubchenco, 1983) or non-native invaders (Aplet *et al.*, 1991).

Invertebrates

Insect herbivores can also have variable effects on succession. In bare sandy patches of grasslands, belowground herbivores feeding on plant roots increased the rate of succession, whereas aboveground foliar-feeding insects decreased it (Brown & Gange, 1992). Bach (1994) studied the role of invertebrates in early primary succession on dunes near Lake Huron (U.S.A.) using exclosure experiments. A willow-specialist flea beetle (*Altica*) directly impacted the density, size and survival of *Salix cordata*. Of greater interest was the response of associated species. Under some circumstances, herbs (*Juncus*, *Carex*, *Aster*, *Artemisia*) and other species of *Salix* were suppressed by *S. cordata* if beetles were excluded. These results demonstrated that some plant species benefited from the reduced competition from *S. cordata* provided by the presence of the beetles. In an analogous example from a rocky shoreline, removal of limpet grazers promoted establishment of later over earlier successional algae, thereby facilitating successional change (Benedetti-Cecchi, 2000).

Insect herbivory can also alter facilitative interactions. On a New Zealand volcano, the dominant *Coriaria* is a nitrogen-fixer that is believed to facilitate later successional trees such as *Griselinia* and *Weinmannia* (Clarkson & Clarkson, 1995; see also Bellingham *et al.*, 2001). However, introduced deer, rabbits and possums preferentially browse *Weinmannia*, reducing its role in any future forest. On Mount St. Helens, *Lupinus* herbs are key pioneer plants and the only N-fixer on barren pyroclastic flows and pumice (del Moral & Wood, 1993). Its presence facilitates the establishment and population growth of other herbs, provided its density is only moderate. Studies by Fagan & Bishop (2000) documented that insects restricted the rate of population growth of *Lupinus* individuals on Mount St. Helens and showed that *Lupinus* mortality was due to stem borers and leaf miners. As a result, the rate of population growth of other plant species was slowed, compared to areas where *Lupinus* was less affected by the herbivores. Thus, the development of herbs in the absence of *Lupinus* was retarded because of a reduction of facilitative effects. However, there was no direct evidence that successional trajectories or species composition were altered.

6.4.3 Parasitism

Parasitism can interact with plants in primary succession and affect both rates and trajectories. Van Andel *et al.* (1993) described a study where *Salix* shrubs were killed by willow rust along the Yukon River in Alaska, possibly accelerating succession to *Picea*. Nematode damage to roots, nodules and mycorrhizae of the N-fixing shrub *Hippophaë* on coastal dunes could accelerate succession to other shrubs (Oremus & Otten, 1981). Similar damage by nematodes to the grass *Ammophila* on coastal dunes appeared to favor growth of the grass *Festuca* (Van der Putten *et al.*, 1993). *Cuscuta*, a parasitic vine, may influence some arid land or salt marsh seres; root hemiparasites such as the herb *Castilleja* occur in primary succession (e.g. within 10 yr of the first plant colonization on pumice on Mount St. Helens). However, we know of no documentation that *Cuscuta* or *Castilleja* alter successional trajectories.

6.5 Interactions between animals

Animal or heterotrophic succession can be in response to changing plant resources, physical conditions or interactions among animals themselves. Animal succession may be more dependent on cover and structure of the vegetation than on plant species composition (Letnic & Fox, 1997a,b; Kritzinger & van Aarde, 1998), resulting in a partial uncoupling of plant and animal succession (Gallé, 1991). Animal succession may also be related to the floristic or structural diversity of physical habitats available (Parmenter *et al.*, 1985) or dispersal abilities of the animals (McCormick *et al.*, 1991; Levin *et al.*, 1996). Majer (1992) found that ant recolonization on mined lands in Brazil was positively correlated with rainfall, but in South Africa was also influenced by the competitiveness of one successful colonist (Majer & de Kock, 1992). Early arrivals may be replaced by later arrivals (relay faunistics) or species may merely accumulate without replacement (initial faunistics; Parmenter *et al.*, 1985). See discussion of the analogous relay floristics or initial floristic composition theories in sections 3.3 on Clements and 3.6 on neo-reductionism.

On new volcanic substrates, the first animals to arrive are borne by wind and feed on other neogeoaeolian fauna such as crickets and spiders. Survivors among these species are probably omnivores, although reproduction among aerially distributed arthropods on Mount St. Helens was limited to microsites where plants had already established (Crawford *et al.*, 1995). Herbivores cannot colonize until plants become available, providing a form of obligatory facilitation. Similarly, predators and parasites must colonize simultaneously or after their prey and hosts (Edwards, 1988). This dependency on habitat variables (food or predator avoidance) has been described by a habitat accommodation model and applied to colonization of mammals after fire in Australia (Fox, 1982), ants on mined lands in Australia (Fox, 1990) and rodents on coastal dunes in South Africa

(Ferreira & van Aarde, 1999). Subsequent changes in vegetation can cause the animal population to go extinct, leading to animal succession where change is due to habitat variables and not species interactions (Ferreira & van Aarde, 1999).

Rocky intertidal habitats (Sousa, 1979, 1984; Paine & Levin, 1981; Turner, 1983; Petraitis & Dudgeon, 1999) and artificial reefs (Carter *et al.*, 1985) have provided a wealth of experimental data on the mechanisms of both plant and animal succession. Bare patches are colonized by readily dispersed green algae that usually inhibit (but sometimes facilitate) later colonists (Turner, 1983). Algal succession on suspended concrete plates varied with depth, but shade-tolerant kelp dominated later stages, inhibiting algae (Hirata, 1986, 1992). Reduction of predation by removal of starfish appeared to increase diversity by allowing more species to survive (Paine, 1966). Inhibition thus appears to be the dominant mechanism in rocky intertidal succession (Sousa, 1979; Dean & Hurd, 1980). Primary succession on soft-bottomed intertidal substrates, however, provides more evidence for facilitation than inhibition because tube builders facilitate the arrival of other organisms (Gallagher *et al.*, 1983). Cages that preclude predation can also act as safe sites or substrates for colonists, so conclusions about successional mechanisms from cage experiments must be carefully interpreted (Hulberg & Oliver, 1980). Succession on benthic substrates in oceans tends to follow patterns of terrestrial succession, with competition becoming more important later in succession (Lu & Wu, 1998). Both facilitation and competition appear to be important in streams, with the relative importance often determined by current speed (Peterson & Stevenson, 1992). Early colonizing bacteria that create biofilms on rock surfaces may facilitate (Korte & Blinn, 1983) or inhibit (Peterson & Stevenson, 1989) subsequent colonization by algae.

Decomposers form a special kind of animal succession where resources diminish over time, ending with the complete utilization of the resource (Schoenly & Reid, 1987). This contrasts with plant succession where some resources (e.g. available soil P or light at the ground surface) also decline with time, but carbon fixation from photosynthesis continues and organic matter accumulates and is recycled. Other processes such as colonization sequences, species interactions and herbivory are just as important in heterotrophic as in autotrophic succession. Therefore, decomposition that starts with a fresh, previously uncolonized substrate provides a useful analogy to primary succession. Wardle *et al.* (1995) studied the development of decomposer food webs for three years in sawdust. They found that food-web structure reached its maximum complexity after

77 days, followed by oscillations in populations of bacteria, fungi and nematodes driven by predator–prey cycles. Their results in this simple system suggest that heterotrophic succession quickly achieves a dynamic equilibrium but is not responsive to gradual increases in nutrient levels.

There is often a sequence during decomposition from mobile bacteria and fungi that decompose simple sugars to less mobile organisms that decompose substrates that are more recalcitrant. Dispersal to a carcass is rapid for flying insects or mobile amphipods (which were the first colonists to reach experimental cetacean carcasses placed at the bottom of the Atlantic Ocean; Jones et al., 1998). Soil-bound organisms are slower to colonize. For example, road kill carcasses (horses, dogs, pigs) on paved roads took four-fold longer to decompose than those that had been moved into the adjacent roadside edges in Puerto Rico (L. R. Walker, pers. obs.). Temperature has a direct positive effect on the rate of carcass decomposition (by increasing the rate of fly larval development) and alters successional patterns of decomposers (Richards & Goff, 1997). On wood (Torres, 1994) or carrion (Goddard & Lago, 1985; Early & Goff, 1986), motility, size of mouthparts and diet are partly responsible for determining the sequence of species. Carrion flies can be attracted to carcasses by metabolites from specific bacteria, and may aggregate in clusters that reduce interspecific competition (Ives, 1991), thereby promoting diversity of decomposer flies. Aggregation did not deter other insect or vertebrate decomposers, so it may not have any significant successional implications. Decomposition is dominated by a continuous change of organisms that rarely separate into distinct stages (Schoenly & Reid, 1987).

Plants influence animal succession by altering habitats or providing food. For example, during beetle succession on mined lands in the Czech Republic, open habitats favored beetles with larval stages whereas closed habitats favored ones lacking larval stages (Hejkal, 1985). Beetle succession on mined lands in Wyoming (U.S.A.) was also closely linked to the vegetation, with herbivorous beetles more abundant in areas of high plant diversity (Parmenter & MacMahon, 1987). Decomposer succession can be facilitated by herbivores that preferentially graze on early successional fungi (see section 4.4.2 on soil microbes). Decomposition of dung is inhibited by mites that feed on fly and beetle larvae (Cicolani, 1992).

6.6 Net effects of interactions

Facilitation, competition, mutualism, herbivory and parasitism can each be important at any given moment in any stage of a primary sere. The

challenge for a better understanding of primary succession is to determine the relative importance of each of these variables (Walker & Chapin,
1987; Brooker & Callaghan, 1998; Levine, 2000) and their net effect on
successional change (Callaway & Walker, 1997), particularly in light of
changes in the physical environment (Sharitz & McCormick, 1973; Shure
& Ragsdale, 1977; Duncan, 1993). One way is to examine facilitation
and inhibition at community as well as individual levels (Goldberg et al.,
1999). Poulson (1999) suggested that the accumulation of humus had a
facilitative effect on dune succession in Indiana (U.S.A.), but that plant
species that promoted fires by being readily flammable indirectly inhibited
invasion by later successional species. Both of these processes operated at
the community scale. Poulson also found evidence for indirect facilitation
and both direct and indirect inhibition at the individual level. Interactions between species also change during the life history stages of each
species. Nurse plants facilitate seedlings of other species, but may then be
out-competed as the colonist overgrows the nurse plant (Yeaton, 1978;
Smathers & Gardner, 1979; Vandermeer, 1980). Nurse plants may themselves be invaders of patches of earlier colonists (Griggs, 1956; Walker
et al., 2002). These interactions can be explained by the asymmetric
competition that occurs between large and small individuals. Alternatively, the nurse plant or thicket permits germination but not growth of
the later successional species (see section 6.3.2 on thickets). A nurse plant
can be negatively affected even when the colonist does not overgrow it.
Desert annuals in California had negative effects on shrub water status and
growth, even as the shrubs facilitated growth of the annuals (Holzapfel &
Mahall, 1999). These bi-directional relationships were most pronounced
early in the growing season. Finally, a thicket can have multiple effects
on each of several life history stages of another species. The impacts of
the invasive Myrica on germination of the native Metrosideros on tephra
in Hawaii were positive (shade), negative (leaf litter) and neutral (leaf
leachates) (Walker & Vitousek, 1991). Similar complexity was found for
effects on seedling growth and survivorship and tree growth (Table 6.4).

Each process may also affect short-term species changes and/or have
longer-term indirect influences on succession. For example, Salix shrubs
on a glacial moraine in Washington (U.S.A.) had a short-term neutral or
negative effect on seedlings of other species, but soils from under the Salix
improved growth of transplanted Pinus seedlings, which suggested long-
term indirect facilitation through soil amelioration (Jumpponen et al.,
1998b). Interactions with herbivores and mycorrhizal associations further
modify species interactions. Addition of mycorrhizal inoculum to mined

Table 6.4. *Effects of the invasive tree* Myrica *on the native tree* Metrosideros *in Hawaii*

Symbols: +, positive effect of *Myrica* on *Metrosideros*; −, negative, 0, neutral, ( ), non-significant trend

Metrosideros Life Stage	*Myrica* Factor	Effect on *Metrosideros*	Source of Data
Germination	Litter	−	Seeds sown in field and greenhouse experiments
	Shade	+	Greenhouse experiments
	Leaf leachate	(−)	Greenhouse experiments
Seedling growth	Litter	(+)	Greenhouse experiments
	Shade	(−)	Transplants into gaps and under shade cloths
	Roots	−	Transplants
	Soil	+	Transplants and competition experiment
Seedling survivorship	Shade	+	Transplants
	Roots	−	Transplants
	Soil	0	Transplants
Tree growth	Entire plant	0	Dendrometers
		−	Observation of *Myrica* forest without native species

Data from Burton (1982), Burton & Mueller-Dombois (1984), Vitousek & Walker (1989) and Walker & Vitousek (1991).

lands reduced growth of the dominant non-mycotrophic herb *Salsola* but longer-term successional implications (for perennial mycotrophic shrubs) were not clear (Allen & Allen, 1988), in part because of the inhibitory effect of *Salsola* on the shrubs.

Strictly defined, facilitation and inhibition alter successional rates. These processes can be distinguished from simple positive or negative interactions that do not. Yet successional consequences of an interaction are not always obvious, so facilitation and inhibition have gradually taken on the broader connotations of any positive or negative interaction. Care should be taken not to imply that such interactions have successional consequences unless changes in successional rates are demonstrated.

A plant may alter the environment in a way that facilitates establishment of one species and inhibits another. On volcanic cinder in Hawaii, the inhibited species was an earlier successional species that established best in open habitats (Wright & Mueller-Dombois, 1988), so succession advanced. Inhibition of a species characteristic of later succession slows succession. Differential facilitation of several species by a dominant species can maintain species diversity. For example, the N-fixing shrub *Carmichaelia* had differential positive effects on three tree species on landslides and adjacent floodplains in New Zealand (Bellingham *et al.*, 2001). Future forest species composition in the area and spatial patterning of species in mixed stands of the three tree species will be determined in part by the variable interactions of each species with *Carmichaelia*.

Conclusions about species interactions may also vary among studies of the same species. *Myrica* shrubs inhibited neighboring plants on dunes in New Jersey, U.S.A. (Collins & Quinn, 1982) but had a net facilitative effect on dune succession in Massachusetts (Shumway, 2000). The invasive annual *Salsola* inhibited later successional plants on disked fields in Wyoming, U.S.A. (secondary succession) but facilitated grass colonization of mined lands (Allen & Allen, 1988). Facilitative effects of *Salsola* on the mined lands also varied with time.

The relative importance of facilitation and inhibition may vary along stress and productivity gradients, which often parallel temporal (successional) gradients. Bertness & Callaway (1994) proposed that facilitation was most important in extremely stressful habitats (for habitat amelioration) and in areas with high consumer pressure (for protection from herbivores; Fig. 6.12). Hacker & Gaines (1997) suggested that facilitative effects could provide a variation of the intermediate disturbance hypothesis (Connell, 1978) whereby species diversity both increases and extends further into stressful habitats when a facilitator is present (Fig. 6.13). These models both imply that indirect facilitation should be important in early primary succession.

Grime (1977) proposed that competition intensity increases with productivity, and Walker & Chapin (1987) suggested that the importance of competition for species change increases with succession (see Fig. 3.7). Tilman (1988) proposed that the ratio of aboveground competition for light to belowground competition for nutrients increases during succession but that overall competition intensity remains constant. This apparent discrepancy could be due to an emphasis on absolute versus relative competition (Grace, 1993) or competition intensity versus competition importance (Welden & Slauson, 1986). Absolute competition or competition intensity, measured as the degree of reduction of a plant's success due to the presence of its neighbor, can be unrelated to site productivity. Instead, competition intensity is measured as a function of the carrying capacity of a site and increases as carrying capacity and resource

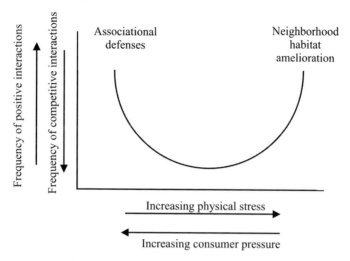

Fig. 6.12. The balance of facilitative and competitive interactions changes as a function of physical stress and consumer pressure. Facilitative interactions are expected to be most important in favorable environments where species aggregations may offer protection from herbivores or in harsh physical environments where species ameliorate the environment. Reprinted from Bertness & Callaway (1994), with permission from Elsevier Science.

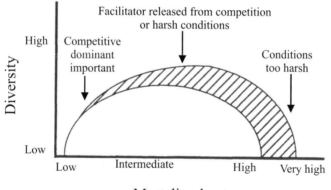

Fig. 6.13. A variation of the intermediate disturbance hypothesis, where a facilitator (hatched bars) extends and heightens the curve of species diversity under conditions of intermediate to high levels of mortality. Because the facilitator is released from competition, species diversity is directly and indirectly enhanced through the creation of new interaction webs. From Hacker & Gaines (1997).

limitations increase (Taylor *et al.*, 1990). Therefore, in the severe habitats of early primary succession, the perceived importance of inhibition to species interactions may depend on one's emphasis on absolute or relative competition.

Goldberg *et al.* (1999) present a third possibility: competition intensity declines with increasing productivity. They cautioned that their conclusion, based on meta-analysis, was preliminary because of the problems inherent in extrapolating from individuals to communities. A possible explanation of their results lies in their observation that herbivore consumption may go up with productivity. Alternatively, a shifting balance between root and shoot competition may be involved, as well as interactions with precipitation (Coomes & Grubb, 2000). Walker & Chapin (1987) suggested that the importance of facilitation for species change is most important early in primary succession. Goldberg *et al.* (1999), using data from relatively fertile sites (not primary succession) found that facilitation of growth declined with increasing productivity (cf. Bertness & Callaway, 1994), but facilitation of survival increased. Comparisons of the importance of inhibition and facilitation for species change in primary seres need to be explicitly made along productivity gradients, as Walker (1993) has contrasted the role of N-fixers in primary succession across N gradients.

The successional implications of species interactions are regulation and alteration of rates and trajectories of succession. In Chapter 7 we explore how life history traits and species interactions shape successional pathways.

7 · *Successional patterns*

In this chapter, we describe how primary successions unfold. We discuss both the trajectories and the rates of development. We explore how biodiversity and biomass change and how major environmental factors control primary succession. Although we believe that community assembly has a large stochastic element, predictable structural and functional patterns often emerge. Trajectories may not be predictable in floristic detail, but relative rates of change and the ultimate vegetation structure can be predicted (cf. Samuels & Drake, 1997). The rates and pathways of primary succession are also controlled by the landscape context and size of the recovering area. Barriers and isolation from sources of propagules limit the number and kinds of species that reach a site (see section 5.3.3 on barriers). Interactions with later arriving plants, animals and microbes (see Chapter 6) all modify the rates and trajectories. The degree of environmental stress affects the rates of biomass accumulation that in turn affects biodiversity, biomass accumulation and successional trajectories. Here we explore the longer-term consequences of these early events. We will examine community patterns that result from the trajectories of succession and explore the forces that may lead to community convergence.

7.1 Types of trajectory

Successional trajectories describe changes in species composition through time. Trajectories may follow definitive pathways, moving through well-defined 'seres' (stages) that lead to a predictable, stable association. In this scenario, multiple starting points converge in terms of dominant species composition. Alternatively, the course of species assembly may be less predictable without convergence. With this second scenario, diversity, biomass and to some degree the presence of functional groups of species are generally more predictable than species composition.

Assessing the nature of a set of trajectories that commence after a major disturbance is problematic. The traditional approach to describing and understanding the nature of a trajectory is to use a chronosequence. In this analysis, vegetation changes observed on the landscape are assumed to represent changes in time (Pickett, 1989). This method, often referred to as a space-for-time substitution, is useful, particularly for longer seres. However, long-term permanent plots offer substantial advantages for the analysis of mechanisms of temporal dynamics (Bakker et al., 1996). Several methods can be used to analyze trajectories. Species can be assigned to functional types, life forms or growth forms, and changes in their proportions can be analyzed through time (Willems, 1985; Olff et al., 1993). Transition probabilities between community types offer a probabilistic approach to predicting stages in succession (van Noordwijk-Puijk et al., 1979), but this approach may fail when environmental conditions change as the sere matures or when adjacent sites affect the trajectory. Synthetic chronosequence approaches, in which associations are constructed from multiple field examples that may be spatially disjunct, are constructed by using dates determined from historical records or direct observations. Initiation dates may also be inferred from such evidence as tree-rings, lichenometry or ecosystem features including soil depth (Poli Marchese & Grillo, 2000). Ordination describes a suite of methods that place samples into a temporal order based on floristic relationships. This method works best when applied to successive measurements of permanent plots because it demonstrates directly both the direction and the rate of succession (Wiser et al., 1996). When applied to chronosequence data, ordination methods help to reveal the overall patterns of the trajectory. Individual species dynamics can be explored using non-linear regressions of species abundance through time (Olff et al., 1993). The analyses of species patterns through succession can lead to better understanding of competitive interactions and ultimately provide a better basis for rehabilitation of degraded lands (see Chapter 8).

Trajectories of assembling plant communities follow one of several models (Fig. 7.1). In *convergence*, vegetation that initially varies in composition is eventually dominated by the same suite of species found in mature vegetation. In *divergence*, the initial composition may be homogeneous or variable, but composition then becomes increasingly heterogeneous because intrinsic environmental heterogeneity permits patchy dominance that accentuates any initial floristic differences. The divergent trajectories then result in multiple stable associations that are distinct from surrounding mature vegetation. There are many reasons to expect that divergence is common, including priority (or sequence) effects, sensitivity

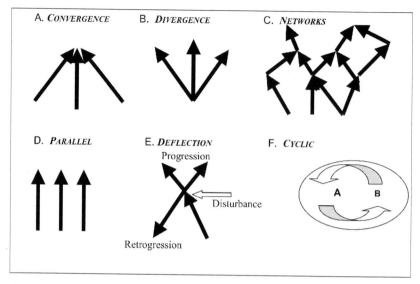

Fig. 7.1. Potential successional trajectories, based on species composition.
(A) Convergence; (B) divergence; (C) networks; (D) parallel; (E) deflection;
(F) cyclic. (Retrogression may be a result of disturbance, a special case of
deflection, or part of a cyclic sere.)

to minor differences in initial conditions (chaos effect) and stochastic
effects.

  Networks occur when one stage develops into two or more stages.
There may be alternative pathways to the same endpoint. *Parallel* trajectories are a special form of divergence that occur when different initial
structural conditions persist, develop independently, but retain the magnitude of their original differences. Subsequent disturbance may reinitiate
succession. In some cases, the vegetation is destroyed and the succession
recommences with a different biota that depends on chance, landscape
factors and any residual species. Alternatively, a disturbance may only
deflect the trajectory, often by selectively impacting a particular species.
Once initiated, *cyclic* seres are a type of secondary succession controlled
largely by autogenic processes where species of each sequential stage cannot reproduce, but so alter conditions that the next stage invades and
succession returns to the original stage. *Retrogressive* seres result from
chronic disturbance, but they also may be caused by gradual degradation
resulting from air pollution, leaching or erosion. Retrogression can also
describe a portion of a cyclic sere.

7.1.1 Converging trajectories

What is the evidence for floristic convergence in primary succession? Complete convergence of species composition occurs when the trajectories of initially heterogeneous communities terminate in a single endpoint, the 'climax'. Given that initial landscape and environmental conditions may vary, convergence requires biologically based processes with low stochasticity. Communities do usually develop similar growth-form spectra after sufficient time under similar environments (Raunkiaer, 1937; Nilsson & Wilson, 1991), but in an increasingly fragmented landscape, this result may occur less often. Profound dispersal barriers and toxic substrates can preclude even life form convergence. Wilson *et al.* (1995) found strong growth-form convergence in understories of temperate rainforests (New Zealand) that suggested functional assembly rules, but they did not find species-based rules. Evidence for convergence in plant guilds also was found by Wilson & Whittaker (1995) working in a salt marsh (Wales), from which they inferred assembly rules based on biotic interactions. These rules predicted species composition due to the limited diversity of the system. Other studies have shown little (see, for example, McCune & Allen, 1985) or variable (see, for example, Inouye & Tilman, 1995) convergence at the species level. Convergence of species composition during primary succession appears to be less likely than generally asserted.

When a biological legacy survives disturbance, convergence is more likely and predictability greater than when there is little or no legacy. Secondary successions are initiated by species from the previous sere and therefore do not rely as much on dispersal, perhaps the most stochastic element in succession. Margalef (1968a,b) emphasized chance early in succession, although he described strong deterministic processes later in succession that caused convergence.

Measurement

Measuring changes in floristic similarity can help to determine whether convergence is important in primary succession. If there is a target (i.e. a local climax), then the species composition can be compared to that of the target by using many similarity measures. During community development, species changes in space (beta-diversity *sensu* Whittaker, 1975) will become reduced if convergence is occurring. Alternatively, increasing beta-diversity along a chronosequence implies divergence, whereas no change implies that the initial conditions still control species patterns

(Egler, 1954). Sample heterogeneity in a sere at a given time may be decreasing, which suggests that convergence is occurring. However, at the same time, similarity to the target may be increasing, implying that the sere is on a different trajectory.

Assembly

The number of species that accumulate during a sere is related to dispersal (influenced by landscape factors), local extinction and interactions between colonists and later invaders. Consistent patterns in the number and kinds of species found in multiple examples of a sere would suggest biologically based rules (Weiher & Keddy, 1995; Wilson & Whittaker, 1995). However, it is unclear that rules operate at the species level in plants or where dispersal limitations are strong. During establishment, which is controlled primarily by physical processes, any assembly rules will be trivial, or confined to higher levels of organization such as functional types. Later, predictable competitive interactions may override variation that results from dispersal and environmental heterogeneity. It is possible that these biological processes create recurring species assemblies that have non-random spatial patterns and which are predictable. However, it is more likely that there are so many exceptions in plants, owing to stochastic processes, that any rules are weak generalities (del Moral *et al.*, 1995; Lockwood, 1997; Walker, 1997; Belyea & Lancaster, 1999). In more labile species that interact strongly with one another, assembly rules may exist and be useful. Kaufmann (2001) examined invertebrate succession on glacier forelands (Austria) and found that it proceeded more rapidly than the associated floral succession and that it was both predictable and deterministic. In contrast, Honnay *et al.* (2001) found that stochastic events combined with weak deterministic factors to structure riparian vegetation (Belgium). Initial colonization was stochastic, such that the upper reaches of six small streams were environmentally homogeneous, but floristically and compositionally distinct. Subsequently, deterministic processes, notably downstream dispersal by water, led to increasingly diverse vegetation in the lower portions of the stream. Dispersal compensated for local extinctions, but floristic assembly led to floristic divergence. No deterministic assembly rule could be inferred from their results.

Priority (or sequence) effects related to the order of colonization do exist and can be powerful. Drake (1991) suggested that the interaction between existing plants and invaders dictates successional trajectories and that normally it was the established plant that would prevail. These priority effects imply that inhibition (see section 6.3.2 on inhibition) is dominant once some species become established. Malanson & Butler (1991) showed that priority was crucial in determining composition and development of gravel bar vegetation in Montana, U.S.A. They hypothesized that chance and variable dispersal events produced different initial compositions that each produced a unique sere. Thus, strong priority effects due to stochastic founders overrode any but the most trivial assembly rules.

Types of convergence
There are two aspects of convergence. The first is vegetation heterogeneity, which nearly always declines from initially high levels during succession. The second, and more interesting, is that individual trajectories come to resemble one vegetation type. Del Moral & Jones (2002) studied a grid of permanent plots on young pumice at Mount St. Helens between 1989 and 1999, using a sampling method devised by Wood & del Moral (1988). They formed composite samples of spatially contiguous plots and investigated the trajectories by detrended correspondence analysis (DCA). The two axes account for over 50% of the variation. In Fig. 7.2, all trajectories move from right to left between 1989 and 1999. For clarity, composite sample positions are shown at 3 yr intervals, plus the final year. DCA provides a good estimate of beta-diversity, measured in half-changes. Beta-diversity of the 11 samples declined from 0.97 to 0.56 in the first axis. Fig. 7.2A shows the trajectories of three samples that converged toward overall similarity. The samples are beginning to converge, but are far from approaching the composition of regional zonal vegetation, which is dominated by mesophytic shrubs and conifers. This vegetation remains too immature to expect strong convergence. Because many primary seres require very long time spans to unfold, they remain distinct from surrounding vegetation for centuries. Close convergence may never occur owing to profound climatic changes (cf. Whitlock, 1992; Whitlock & Bartlein, 1997), further disturbances, or the introduction of new species. In the example, there was a suggestion of a floristic convergence among some plots. The other plots either diverged (Fig. 7.2B) or developed parallel trajectories (Fig. 7.2C) and are discussed below.

Convergence often does occur and is well illustrated in long-term trajectories. Mount Etna has a long and extremely well documented history of eruptions that formed lavas of different ages with similar geochemistry. Flows of several ages are in close juxtaposition, so permitting the use of the chronosequence method. Lavas dating from AD812 to 1983 were

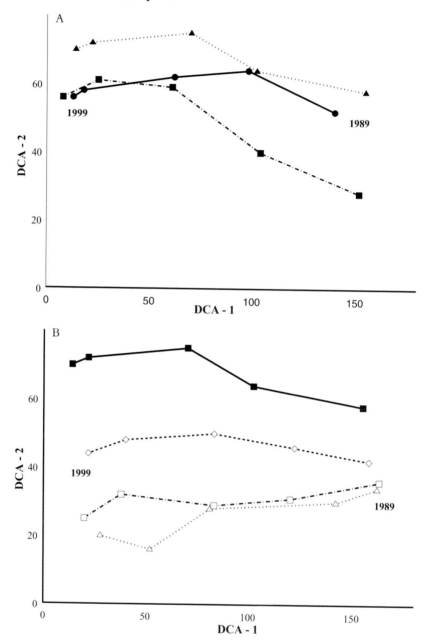

Fig. 7.2. Trajectories of composite samples on young pumice, Mount St. Helens, demonstrating (A) *convergence* among three representative plots (filled); (B) *divergence* of three samples (open) from each other and from a reference stand (filled); and (C) *parallel* trajectories of two plots (crosses) that diverge from a reference plot (filled). Time goes from right to left between 1989 and 1999.

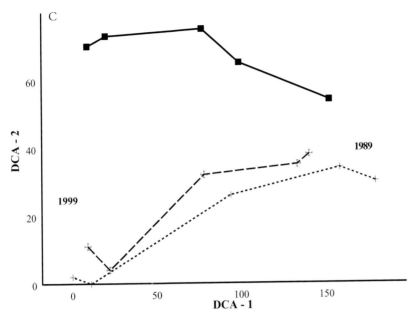

Fig. 7.2. (cont.)

sampled by Poli Marchese & Grillo (2000) on the south slope between 700 m and 1100 m. Although several factors other than age determined the stage of vegetation (e.g. surface morphology, microclimate, dispersal and reproductive success), there was convergence from several pioneer stages that were initiated in different surface types, to shrub stages dominated by three shrub species, to woodlands dominated by either of two *Quercus* species, and putatively to a single woodland dominated by *Q. ilex*.

Where there are barriers to invasion that limit the potential colonists, floristic convergence is common. Rydin & Borgegård (1991) studied small, recently formed islands in a Swedish lake. After 100 yr, the vegetation had not yet developed characteristics of mature forests, but it was becoming more similar among the islands owing to the emerging dominance of wind-dispersed shrub species. It remains unclear whether the larger, less isolated islands will develop a mature vegetation structure similar to that of the surroundings or remain dominated by shrub thickets due to the failure of tree species to invade. Avis & Lubke (1996) described a toposequence on South African dunes that they interpreted as a primary chronosequence. Population dynamics, strong deterministic factors, few dispersal barriers and stressful environmental conditions combined to force structural and floristic convergence in this sere. Graded

Wyoming (U.S.A.) mine tailings form a homogeneous surface. Hatton & West (1987) applied several treatments over 4 yr. The trajectories of each treatment converged. Planting shrubs typical of mature vegetation accelerated succession. This experiment was too short to draw strong conclusions, but they noted that early assembly was stochastic and that sites would become uniform. This study suggested that convergence is most common in species-poor, stressful environments or where interference is intense. The available evidence suggests that floristic convergence is most likely where dispersal limitations are few, where initial conditions are relatively homogeneous and when the substrates permit the rapid development of dense vegetation. Under these conditions, stochastic dispersal events become less important, environmental factors do not lead to alternative trajectories and deterministic factors such as competition can guide the developing trajectory into one channel.

7.1.2 Diverging trajectories

Divergence occurs when different parts of an impacted landscape become less similar as the vegetation matures. In primary succession, divergence can result from priority effects, local edaphic heterogeneity, environmental differences that develop after the initiating disturbance or from vegetation mosaics that develop from the sporadic establishment of strong dominants. Matthews (1999) discussed divergence in alpine glacial moraines. He found that the limited array of species resulted in pioneer communities that were relatively homogeneous, whereas within late-successional shrub communities, floristic variation was greater and the communities were more distinct than were earlier communities. Prach (1994b) found early successional floodplain communities to be similar across a wide geographic range in the Himalayas, probably owing to comparable water table dynamics. However, later stages diverged, reflecting the variety of precipitation regimes that supported vegetation as diverse as forests and semi-deserts. Prach's data support the concept that convergence and divergence can both be found depending on the scale of analysis (Lepš & Rejmánek, 1991). There may be divergence on a small scale because different species achieve dominance, whereas at a larger scale the community begins to resemble adjacent ones.

Local heterogeneity
Initial heterogeneity may become exaggerated as a closer tie between species and their environment develops. Londo (1974) studied maps of

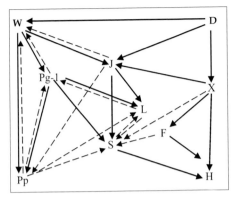

Fig. 7.3. Succession scheme for slack dune vegetation change over 12 yr. Solid lines are major transitions; dashed lines are minor ones. W, water; D, sand lacking vegetation; X, sparse, xeric vegetation; Pp, *Phragmites australis*; Pg-1, mixed sun-loving species including *Eleocharis* and *Scirpus*; L, *Lycopus* and *Lythrum* communities; J, pioneers with *Juncus* spp.; S, *Salix* communities; F, mesophytic grasslands; H, scrub of *Hippophaë* and *Salix*, with *Populus*. This scheme illustrates a complex network of transitions as well as the potential for occasional retrogression. Based on changes in maps between 1956 and 1968. Modified from Londo (1974), with kind permission from Kluwer Academic Publishers.

Dutch dune slacks and found that community divergence, with elements of networks, was based on local heterogeneity (Fig. 7.3). Open water (W) and dry sand (D) could develop into five communities, three of which could also regress. During his study, some communities disappeared (e.g. X) or waned (F). Dry sites could develop into dominance by *Hippophaë* (H) or *Salix* (S) shrubs or other semi-stable communities. Open water could develop into *Phragmites* grasses or *Salix*, from which *Hippophaë* also could emerge.

Subsequent disturbances
Matthews & Whittaker (1987) described the pattern of early succession in a stressful habitat subject to recurrent disturbance. On a glacial moraine, they found that lower and higher elevation sites diverged from a similar first stage. Species composition and rates of change were influenced by terrain age and by topographic position. Succession was initially progressive, but retrogression due to continued disturbances was also common after 50 yr. Vegetation heterogeneity declined on sites up to 50 yr old and then increased on terrain over 200 yr. This resulted because alternative mature communities developed despite similar initial conditions.

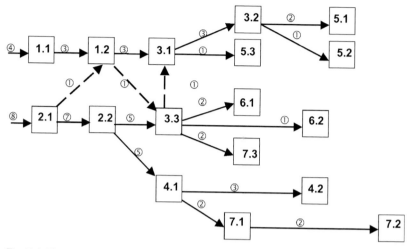

Fig. 7.4. Transition diagram demonstrating divergence in reclaimed tidal flats. Circled numbers are the number of transitions to the next stage; numbers in boxes refer to cluster types and subtypes (e.g. 1.1 = cluster type 1, subtype 1). The patterns indicate some elements of a network and of divergence. Modified from van Noordwijk-Puijk *et al.* (1979), with kind permission from Kluwer Academic Publishers.

An initially homogeneous habitat may develop heterogeneity from erosive forces and support vegetation that diverges from other plots. In the Mount St. Helens pumice example, plots that were initially moderately similar to the others diverged floristically through time from the other plots, and have become more similar to each other (Fig. 7.2B). These changes were due to surface changes driven by wind erosion, which permitted mosses to develop in gaps among the residual pebbles.

Mosaics

Mosaics that reflect land use patterns, landscape effects and human disturbance are common. When tidal flats are reclaimed, desalinization, desiccation, loss of nutrients and oxidation of sulfides occur. Over large areas, these processes do not occur at the same rate, so vegetation becomes a temporal and spatial mosaic (Fig. 7.4). Van Noordwijk-Puijk *et al.* (1979) mapped vegetation that developed between 1963 and 1973 to determine successional trajectories. From year to year, a plot either remained static or changed into more mature vegetation. However, plots of one type could change into alternative types. The fifteen plots were initially grouped into two clusters; ultimately there were four clusters with variants and seven clusters occurred over the 11 yr.

Recurring natural disturbances also create mosaics. Because recurring disturbances do not impact the same areas with similar intensities, mosaics are likely (Motzkin *et al.*, 1999). Subsequent disturbances may be similar to or very different from the original, so the trajectory of seres will be deflected. Alternative trajectories can arise because new conditions are created by the disturbance, or because the competitive or predatory balances are changed, leading the sere into a deflected trajectory (Law & Morton, 1993). One implication is that present conditions are 'ghosts' (cf. Connell, 1990) of past interactions mediated by disturbances. Petraitis & Dudgeon (1999) point out that transient disturbances can still have profound effects. Their experimental work on mussels and algae demonstrated how disturbance can dictate which group will dominate and also how disturbance size is crucial to further development.

7.1.3 Trajectory networks

Some seres studied with chronosequence methods suggest that trajectories form a network (e.g. Michigan dunes; Olson, 1958; Boerner, 1985; Packham & Willis, 1997). Networks result when initial conditions differ in space or time, the environment does not become homogeneous, stochastic factors control some transitions and dispersal is stochastic, thus permitting the invasion of different pioneer species. When there are a few different dominant invaders, each invader can initiate a different trajectory, even if the habitat is homogeneous. Fastie (1995) noted three sequences at Glacier Bay. Conifers quickly invaded sites released by the glacier before 1840, but *Alnus* persisted in dense thickets at sites initiated more recently. Regardless of age, *Populus* dominated sites that were closest to existing *Populus* stands. Networks of trajectories were formed in which the initial woody associations could develop into each of the other seres. Other networks form through the interactions of habitat heterogeneity and competition. Bliss & Gold (1994; Bliss, 2000) described a network of successional types on coastal Devon Island, Canada (Fig. 7.5). Initial surfaces could be concave, flat or convex, and both convex and concave sites could lead to *Puccinellia* marshes if there was an input of organic matter. These marshes would diverge depending on water status. Hummocks could further diverge depending on salt spray. *Dupontia* meadows develop in a complex way depending on the interactions of salt

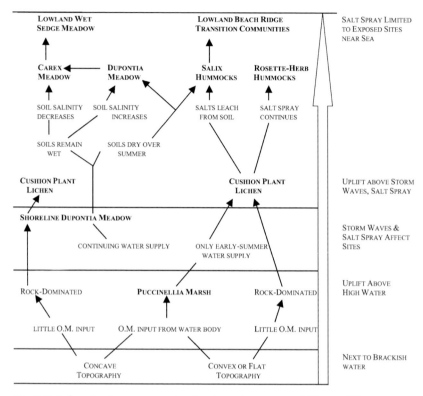

Fig. 7.5. Inferred primary successional trajectories on coastal fringe of Devon Island. Plant communities in bold, processes in normal font. Communities with no further arrows are stable. The broad arrow represents successional time. O.M., organic matter. Modified from Bliss & Gold (1994).

and competitive interactions. This network has only a few decision points and one of the transition choices is much more likely in each case, so it is more predictable than many other cases. Simple trajectories described on the basis of significant structural or functional types are more likely to be successfully modeled than are complex networks in which associations are based on floristic differences alone.

Networks have been described for several dune systems (Cooper, 1919; McBride & Stone, 1976). However, no evidence exists for the dynamic conversion of, for example, dune scrub to either chaparral or pine–oak scrub. Rather, such a chronosequence may be merely a toposequence, with transitions unlikely. The problem of confounding a chronosequence with a toposequence is common, but is particularly problematic in primary succession. Dunes in other situations (cf. van der Maarel, 1998;

Lichter, 2000) have also been treated as networks, but some stages or examples of a stage may be stable and not be transformed further.

7.1.4 Parallel trajectories

Strong environmental determinism can prevent initially established communities from converging. Like runners confined to their own lanes, these communities are unable to converge. Alternatively, strong competitive interactions may preclude invasions that would facilitate increasing floristic similarity. If the environment is initially patchy and patches strongly influence the initial colonists, then the vegetation of each patch may develop independently and differences persist. Schubiger-Bossard (1988) described such a case in alpine glacier moraines (Switzerland). There, habitat heterogeneity imposed constraints that precluded either convergence or divergence among the trajectories. Salt marshes can also display parallel development. Roozen & Westhoff (1985) monitored permanent plots in Dutch salt marshes for 27 yr to follow community transitions. There were four independent parallel seres, each responding to local conditions. These discrete seres were maintained by different initial conditions and variations in plant modifications of these conditions.

Forest systems might be expected to converge because trees can impose strong constraints on invading species. However, Elgersma (1998) found two parallel forest successions occurring on sandy soils in The Netherlands. Sites of similar age, but with distinctly different edaphic conditions maintained different and non-intersecting trajectories. The infertility of the substrate implied that succession was slow and that habitat heterogeneity has not been overcome by biotic factors. The biodiversity of the landscape was maintained by the inability of forest vegetation to reduce site heterogeneity. In fact, the forests became more diverse as similar pioneer communities diverged into several woodland communities.

Several plots of the Mount St. Helens pumice example (Fig. 7.2C) changed in parallel. Each plot occupied a different combination of environmental factors that retained their identity. These parallel trajectories are an interim situation because convergence in this small area is likely to occur when conifers become established. However, if the samples were physically isolated, trajectories could retain their identities because the spread of species would be restricted. McCune & Allen (1985) investigated mature forests in Montana, U.S.A. where succession was initiated by fire. They were environmentally similar without evidence of convergence. Priority effects resulting from chance and landscape factors permitted unique communities to develop without convergence (cf. Facelli & D'Anela, 1990).

Convergent, divergent and parallel seres all occur. Convergence to a single, stable state is more likely when conditions are stressful and bio-diversity is low. This combination of conditions reduces options and the importance of chance. Divergent seres are common when stochastic processes permit alternative species to become locally dominant and when subsequent biotic interactions accentuate these differences. Divergent trajectories have been documented, but it remains unclear to what degree they result in stable assemblages that resist subsequent convergence. Where seres are isolated or in sites that contrast sharply, vegetation development may be parallel, with similar rates of species turnover, but no changes in their degree of similarity. Parallel trajectories have been documented directly over a few decades and indirectly for a few centuries. In such cases, the differences are floristic, not structural. These three cases assume that no further significant disturbances occur.

7.1.5 Deflected trajectories

Deflections of primary successional trajectories result from later disturbances or the pre-emptive invasion of unusual species (Whittaker *et al.*, 1989). Landscape effects and chance can result in unusual or exotic species establishing rapidly and deflecting the sere from the local norm (Tagawa, 1992). Deflections may lead to convergence, divergence or networks of trajectories. Primary seres are subject to disturbances that can obliterate all vegetation, resetting the process to commence under potentially different conditions. Alternatively, disturbance may merely retard development (cf. Ramsey & Wilson, 1997). Studies of Krakatau Volcano (Indonesia) demonstrated divergence caused by subsequent disturbance. Panjang, Rakata and Sertung, three remnants of Krakatau, had similar vegetation after 50 yr. After another 50 yr, several teams found that Rakata was distinct from the others (see, for example, Bush *et al.*, 1992). Anak Krakatau erupted in 1932 and impacted Panjang and Sertung, but not Rakata. As a result, the vegetation of Rakata is older and more developed than that of the others. Disturbances after an initiating trauma are a major way to create divergent seres. Restoration ecologists must consider the effects of post-trauma disturbances.

Multiple disturbances can interact to deflect a sere and create complex landscape mosaics (cf. Castillo *et al.*, 1991). In the Serengeti, savanna and woodlands are in a dynamic balance. Fire alone can deflect a primary woodland trajectory to savanna, and browsing can maintain the new state (Dublin *et al.*, 1990). Extreme flooding along rivers in Colorado (U.S.A.) (Baker & Walford, 1995) resulted in divergent seres with little predictable pattern. Rather, disturbances led to many alternative stable states in these dynamic riparian systems. Floodplains are very dynamic (Malanson, 1993) and initiation of succession depends on several sets of stochastic factors (e.g. disturbance interval, disturbance intensity and dispersal).

Disturbances may be so intense that recovery options are limited, leading to homogeneous vegetation distinct from that of the surroundings. An initially variable system can converge toward one vegetation type. Although only 2 km apart, the two pumice habitats on Mount St. Helens differ in the degree of stress. The Pumice Plain, described previously, retains substantial variation after 20 yr, whereas the other (Abraham Plain) is homogeneous. The latter is drier, at higher elevation and blocked from prevailing winds that can transport seeds. Cluster analyses of each showed substantially greater floristic variation on the less stressful site (R. del Moral, unpublished data). Stress limits floristic options, produces greater homogeneity and increases the chance of convergence.

Another extreme disturbance is exemplified by andesite lava flows formed during recent eruptive episodes on Mount Tolbachik (Kamchatka, Russian Far East). Lava that was 25 yr old had little vegetation except where scoria had accumulated (P. Krestov, pers. comm.). On scoria, but not on lava, a flora dominated by the grasses *Leymus* and *Poa*, the forb *Chamerion*, mosses and lichens occurred. Nearby lavas are over four centuries old. They had developed a woodland of stunted *Populus*, *Larix* and *Pinus* trees. Scoria deposits deeper than 1 m destroyed this vegetation. Succession on scoria deposited on lava depended on the degree of survivorship, whether any dead trees remained standing and whether subsequent erosion removed scoria. A complex recovery pattern of primary and secondary succession resulted. This complex vegetation mosaic combines differential survival and different substrates. The subalpine forest that had developed on old lavas is now undergoing primary succession dominated by alpine herbs and mosses. Landscapes subject to repeated disturbances with different effects would support complex mosaics of recovering vegetation (Fig. 7.6).

Stable dune vegetation may change when an alien species invades. Hodgkin (1984) described colonization by the alien shrub *Crataegus* in Great Britain that began in 1954 because of the decimation of rabbits

Fig. 7.6. Succession on Tolbachik Volcano (Russia). Lava is 400 yr old, with a dense layer of scoria 25 yr old. Many *Populus*, *Larix* and *Pinus* survived, but the ground layer was destroyed in most places, leading to an interesting mixture of primary and secondary succession.

by myxomatosis and the subsequent release from browsing. *Crataegus* advanced from woodland onto dunes at a rate of from 170 to 270 m per year, altering the soil by increasing nitrogen, phosphorus, soil organic matter and moisture. As a result, *Ammophila* grasses declined while grasses such as *Poa* and *Festuca* and woody species including *Pinus* and *Ulex* all increased. These invasions strongly altered the nature of the vegetation on these fragile dunes and increased the erosion rates. Unlike *Ammophila*, the newly dominant species could not bind the loose sand. Similar invasions of dunes by seemingly innocuous species lead to impoverishment and degradation of pristine dunes. The subshrub *Lupinus arboreus* has invaded many California dune systems, leading to enriched soils, invasion of many European grasses and the decline of native species (Maron & Jeffries, 1998). Alien species are an increasingly important part of seral dynamics and a major force in deflecting seres. Restoration ecologists now recognize that many invasive alien species, particularly those that enrich the soil, can trigger massive structural changes. The control and eradication of such species has become a major priority in many sensitive habitats (Pickart *et al.*, 1998), but both manual and mechanical approaches can be counterproductive if they stimulate other undesirable species (see section 8.5.1 on increasing restoration rates).

7.1.6 Cyclic patterns and fluctuations

Cyclic patterns can develop late in primary succession but they are more common in secondary succession and on shorter time scales than most of the seres we consider in studies of primary succession (see section 1.2 on definitions). Fluctuations are temporary shifts in species composition or abundance and might occur on even shorter time scales (i.e. seasonal fluctuations). Cycles and fluctuations are similar because the vegetation returns to a previous state, yet they are not easily distinguished from directional succession until after they have occurred (Glenn-Lewin *et al.*, 1992). Cycles are often described in response to recurrent fires (cf. Arianoutsou, 1998), flooding (Drury, 1956), sand dune movement (Ranwell, 1960) or other disturbances, but this pattern of response to allogenic disturbance could simply be considered a resetting of succession. Autogenic cycles are more distinctive (Watt, 1947, 1955). For example, in wet boreal regions, *Picea* forests often develop on *Sphagnum* bogs. As a result, the bog desiccates and *Sphagnum* is reduced. However, *Picea* cannot regenerate in the dark, infertile conditions and, as the stand degenerates, *Sphagnum* returns. Eventually, *Picea* invades again and the cycle continues (Klinger, 1996). Another example from the Chihuahuan Desert in Texas (U.S.A.; Yeaton, 1978) involves an allogenic disturbance (wind erosion). The common desert shrub *Larrea* colonizes open areas but is then invaded by the cactus *Opuntia*. *Opuntia* displaces *Larrea* by being a better competitor for water but cannot withstand wind erosion and eventually dies. *Larrea* seedlings colonize the newly opened habitat. This pattern has been widely observed in other deserts (Yeaton & Esler, 1990; Valiente-Banuet & Ezcurra, 1991; Eccles *et al.*, 1999). Many cycles have an upgrade phase of developing biomass and complexity and a downgrade phase of disintegrating structure (Watt, 1947). Yet many apparently directional seres actually encompass cycles of species change as well as cycles in other characteristics such as spatial heterogeneity (Armesto *et al.*, 1991) and herbivory (Bryant & Chapin, 1986; Krebs *et al.*, 2001). Therefore, cyclic succession is most often considered a localized phenomenon within the broader temporal and spatial scales of succession.

7.1.7 Retrogressive trajectories

Chronic erosion and grazing are two forces that can cause retrogression, i.e. a gradual change towards structurally less developed and less diverse vegetation. Retrogression is a special case of deflection. Guerrero-Campo & Montserrat-Marti (2000) studied degraded sites in the foothills of the Spanish Pyrenees. Vascular plant cover and diversity decreased with increased soil erosion. Under the most severe erosion regime there was no further species losses, only reduced abundance. Hobbs (1999) suggested that a degradation threshold exists below which full restoration is impossible. Under these conditions, a site might be rehabilitated, but it could not be coaxed to sustain natural vegetation.

Seres that develop on glacial moraines are subject to repeated disturbances that often result in retrogression. Storbreen Glacier in Norway presented a complex set of patterns (Matthews & Whittaker, 1987) in which trajectories were affected by terrain age, topography and disturbance. Young sites retrogressed during the 11 yr study in response to disturbances such as water erosion.

Alien vertebrates can initiate retrogression. This is evident in Hawaii, where pigs, introduced by James Cook in the 1770s, have run amok. Feral pigs alter the species composition to favor alien over native plants, but because many alien plants dominate in the altered environment, removal of pigs alone will not encourage recovery of native vegetation (Aplet et al., 1991). The alien flora will simply readjust to the new disturbance regime.

Retrogression can also occur on much longer time scales, as soil fertility declines from extensive weathering and leaching of minerals (Wardle et al., 1997). Soil microbes tend to shift from dominance by bacteria to fungi (Bardgett, 2000; see section 4.4.2 on soil microbes). Hardpans may develop under arid climates that restrict rooting depth, and declines in plant size and primary productivity usually accompany the soil changes. Although such changes can occur within several centuries, studies of ancient lava surfaces in Hawaii (Crews et al., 1995) and dunes in Australia (J. Walker et al., 1981, 2001) suggest that changes can continue for millennia (see sections 4.7 on soils and 5.6.2 on under-saturated communities).

7.1.8 Arrested trajectories

Arrested trajectories occur when there is a cessation in species turnover before the predicted stable state is reached. There are many examples of vegetation that fails to develop even though it is free from disturbances such as grazing. Allelopathy is involved in some cases of arrested succession, particularly in secondary succession (Rice, 1984). Perhaps the best-known example of arrested succession resulted from the purposeful manipulation of power transmission corridors in the northeastern U.S.A. (Niering *et al.*, 1986). Invading trees were killed by herbicides in 1954 and the resulting vegetation, dominated by the shrub *Viburnum*, remained stable without further intervention. Invasion of tree species was precluded by inhibition, although whether the inhibition was due to competition or allelopathy is uncertain (cf. Inderjit & del Moral, 1997). Where stressful conditions prevail, developmental change may require substantial ecosystem development. Imbert & Houle (2000) described how the grass *Leymus* and the forb *Lathyrus*, pioneers in sub-Arctic dunes (Canada), persist indefinitely. Arrested seres may recommence successional development if an arresting dominant species becomes senescent. Further examples of arrested trajectories are provided in section 6.3.2 on inhibition.

7.1.9 Trajectory summary

Vegetation stature, cover and persistence all usually increase during primary succession. However, change in species composition relative to surrounding vegetation is not predictable because it is affected by proximity to colonists, priority effects and local dominance patterns. Convergence and divergence are both common. Primary seres usually become more homogeneous, but the result may not approach the composition of some particular mature vegetation. If left undisturbed long enough, convergence to the local vegetation will usually occur, provided that the dominant species invade. Stressful conditions, a limited flora and strong autogenic factors enhance the probability of convergence. Over time, deterministic factors reduce variation and species may become more closely tied to the local environment, yet even if vegetation is structurally similar to its surroundings, it still may differ floristically owing to chance or historical factors.

Divergence is more likely when an initial set of invading species, dominated by those with good dispersal powers, is rearranged in response to strong local habitat variation. This divergence is allogenic, but autogenic processes can reinforce it. For example, if a tree can only establish in lava cracks, then the vegetation in that site will develop differently from the surrounding smooth sites where trees do not establish. Different trajectories result from both the intrinsic site differences and the environmental changes caused by the tree. Parallel successional trajectories may persist, but it is more likely that they will eventually diverge, converge under the control of biotic forces and greater habitat differentiation, or develop a more complex combination of both to form a network.

Finally, most primary seres will be disturbed repeatedly. These disturbances may initiate new primary seres but, more likely, the successional trajectory is merely altered. The rate of development can be retarded through fire, grazing or other forms of biomass loss, accelerated through facilitation of mineral cycling, or altered by the introduction of new species. In the next section, we investigate community patterns in primary succession.

7.2 Temporal dynamics

The rate of succession, or species turnover, is crucial because it determines when a landscape will return to its 'normal' functions. Restorationists wish to know natural rates of successional development to judge their efforts at accelerating the process. Wildlife biologists may wish to project the rate of recovery for such species as elk to better prepare management plans. Agronomists want to predict a return to useful agriculture. Knowing typical rates for a region and a particular type of succession may also help to detect unusual situations. For example, a survey of recovering sites that estimate the recovery rates may help to identify sites amenable to rehabilitation efforts.

Determining rates for long-term processes is problematic. Many permanent plot studies now exist that permit us to calculate rates during succession. These studies are distributed over many types of succession in many climates. No rules have been developed to predict rates, but general principles can be summarized. For later stages, chronosequence approaches are required. These are subject to several problems, including when to make the decision that succession has ceased. Here we emphasize studies that address rates of early primary succession in which permanent plots or reliable chronosequence approaches have been applied. The rates of succession vary with climate, soil conditions and landscape factors. Rates vary during a sere in several ways. Early primary succession is rapid if the potential for biomass accumulation is large. In such cases, the pioneers may sequester nutrient resources leading to significant deceleration and landscape effects or competition may retard further species turnover. In contrast, early rates may be virtually imperceptible, as on smooth lava or dunes (van der Maarel, 1998), and then accelerate when a few plants become established. On many dunes, development is spatially variable and rates differ strongly over short distances (Martínez et al., 2001). Despite variation early in a sere, rates of succession generally slow as the longevity of dominant species increases.

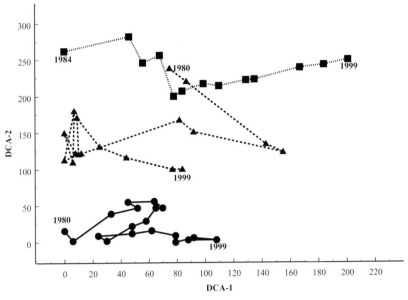

Fig. 7.7. Detrended correspondence analyses of three permanent plots sampled annually on Mount St. Helens. DCA-2 scores were displaced 100 units (one half-change) to clearly display the individual trajectories. First and last year of each study shown for each vector; points are individual years, with 1991 missing for Butte Camp (circles), 1993 and 1997 missing for Pine Creek (triangles) and 1988, 1993 and 1996 missing for Studebaker Ridge (squares). Variation is least in Butte Camp and greatest on Studebaker Ridge; after the first few years, changes in each plot decline.

7.2.1 Definitions

Succession rates are measured in terms of changes in species composition. Whittaker (1974) analyzed vegetation gradients in space in terms of *beta-diversity*, a measure of species turnover along a gradient. The concept can be extended to gradients in time (Rydin & Borgegård, 1988b). An elegant way to determine beta-diversity is with detrended correspondence analysis (DCA). Since its introduction (Hill, 1979), DCA has been used in many succession studies to estimate rates of succession. Using changes in DCA space works best when time is related to an axis (Avis & Lubke, 1996). This method is appealing because DCA units are scaled in terms of floristic change and can be used to compare different seres. For example, a moist site recovering from a volcanic eruption may be found to achieve 50% species turnover more quickly than a comparable site under dry conditions. As an example, Fig. 7.7 shows data from three permanent plots on Mount St. Helens, tracking DCA changes through time. Each set was run independently, but plotted together. The plots are nearly undisturbed (Butte Camp), severely scoured with survivors (Pine Creek) and a true primary succession resulting from the directed blast (Studebaker Ridge). A strong directional trend occurs in the primary succession, and no strong trend occurs at Butte Camp. Pine Creek recovered quickly, stabilized and then moved (DCA-2) because of a strong moss invasion.

7.2.2 Methods of measuring rates

Several suites of methods can be used to determine succession rates. The most simple is the time required to reach a terminal stage, or some specific target. Rates are often expressed in general ways without reference to absolute time. For example, the results of competition, seed predation or grazing are said to accelerate or retard succession (cf. Bach, 2001), but the time required to reach a given stage is rarely given. Other studies may describe changes at several stages determined from chronosequences and may estimate rates. However, these studies do not usually characterize the internal mechanisms of change, and can be in serious error if dating is inaccurate or if retrogression and deflections have occurred. Rates determined with these handicaps remain useful for comparisons among succession types and to determine the effects of environmental factors within types (see, for example, Bornkamm, 1984, 1985).

Rates can be determined by several methods. Chronosequence approaches are useful for comparisons of ecosystem functions such as productivity changes during succession because climatic effects are minimized and differences can be attributed to age. Space-for-time substitutions permit direct (if cautious) rate determinations (see section 7.2.1 on definitions). A special case of this chronosequence approach is the experimental study, where several treatments are compared to determine the effects of such factors as competitors, grazers or predators. Rates are rarely explicitly calculated in these studies.

Permanent plots established at the onset of a trajectory and monitored frequently eliminate historical and chance effects and can be used to study species replacements and competition. These studies can provide better answers to questions about competitive dominance than short-term removal experiments (Herben, 1996) because permanent plots eliminate uncertainties inherent in the experimental procedure. Density manipulations monitored for extended periods may prove to provide the most complete analyses. By selectively removing dominants or introducing

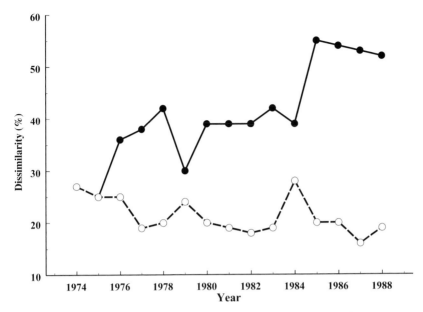

Fig. 7.8. Dissimilarity during the development of pasture vegetation after cessation of fertilization. The rate of change is declining, but by either comparisons with the previous year (filled) or comparisons with the mean of the previous three years (open), the vegetation had not changed dramatically. Modified from Olff & Bakker (1991).

species that typically invade later in a trajectory, the effects of competition and habitat alterations can be determined directly. Permanent plot studies also can lead to testable hypotheses, constrain possible replacement mechanisms and verify predictions based on experiments.

Most quantitative methods measure some aspect of species turnover. Species turnover can be used to determine the dynamics of colonization and extinction at different scales and over different time intervals. As plot size increases, intervals between samples should increase. Fröborg & Eriksson (1997) measured turnover in large permanent plots over decades, whereas del Moral (2000b) measured colonization and extinction rates in small plots over 3 yr intervals. These species-based methods can be used to estimate rates during any interval of the sere.

The similarity (or dissimilarity) between successive samples of permanent plots is a good way to calculate rates. Because of annual fluctuations in plant cover, it may be desirable to make comparisons to a running average rather than to a single season. Fig. 7.8 is summarized from Olff & Bakker (1991), who studied secondary succession in grasslands in The

Netherlands. The solid line shows the difference between a given year and the first year. It demonstrates that dissimilarity has generally increased in comparison to the baseline year. The dotted line indicates dissimilarity to the previous 3 yr and shows that changes between years are declining. A gradual decline in the rate of change appears to be a general pattern in primary succession as well.

Alternatively, the dissimilarity between endpoints can be calculated to obtain an overall rate. Del Moral (1998 and unpublished data) used percent similarity of the first study and each subsequent year of primary succession to track succession rates (Fig. 7.9). The data are from two grids that sampled lahar vegetation cover from 1987 to 2000. Percent similarity used the cover index scores. Each line represents the percent similarity of one year compared to all subsequent years. Lahar 1 (Fig. 7.9A) developed more quickly owing to proximity to adjacent vegetation. Therefore, comparisons of successive years to 1987 show greater percent similarity than do the comparable curves for the more isolated Lahar 2 (Fig. 7.9B). However, by 1992, bi-annual changes in similarity on the two lahars were small and similar. Similarity among plots in a given year increased through time, suggesting greater homogeneity. As the time between comparisons increased, similarity decreased, a result of both accumulation of species and increasing cover. The slopes of the curves are steeper in Fig. 7.9A than in Fig. 7.9B, indicating that the rate of development and turnover is more rapid where there is a proximate pool of potential colonizers.

The rate of succession can be investigated directly by looking at cover (or biomass) changes through time, summed for individual species. In short-term studies with rapid changes, this can readily be accomplished. Prach et al. (1993) constructed a measure based on summing absolute cover changes of dominant species between two sample points and used the results to compare rate changes in one trajectory and between trajectories. This is a less ambiguous and more readily interpreted estimate of rates than is statistical indices such as percent similarity and is less dependent on sample sizes. Succession rates may also be inferred from other data. Donnegan & Rebertus (1999) developed a chronology based on tree ring analyses. From these they used similarity measures to track succession rates. They concluded that environmental stress governed the rate of succession, in this case development of subalpine vegetation after fire. Foster & Tilman (2000) used annual changes in the proportions of perennials and natives to calibrate secondary succession rates. Their approach, pioneered by Austin (1981), was to use repeat measure studies of a transect. Many others have used life-form changes in a similar way,

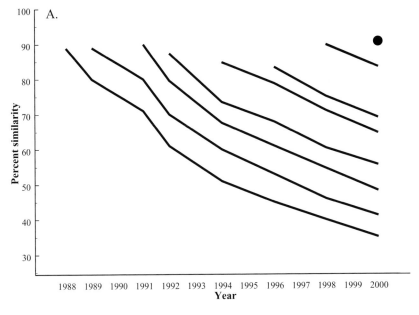

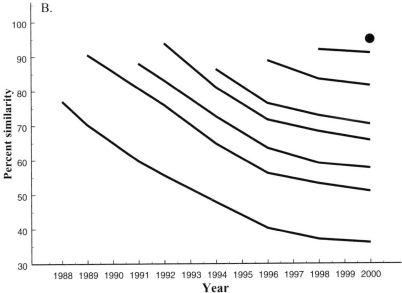

Fig. 7.9. Percent similarity among permanent plots on a lahar. Each curve is the percent similarity of later years compared to 1987, the first year of study. Similarity of lahar plots to each other increased through time, so the curves all increase on the y axis. Similarity of plots relative to the 1987 baseline declined as the comparison years became more distant. (A) Lahar adjacent to intact vegetation; (B) lahar distant from intact vegetation.

although such studies normally involve chronosequences and therefore cannot accurately be related to trajectories.

7.3 Changes in biodiversity and biomass

Early in primary succession, biodiversity of course increases. During the assembly phase, the number of species increases at variable rates, but few established species are replaced. Turnover is limited until the physical environment changes sufficiently to preclude some colonists, or, as is more likely, biotic effects lead to the exclusion of some species. The subsequent course of biodiversity then becomes uncertain. Likewise, biomass must increase for a substantial period, but eventually the mature community ceases to accumulate biomass, although the store of organic matter in the soil and in standing dead material may increase. Eventually, the community begins to senesce owing to nutrient limitations that lead to changes in biotic interactions.

7.3.1 Biodiversity

The most prominent feature of development on newly formed surfaces is the accumulation of species. Species continue to accumulate until a temporary balance occurs between long-distance dispersal rates, local expansion of early colonists and local extinction. When major changes in the growth form of dominance occur, changes in biodiversity, either positive or negative, ensue. On Mount St. Helens, mean species richness in 100 m^2 plots increased significantly in the first decade after the eruption, but after 20 yr, there was little further increase in species numbers. The harsh, isolated Abraham Plain was barren in 1988, with only 0.4 species per 100 m^2; by 2000, there were 12.0 species per 100 m^2 (R. del Moral, unpublished data). Less isolated pumice had 4.3 species per 100 m^2 in 1989, and had stabilized by 2000 at 16 species per 100 m^2. A third isolated surface, on a lahar, increased from 7 to 17 species per 100 m^2 between 1987 and 2000. Similar increases during early succession were noted by Odland (1997) on newly formed lake margins, where richness increased annually between 1985 and 1997. Andersen (1995) studied coastal mud flats formed by dikes. Permanent plots showed that species richness increased from 23 to 91 species between 1980 and 1993. He expected further increases, but speculated that richness would decrease when woody species invaded. This pattern of increasing, then decreasing richness is a general rule when primary succession moves beyond the pioneer stage.

7.3.2 Stability

Stability may be expressed as the persistence of species in the absence of further disturbance. Therefore successional systems are intrinsically unstable. Most plant communities develop greater stability through time because they become dominated by long-lived species (see section 5.5.1 on persistence). This reduces the rate of species turnover. Species turnover is scale-dependent and even stable systems have significant species turnover at several scales (Fröborg & Eriksson, 1997). This feature was termed a 'carousel' by van der Maarel & Sykes (1993, 1997). The carousel model proposes that even in stable systems species turnover on a local scale is common, with individual replacement being stochastic. We are unaware of any studies that have investigated the relationship between turnover rates and the stage of succession, although del Moral (2000a) showed that turnover rate in early primary succession was higher than that in either secondary succession or stable sites.

Of more interest is the degree to which a sere resists disturbance and the degree to which it can recover. Later successional communities are usually more diverse than younger ones, but outside the realm of trophic interactions where alternative prey may confer population stability to an herbivore or predator, we find no evidence that the increase in the number of species conveys enhanced stability in the face of intense disturbance. In this regard, a primary sere, once it is relatively mature, is likely to behave like mature systems with similar structure. However, this assumption has not yet been tested.

7.3.3 Biomass and allocation

Most functional aspects of primary succession are driven by the rate of biomass accumulation (Goldberg et al., 1999). Grime (1979) proposed that early primary succession is dominated by slow-growing, stress-tolerant species. The combination of stressful conditions with species incapable of utilizing ephemeral pulses of nutrients or water result in initially slow biomass accumulation. Waide et al. (1999) surveyed the relationships between productivity and species richness in mature communities and found unimodal, positive, negative and no relationships, with no relationships representing 40% of the studies of terrestrial plants they reviewed. There does not appear to be any mechanism during early primary succession by which biodiversity and the rate of biomass accumulation are related. Instead, this rate is governed by other factors such as temperature, moisture, grazing and nutrient levels, as well as the type of ecosytem and the kinds of species dominating a particular system. As a system matures, biomass may accumulate rapidly if growing conditions are favorable, so it is expected that succession will proceed more rapidly under moist tropical conditions than in the Arctic (see section 5.4.2 on early plant growth). Ultimately, large biomass accumulations that usurp soil fertility may arrest succession by precluding the invasion of additional species. The resistance to invasion may result from direct competition or from the absence of species sufficiently tolerant to invade dense communities with limited nutrient availability. In contrast, under stressful conditions, any residual soil nutrients will accelerate succession. Harper & Kershaw (1997) used soil organic matter to compare recovery on denuded sites (borrow pits) and vehicle tracks of the same age with undisturbed sites in the Canadian low Arctic. Organic matter in vehicle tracks was much greater than the borrow pits which lacked an organic layer. After 50 yr, the borrow pits had scant vegetative cover.

Of course, biomass will increase for most of the primary sere until subsequent disturbance removes biomass. Fire, browsing, storms or a return of the initial disturbance all reduce biomass. Other factors related to biomass increases are of greater interest. Soil organic matter is a direct corollary to biomass accumulation, but it is affected by the decomposition regime. In cold environments, Chapin et al. (1994) found a dramatic increase in soil organic matter in glacial moraines in a sere that included nitrogen-fixing plants. In warmer climates, decomposition may outstrip biomass accumulation so that organic matter increases slowly, and the lack of available nutrients can restrict productivity (see section 4.5.2 on organic matter).

De Kovel et al. (2000) used a chronosequence approach to study development to pine forests on five similar sites on inland drift sand areas in The Netherlands: barren (e.g. no vegetation), very young with only the grass Corynophorus and algae present or Pinus stands that were 10, 43 and 121 yr old. Vascular plant cover increased dramatically as trees (Pinus, Betula, Quercus) and shrubs (Empetrum and Vaccinium) invaded, thus greatly increasing the shoot : root ratio. Biomass accumulation aboveground increased from 2.7 to 9700 g m^{-2}, while roots increased from 8.8 to 5734 g m^{-2}. The shoot : root ratio shifted from 0.32 to 1.7 as dominance by trees developed.

As the vegetation develops, changes in shoot : root ratios may occur but neither the patterns of these changes nor their causes have been widely investigated. Early in many primary seres, nutrients are limiting and plant density is low. These habitats are invaded by light-responsive species with a greater allocation to photosynthetic tissue. However, more nutrient-efficient species with a higher allocation to roots have a competitive advantage in such situations, so they will eventually dominate (cf. Rees & Bergelson, 1997). Wood & Morris (1990) confirmed that competition for soil nutrients increased with system productivity, resulting in greater allocation of resources to roots. If nutrients are not limiting, as in wetlands that develop after draw-down, then there may be a greater allocation to shoots, although the level of disturbance may affect this pattern.

After colonization, a successful population may develop by vegetative expansion, seed production or combinations of the two. Data are scant with respect to changes in reproductive allocation during primary succession. Seedling establishment will be crucial if the system is developing rapidly, as in areas where biomass can accumulate quickly. In contrast, if seedling establishment is difficult, slow vegetative expansion may be more successful. Seres dominated initially by annuals and short-lived perennials will have more resources allocated to reproduction than will those in which long-lived perennials invade. Pyšek (1992) demonstrated that as *Deschampsia* replaced *Calamagrostis* on barren sites, reproductive allocation declined in both grass species. Within any sere, the pattern will depend on local circumstances. Infertile, open systems will develop gradually and we expect that reproductive effort will increase until multiple strata develop. As long-lived woody species come to dominate, no particular pattern appears inevitable. Later in succession, mutualisms with organisms such as mycorrhizae and pollinators may augment diversity as biomass and structure develop.

7.4 Environmental feedback

The initiation of any succession depends on the initiating disturbance, but thereafter the local environmental and landscape factors come into play. Stress is the major factor that governs successional rates. Grime (1977) defined stress as any external constraint that limits the rate of dry matter production of all or part of the vegetation. In early primary succession, stresses typically involve infertility, drought, toxicity, or sometimes grazing effects. Once plant species commence their invasion, a process strongly affected by landscape factors, conditions normally start to ameliorate and an interaction with the local environment commences. Here we investigate how the local environmental context affects the rate and direction of succession. Any environmental factor that affects the rate of biomass accumulation will affect the rate of succession. Moisture, temperature, nutrients, salinity, landscape effects and disturbance are the principal factors of concern.

Once established, a seedling must develop in the face of limitations such as drought, nutrient stress, toxicity and low temperatures. Growth will proceed in accordance with the degree of stress such that the rate of succession may be different in different parts of the same disturbance. For example, rates of succession on Mauna Loa, Hawaii, were greater on wet sites than dry and greater in warmer sites than in cold ones (Aplet *et al.*, 1998) and different parts of a desiccating mud flat developed at different rates in response to differential drying and reduced salinity (van Noordwijk-Puijk *et al.*, 1979). These disparate examples demonstrate that succession may proceed at different rates in adjacent habitats in response to local environmental conditions. The result is often a mosaic of seres that ultimately can interact with adjacent mosaics. The tendency for subtle trajectory responses confounds modeling attempts that do not include spatial components.

7.4.1 Moisture

The rate of recovery early in primary succession is correlated with moisture unless some other factor exerts greater control. Moisture varies with elevation, topography and proximity to bodies of water among other factors. Biomass accumulation is low at both moisture extremes and peaks at intermediate moisture, while biodiversity varies in complex ways.

Primary seres frequently develop on xeric sites. Many volcanic effluents yield initially xeric substrates. The rate of development will be determined largely by how quickly these sites weather to permit the invasion of higher plants. Typically, young lava substrates fracture, creating the opportunity for cryptogamic seres on the surface and vascular plant seres in cracks, crevices or in small depressions that collect moisture. We have observed such 'two-track' seres on volcanoes as different as Mount Tolbachik (Kamchatka, Russia), Mount Etna (Sicily), Mount Lassen (California, U.S.A.), Mount Sakurajima (Kyushu, Japan) and Kilauea (Hawaii, U.S.A.). Less well recognized is the effect of moisture on lavas of similar age and composition. Available moisture will affect weathering and leaching and hence soil development. Therefore, wetter sites normally develop more quickly than dry sites of similar age. Kitayama & Mueller-Dombois (1995a) found that species turnover on moist lava in Hawaii was substantially greater than on drier sites.

The rate of succession can also be affected by moisture indirectly. Under dry conditions, a dominant may develop that, when combined with periodic drought, can inhibit subsequent invasions. *Stereocaulon*, a N-fixing lichen that usually colonizes Hawaiian lavas (Kurina & Vitousek, 1999), invaded more rapidly under warm, moist conditions, but persisted longer under xeric conditions. This persistence inhibited colonization of vascular plants. Under moist conditions, vascular plants could establish and shade out the lichen. Under drier conditions, lichens decline appeared to be related to nutrient limitations on lavas whose surface nutrient supplies (especially P) had been exhausted. This study hinted at complex interactions between moisture levels, nutrient supply and competition as they govern succession rates (see section 6.2 on plant–soil interactions). Lichen deserts also occur on pumice. Near Mount Ksudach (Kamchatka), surfaces are homogeneous and dry during much of the summer. In contrast to lava, there are no safe-sites for the invasion of vascular plants, and lichens have resisted vascular plant invasion for over 90 yr (Grishin *et al.*, 1996). Salt marsh successions are problematic, combining as they do physiological drought stress, chronic disturbance, fluctuating gradients, and steep environmental gradients. In addition, other disturbances such as grazing can alter patterns. Olff *et al.* (1997) conducted an excellent study of how these factors interacted over two centuries in a Dutch coastal salt marsh. The salt marshes formed because of sedimentation derived from sand dunes on a coastal bar island. Species were separated on response surfaces that integrated surface age and elevation. These factors were strongly correlated with the development of forests. Geese appeared to accelerate the succession because they differentially grazed on pioneer species common to the lowest elevations, thus permitting species from later seres to develop. Forage quality declined as species such as the grass *Elymus* displaced succulent species. Cattle grazing on *Elymus* on old marshes caused retrogression as succulent halophytes and geese returned. Grazing by birds altered succession and grazing by cattle resulted in retrogression, which permitted the return of geese (see section 6.4.2 on herbivores).

Local environmental factors can affect the rate and pattern of succession within a homogeneous system. Bunting & Warner (1998) found that the rate of succession differed greatly along different lake margins within similar climates. They attributed the differences to chance, not to moisture. Small dune slack ponds near Lake Michigan (Jackson *et al.*, 1988) appeared to reflect a hydrarch succession, with the topographic positions strongly suggesting a chronosequence based on gradually filling lake margins and decreasing moisture levels. Ponds close to the lake had submerged and floating macrophytes, whereas the most distant ponds were dominated by emergents, especially *Typha*. A 3,000 yr pollen and macrofossil record indicated that all these ponds had been stable and similar to the ponds closest to the lake until 150 yr before the study, and then they shifted to their current stage. The recent pattern that mimicked a chronosequence was actually the result of anthropogenic alterations of the hydrological regime and was not an autogenic succession. This study shows problems of inference from space to time, and that succession rates may be poorly related to local environmental conditions.

In contrast, spatial changes may well be related directly to temporal changes in moisture the affect the sere. Ground water depth controls vegetation development on dune slacks next to the Baltic Sea (Poland) (Zoladeski, 1991). The sere develops when a dune, colonized by *Ammophila* grass, blows out to reveal the ground water. The grass *Agrostis* then establishes and initiates development that culminates within 80 yr in a *Pinus* scrub. Heath species such as *Empetrum* and *Vaccinium* hasten the acidification of the soil. Chronosequences inferred from topography often do accurately reflect the development of vegetation, but each study that asserts that the analogy is correct must be carefully examined for hidden assumptions and confounding processes.

7.4.2 Temperature

Temperature controls most ecosystem functions and therefore can also regulate the rate of succession. There are broad latitudinal gradients in temperature and it is well known that tropical seres are more rapid than temperate ones and that polar seres are slow. Similarly, succession rates slow with elevation (cf. Dlugosch & del Moral, 1999). In temperate zones, factors other than temperature usually are more significant. However, under conditions of extreme cold or heat, temperature assumes considerable importance.

Tropical
Raich *et al.* (1997) conducted a study of development in *Metrosideros* stands on Mauna Loa, Hawaii, as a function of elevation. Sites were on lavas that ranged from 111 to 136 yr in age and were comparable in all measurable traits except that they ranged from 290 to 1,660 m in elevation. Two 3,400 yr old sites were paired with the 700 m and the 1,660 m

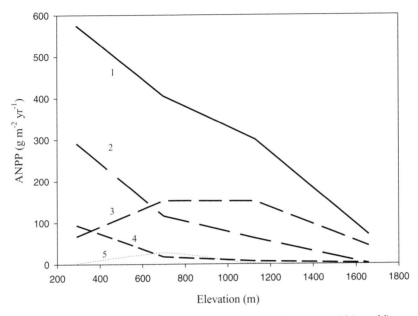

Fig. 7.10. Annual net primary production on young lava flows (110–136 yr old) on Mauna Loa, Hawaii. There is a strong negative correlation between elevation (which accurately indexes temperature) and community productivity. Line 1, total production; line 2, *Dicranopteris* fern; line 3, herbs; line 4, trees; line 5, grasses. Modified from Raich *et al.* (1997).

sites. Ecosystems developed more rapidly at lower elevations (warmer) than higher elevations, and there was a linear relation between annual net primary production and elevation in young sites (Fig. 7.10). Both nitrogen and phosphorus uptake were correlated with elevation as well, suggesting that nutrient limitations were a major factor in productivity on these flows. The old site at 700 m elevation was more productive than the old site at 1,660 m. Temperature affected decomposition, mineralization and nutrient uptake.

Deserts

Desert soils become very hot, enhancing the importance of safe-sites to succession. Cracks in dried mud and mounds of coarse soil offer two desert habitats in which new seedlings can survive (Vasek & Lund, 1980). High temperatures restrict the rate of succession, but usually this is a result of low productivity and drought associated with high temperature.

Boreal forests

Short growing seasons, low temperatures and adverse hydrological regimes slow succession in boreal habitats. Grishin *et al.* (1996) demonstrated that complete vegetation recovery of *Betula* forests on pumice in southern Kamchatka required at least 1,000 yr and that the soil development took much longer. Because of the large number of volcanoes and frequency of their eruptions in the Kamchatka Peninsula – Kuril Islands chain, it is unlikely that there is any fully mature landform to be found there.

Arctic and Antarctic systems

Tundra succession is very slow, owing in part to low temperature and limitations in biomass accumulation. Bliss & Gold (1994) described primary succession in the high Arctic of Devon Island, Canada. Based on related studies (Bliss, 2000), succession from lichens to *Dryas* may take over 3,000 yr, and succession from newly formed salt marshes also required over 3,000 yr in these habitats with a very short growing season. Temperature-dependent processes in these systems include leaching of salts, decomposition and productivity. Each of these affects the rate of succession. Salt leaching permits the invasion of upland species, decomposition improves nutrient availability and productivity increases facilitation to make seedling establishment more likely. Under extreme temperature conditions, little else affects succession. Davey & Rothery (1993) studied colonization by algae in soil polygons in Antarctic fell fields. Under these extremely cold conditions, there was little evidence of competitive reassortment, only limited accumulation of species over time. Differences among the communities appeared to be due to chance.

7.4.3 Nutrients

Nutrient levels are affected by moisture and temperature, but nutrients are often the proximate limiting factor in succession. Seres initiated by similar events will develop at rates consistent with their fertility. Infertile sites may be invaded quickly, but biomass accumulates slowly, thus delaying the development of biotic interactions and invasion of less tolerant species. A general rule is that the rate of succession is positively related to site fertility (Harper & Kershaw, 1997), although under extreme conditions of drought, salinity or cold, fertility may be irrelevant. We discuss the effects of site amelioration by nutrient additions in Chapter 8. Altering the nutrient regime is a cost-effective way to accelerate succession, but doing so can affect trajectories in adverse ways. Below we explore how nutrients affect some typical natural seres.

Old-fields

Early secondary succession under low nutrients can offer insight into primary seres. Nutrient-poor sites are initially dominated by species that allocate more energy to seeds and leaves than to roots but, over time, root biomass increases, suggesting competitive interactions for nitrogen become more important (Gleeson & Tilman, 1990). However, transitional phases in succession may be more complex. In many seres, the transition from herbs to shrubs is crucial, but infertility often prevents seedlings of woody species from establishing before drought or cold kill them. Infertile soils do remain open to invasion, but only large-seeded taxa such as *Quercus* or *Fagus* trees may succeed. Fertile sites often support the development of dense herbs that then preclude or retard the invasion of shrubs or trees. Similarly, well-dispersed taxa (e.g. *Salix* and *Alnus* shrubs) can be expected to predominate early in succession on rich soils because they establish before the vegetation can become dense (Smith & Olff, 1998).

Floodplains

Floodplains and sand bars are usually infertile as well as being dynamically unstable. Clein & Schimel (1995) noted that N turnover and availability dictated the shift from *Alnus* to *Populus* in successions on floodplains. The effects were mediated through the soil microbes (see section 4.4.2 on soil organisms) and appear to be a two-step process. First, the initial saplings of *Populus* decrease N availability to reduce nitrification. Second, continued C input converts the system to N limitation, and stimulates the dominance of *Populus*.

Volcanoes

Volcanic ash is nutrient-rich in many cases. Consequently, where decomposition rates are high, as in the moist tropics, succession on volcanoes is rapid. Nitrogen is a common limiting nutrient; P can also limit development in many cases (Schlesinger *et al.*, 1998). In contrast, pumice in temperate zones is strongly N-limited and often P-limited, so succession proceeds slowly (del Moral & Bliss, 1993). These systems depend on external nutrient sources to accelerate vegetation development. Winds deposit materials collected from surrounding sites to provide allogenic amelioration that is crucial in starting succession in severely nutrient-deficient systems (Howarth, 1987). Similar processes operate on dunes and glacial moraines.

Dunes

Among the best studied of all primary seres are sand dunes. Classic studies by Cowles (1901) and by Olson (1958) set forth many principles of primary succession that are still accepted (see section 3.9 on models). These dynamic systems are often conveniently located and sometimes result from deliberate manipulations. Because they are also fragile and easily disrupted, they have attracted intense scrutiny. Succession on most dunes is controlled by nutrients, in combination with moisture and substrate stability (Tielbörger, 1997).

Sands shift and bury vegetation, but some species survive to reinitiate development (Lichter, 2000), so some dune seres are technically secondary succession. Wind or trampling often initiate blowouts that require long periods to recover. Because of instability, infertility and drought, dunes are among the most difficult habitats to restore adequately. Understanding the limitations in particular cases is needed to plan successful rehabilitation programs. By combining a chronosequence approach with several ordination methods, Henriques & Hay (1998) determined that succession along coastal dunes in southeast Brazil was primarily in response to infertility (low Ca and Mg). Biomass declined in older sites owing to leaching of these limiting nutrients. The chronosequence approach was applied successfully to a study of drifting sands in The Netherlands (De Kovel *et al.*, 2000). Here, C and N accumulations controlled succession rates as mosses, grasses and finally forests developed during the 121 yr of the chronosequence. The rate of development had accelerated during the last 100 yr due to rapid increases in N deposition from external anthropogenic sources. This is but one of a growing number of studies that demonstrate how pervasive are anthropogenic effects on primary succession (see section 4.3.5 on N accumulation).

Nutrient levels also drove early succession on coastal sand dunes in The Netherlands (Olff *et al.*, 1993). The topographic gradient was controlled by differences in salinity, flooding and moisture. During the 18 yr study, soil N increased dramatically on the dunes and dune slacks. Biomass increased with N levels and light at the surface declined, suggesting that competition for light should increase. In this system, other stress factors remained constant, while nutrients increased as vegetation developed. With biomass accumulation came competition for light. As the sere develops, taller species should colonize and prevail. This study demonstrates the value of permanent transects in exploring spatial and temporal changes.

Recent studies (Lichter, 1998) provided analyses of how nutrients affect long-term succession on dunes, as well as how experimental methods can be used even in very long seres. Lichter's chronosequence along northern Lake Michigan included a long period of herb dominance as well as over three centuries of forest development. Soil C increased at an average rate of 23 g m^{-2} yr^{-1}, whereas N accumulated slowly (0.38 g m^{-2} yr^{-1}) because N-fixers were rare. The dune environment became less stressful as vegetation developed, organic matter accumulated and N and P accumulated. Although nutrients accumulated as dunes stabilized, Lichter (2000) showed that colonization by *Pinus* and *Quercus* was constrained by seed dispersal, seed predation and seedling desiccation and not by low soil nitrogen. Landscape and predation effects were much more important in determining the rate of succession. These studies amply demonstrate that experiments can clarify and elucidate patterns suggested by descriptive studies (Wood & del Moral, 1987) and that experiments often reveal complexities even in simple systems.

Dune slacks

Ponds isolated between high dunes develop gradually in relatively protected and stable environments (see section 7.4.1 on moisture effects on dune slacks). Nutrients become available more quickly than in the surrounding dry dunes and development rates may be limited by other factors. *Carex* meadows among dunes near the Dutch Wadden Sea islands were controlled primarily by moisture and pH variations (Lammerts & Grootjans, 1998; Lammerts *et al.*, 1999). Older vegetation did not develop unless substrates become more acid and chlorine was leached. The rate of succession was retarded if the soil could buffer pH changes. Another Dutch slack dune study points out the importance of local conditions. Dune slack vegetation developed from a beach under unusual soil conditions (Sival, 1996) and was dominated by *Juncus, Glaux* and *Salicornia*. Within 30 yr, diverse vegetation developed on mesotrophic soils, dominated by *Schoenus*, orchids and other rare species. Within 70 yr, these sites were dominated by shrubs such as *Betula, Salix* and *Alnus* that reduced diversity greatly. Soil pH declined, but exchangeable ions increased during the chronosequence. The biologically interesting *Schoenus* (a graminoid) stage persisted twice as long where high soil buffering retarded organic matter build up and the pH remained high.

Glaciers

Receding glaciers yield barren, nutrient-poor substrates. Infertility, drought, dispersal and secondary disturbances are a few of the constraints to succession. The soil microbial community is crucial in ameliorating adverse nutrient conditions (Ohtonen *et al.*, 1999; see section 4.4.2 on soil microbes).

Decomposition

Decomposer successions offer the advantage of being completed in short times. Wardle *et al.* (1995) examined microbial development in sawdust, in which stability was achieved within 2 yr. This model succession did not fit several general concepts defined by Odum (1969). Large organisms dominated the pioneer phase and *K*-selected nematodes were pioneers. This microcosm study demonstrated the intrinsic complexity of apparently simple systems.

Vitousek & Hobbie (2000) showed that over a broad successional range in *Metrosideros* forests in Hawaii, decomposition was limited by high lignin; this in turn limited decomposers and maintained N limitations. Indirectly this study implies that succession may be accelerated by N additions or by adding more readily decomposed C sources.

7.4.4 Salinity

Salt impedes succession rates because of osmotic and toxic effects. The metabolic costs of dealing with salt are large. Many parts of the world are becoming increasingly saline, so that primary succession often involves halophytes expanding into formerly less saline habitats. Rehabilitation of these habitats will be discussed in Chapter 8. Here we explore how salt affects succession.

Salt flats

Diversity normally develops very slowly on salt flats (Vasek & Lund, 1980; but see Andersen, 1995). On a fine-textured dry lake with high sodium chloride and nitrates, succession required safe-sites (see section 5.4.1 on germination) that permitted small shrubs to establish. These shrubs created mounds that were colonized by less salt-tolerant species. With the death of the mound builders, the system eroded to resemble the initial condition. True succession occurred only as soil accumulated and nutrients increased. In some environments, this may never occur because evaporation retains the initial adverse conditions.

Salt marshes

Development in salt marshes must contend with chronic disturbances from the tides and a gradual accretion of sediments. The approximate zonation found in many salt marshes may not result from primary succession in any meaningful time scale, although such primary succession certainly occurs in salt marshes. Rather, it may result from differential tolerances to essentially static conditions. Nutrients are rarely limiting because salt, poor aeration or grazing dominate the system. A special case deserves mention. The upper levels of salt marshes receive substantial amounts of storm wrack at the high tide level (Pennings & Richards, 1998). It kills the vegetation but initiates secondary succession in a nutrient-rich substrate that promotes dominance by the forb *Batis* in the upper marsh.

Levine *et al.* (1998) developed an experimentally based model of the importance of nutrients against a background of physical stress caused by salt. The study supported the hypothesis that competitive displacement occurred with competition for nitrogen. Thus, biotic interactions (see Chapter 6) may generate plant zonations among different graminoids. *Spartina alterniflora* usually occurs in the low marsh, where physical conditions are most stressful. However, as nutrients are increased, it may come to dominate the entire marsh. *Distichlis*, an early colonizer of salt marsh disturbances, is replaced by *Juncus* and *S. patens* in the higher marsh under normal conditions, but it can persist if nutrient levels are high. This study strongly supported the hypothesis that nutrients are very important in controlling plant zonation in salt marshes despite salinity gradients and tides and that nutrient levels have predictable consequences for species patterns.

Mangroves

Mangroves are trees from several plant families that form barriers against the sea and are involved in primary succession. The direction of succession depends on the relative movements of sea levels, siltation rates and nutrients. It can be affected by autogenic factors and local disturbance. No long-term studies exist to document primary succession, and chronosequence approaches may not demonstrate directional change. Chen & Twilley (1998) used a model that challenged beliefs concerning mangrove zonation. Nutrient levels and salinity interact along gradients in ways that appear to control zonation, but suggest stable vegetation if the coastline remains stable. The model integrated demography of three mangrove species with salinity and soil nutrient availability. The model

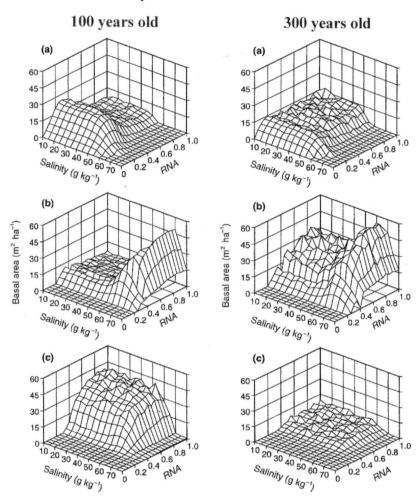

Fig. 7.11. Model distributions of (a) *Rhizophora*; (b) *Avicennia*; and (c) *Laguncularia* after 100 and 300 yr of simulated growth. From Chen & Twilley (1998).

supports the concept that *Rhizophora* could dominate low-nutrient habitats across a spectrum of salinity, and that early in succession *Laguncularia* dominates more nutrient-rich habitats (Fig. 7.11). *Avicennia* dominates hypersaline conditions. With continued development, competition results in *Avicennia* coming to dominance, replacing *Laguncularia*. Under natural conditions, subsequent disturbance resets the clock, and it is rare to find *Avicennia* dominant, perhaps because its recruitment is limited. However, the model has been validated with field data in the Florida Everglades.

7.4.5 Landscape factors

Distance frequently limits the rate of succession, and sites near a source of potential colonists usually develop more rapidly than do isolated sites. Surrounding vegetation can host herbivores that can affect succession. For example, isolated tropical pastures revert to forest more slowly than do pastures near forest remnants. In part, this is because fruit-eating birds seldom wander more than 80 m from the forest (Da Silva *et al.*, 1996). Del Moral (2000b) demonstrated the effect of distance on species richness and cover in four permanent plots on lahars at Mount St. Helens (Fig. 7.12). Plots A and B were within 50 m of an intact conifer forest, while Plots C and D were over 200 m distant. Richness accumulated in a similar fashion, although the isolated plots may be slightly slower. Cover, however, developed significantly more slowly, especially after 1990 when cover estimates from small subsamples are reliable. By 2000, isolated plots had only 50% of the cover of proximate plots (see section 5.4 on establishment).

7.4.6 Chronic disturbance

Chronic disturbance, rather than unpredictable events, occurs in many primary successions. During the early course of primary succession, disturbance nearly always retards the process. In some systems, chronic disturbance leads to a blurred distinction between static zonation and development. Salt marshes are routinely treated as vegetation gradients in equilibrium with a dynamic environment.

Rock outcrops
Outcrops have been treated as either static or dynamic habitats. Ohsawa & Yamane (1988) documented that both the size and depression depth on rock outcrops were important to succession because some depressions in small outcrops simply could not support mature vegetation. Communities dominated by annuals and small biennials persisted in small patches on outcrops owing to frequent disturbance that prevented a robust grass (*Miscanthus*) from invading. The rock outcrop sustained at least three communities, some of which changed cyclically in response to disturbance, others of which were stable due to their small size and extreme stress. Without chronic resetting, the annuals and some biennials would be excluded by competitive inhibition by robust species.

In their study of Oklahoma rock outcrops, Uno & Collins (1987; Collins *et al.*, 1989) found that vegetation patterns were correlated with

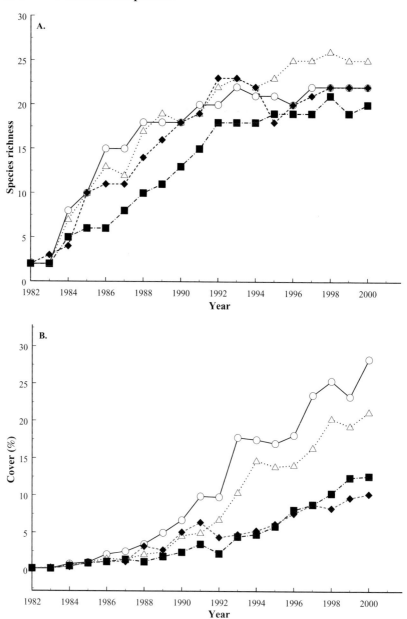

Fig. 7.12. Distance directly affects all aspects of community structure. Here the development of richness and cover are shown on two permanent plots on each of two lahars on Mount St. Helens (U.S.A.). (A) Richness; (B) cover. Open symbols are from plots adjacent to intact vegetation; filled plots are from isolated plots. From del Moral (2000a).

soil depth, size and several edaphic factors. They viewed the patterns as a toposequence, with communities relating to existing depressions, cracks, etc., and to disturbance, not a sere. Soil development, which would drive succession, was slow; erosion may restrict soil formation, precluding succession.

7.4.7 Pollution

Pollution can alter trajectories and may cause mild deflections or significant retrogression. Intense pollution (e.g. radiation or heavy metals) may eliminate all vegetation and set the stage for new primary seres to commence. These effects will be treated more fully in Chapter 8.

Air

Air pollution has subtle, long-term effects that cause retrogression. Newly created sites are no less susceptible to pollution than are other systems. Acid deposition and various oxidants alter physiological processes and when extreme, eliminate particularly sensitive organisms such as orchids or lichens. For example, Hogg *et al.* (1995) described changes around a bog in Yorkshire (U.K.) during 13 yr of observation that resulted from increased acidification and deposition of nitrates and sulfates. *Sphagnum* species declined substantially owing to sulfur deposition and, regionally, several species were eliminated from mires. The ground flora changed dramatically and *Alnus* is now invading. Soil pH declined by over 0.5 units. Because N deposition is a widespread phenomenon in the developed temperate world, the implications of this study are worrisome. Studies by Rose *et al.* (2000) that describe permanent plot studies in English heaths, and by Tybirk & Strandberg (1999) that describe Danish forest responses to nitrogen deposition, are but two that emphasize the seriousness of the aerial deposition effects on the dynamics of vegetation. Enrichment of many developing ecosystems is an aspect of global change that is not widely appreciated, yet it has the potential to deflect and arrest many seres. In addition, nutrient enrichment can cause sharp reductions in biodiversity (Tilman, 1993).

Water

Water pollution effects on vascular plant succession are rarely studied. Eutrophic effects, and effects on fish, bacteria, algae and marine mammals, are all well documented. Waters may be polluted in several ways, and the kinds of pollutants are increasing. Aquatic macrophytes respond

to nutrient levels. In the northern Vosges Valley of France, Thiebaut &
Muller (1995) found four communities that appeared to relate sequentially to the nutrient level. Pollution drove the vegetation to progressively
more eutrophic communities. Lu & Wu (1998) described macroben-
thic succession because of deposition of fish farm effluents that appear
to have caused long-term alterations in community dynamics in a variety of worm species. The arid coasts along the Arabian Gulf have
been polluted by crude-oil tar in many places. Hegazy (1997) studied
a 14 yr chronosequence in Qatar, where tar is piled in marshes. Primary
succession on this unusual substrate none the less followed predictable
trends and was controlled by time since deposition, tar content, moisture, and size. Reclamation of this substrate may be possible using native
species, but bioremediation has yet to be effected. Just as nutrient enrichment and air pollution have impacted the rates and directions of
terrestrial seres, aquatic seres also respond to these impacts. Unlike terrestrial systems, much less is known about the extent and magnitude of
these effects.

7.5 Summary

Primary succession is a variable process that defies exact prediction, although its general outlines usually can be anticipated. All aspects of the
physical and biological landscape affect it. Rates and trajectories vary
with landscape context and can be altered by further disturbances. However, we have identified several broad patterns found in primary seres.
In general, a sere has three broad and overlapping phases: colonization,
development and maturation. A fourth stage, senescence, may occur in
many seres, but usually after centuries have passed. By this time, the
sere behaves like secondary succession (see section 9.3.1 on the end of
primary succession). During colonization, species accumulate and biotic
interactions are less important than are tolerance and expansion by the
pioneering individuals. Site amelioration by the colonists may begin to
affect the success of subsequent invaders. As a sere develops, positive and
negative interactions among plants and herbivory by both invertebrates
and vertebrates become increasingly important. The community eventually reaches maximum biomass, but changes in soils, species composition
and biodiversity continue. For several important structural features, we
predict how they may change through a sere and affect a variety of processes (Fig. 7.13). These patterns will not hold in all primary seres; rather
they are null models against which to test the particular case. Note that

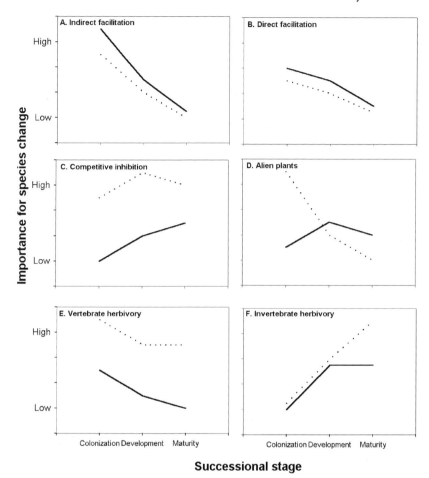

Fig. 7.13. Importance of several processes to species change during three generalized stages of succession (colonization, development and maturity) at two levels of environmental stress (low stress, dotted lines; high stress, solid lines). (A) Indirect facilitation; (B) direct facilitation; (C) competitive inhibition; (D) alien plants; (E) vertebrate herbivory; (F) invertebrate herbivory.

Walker & Chapin (1987; see Fig. 3.7) modeled many of the same processes for all types of succession.

The one factor that most strongly dictates the course of primary succession is the intrinsic quality of the site. Stress includes any physical factor that limits establishment and growth, so it must affect the rate of succession as defined by species turnover. The rate of succession generally slows with the development of a sere because more persistent,

Table 7.1. *Rate of early primary succession, determined from estimated biomass accumulation*

Properties	Rates low if:	Rates high if:	Examples
Environmental			
Moisture	Dry	Moist	Lava (Aplet *et al.*, 1998)
Nutrients	Infertile	Fertile	Sand dunes (Olff *et al.*, 1993)
Growing season	Short	Long	Lahars (Dlugosch & del Moral, 1999)
Substrate			
Stability	Unstable	Stable	Dunes (Avis & Lubke, 1996)
Safe-sites	Few	Many	Pumice (del Moral & Wood, 1993b)
Extremes			
Salinity	Hypersaline	Low	Salt flats (Fort & Richards, 1998)
Acidity	Low pH	Neutral	Mine tailings (Borgegård & Rydin, 1989)
Temperatures	Low or high	Moderate	Deserts and tundra (Danin, 1991; Davey & Rothery, 1993)
Landscape			
Isolation	High	Low	Lahars (del Moral, 1998)
Size	Large	Small	Rock outcrops (Houle, 1990)
Latitude	High latitude	Lowland tropics	Tundra (Beyer *et al.*, 2000)

longer-lived species become dominant. Seres with low stress and high productivity (e.g. wetlands) develop quickly at first compared with high-stress sites (e.g. pumice, sand dunes, toxic mine wastes). However, during development, turnover may cease in productive systems before structural convergence to local mature vegetation. Dense vegetation eliminates the pioneers, and competition or allelopathy precludes the invasion of climax species. The relationship between stress and succession rate can be described more fully in relationship to environmental factors (Table 7.1). Dry conditions lead to stressful conditions and slow development, whereas moist conditions provide opportunities for rapid development. Infertility severely retards development, but nutrient-rich sites develop quickly. Low temperatures and short growing seasons retard the development of a sere. Even moist, fertile sites may fail to develop quickly if they are unstable or if there are few germination sites. Saline, acid and polluted habitats all develop slowly and represent other aspects of stress.

A sere rarely commences development without indirect facilitation (physical amelioration; see Table 6.1). The effects of these processes differ somewhat depending upon the level of stress in the system. In Fig. 7.13A–F we contrast systems with high stress (e.g. limited production, extreme temperatures, toxic soils) with systems developing under more favorable conditions. Most sites require the addition of nutrients from beyond the system. Other processes that ameliorate a site include leaching of toxins, formation of surface heterogeneity and erosion prevention, thereby alleviating the unfavorable conditions listed in Table 7.1. These forms of indirect facilitation (Fig. 7.13A) decline in importance during development as biotic factors come into play and more species become capable of establishment. Direct facilitation (Fig. 7.13B) is less likely to occur in early succession than indirect facilitation, but more likely to occur later when species interactions supersede site amelioration in driving species change. Both types of facilitation have somewhat greater impact on species change in stressful than in productive environments. Competitive inhibition (Fig. 7.13C) is most important in later stages of succession and in less stressful habitats. For example, mature thickets or dense swards of rhizomatous species can inhibit further invasion (see section 6.3.2 on thickets), but these thickets do not form in the initial colonization stage of primary succession.

Introduced plant species (aliens) can foreclose certain trajectories by eliminating other colonists or they can arrest further development by forming dense thickets. Some aliens can alter the nutrient regime and combine with species of mature vegetation to form new, stable assemblages. The ability of alien species to arrest or deflect succession is likely to be greatest in the colonizing stage because many aliens are well adapted to open habitats (Fig. 7.13D). Aliens are usually better adapted to productive conditions, so we expect that their impact will be greatest in low-stress systems. Subsequent invasions by aliens are inhibited by existing species, a process that is more intense in productive habitats. Therefore, the potential role of aliens in productive sites declines sharply as the system develops. Aliens are unlikely to invade mature sites without further disturbance, but if they persist in the vegetation, then their impacts will also persist. High-stress habitats present significant barriers to establishment for native species as well as exotics. However, established exotics may persist in high-stress communities and resist elimination as the vegetation matures. If the native flora is impoverished, as on Hawaii, or if the surroundings are dominated by introduced species, then aliens may dominate the colonization process regardless of conditions.

Herbivory can alter the rate and trajectory of succession in complex ways. Both vertebrates (Fig. 7.13E) and invertebrates (Fig. 7.13F) can have profound impacts at any point in a sere, and unusual events such as an insect plague may be more crucial than routine herbivory. Often, the impacts are contradictory, or operate indirectly. For example, a herbivore may reduce the competitive pressure of a dominant pioneer, allowing invasion by species found at later stages. During early colonization, vertebrates are likely to have a significant impact because there is little biomass and limited grazing can have a large impact. However, the impact of vertebrates may become crucial during development and maturation phases, though less so than early in the sere. Vertebrates can arrest succession at any stage by consuming potential dominants, or accelerate succession by consuming colonizing species that might inhibit succession. As diversity increases, this becomes less likely. Birds and bats may accelerate the invasion of many species after the vegetation develops vertical structure. All of these effects exist to lesser degrees in less productive (more stressful) habitats. In mature communities, the plant species present are those adapted to the herbivore regime, so the potential effects of animals on further succession are reduced.

Invertebrates are more likely to have their impact on a sere through their impact on a key species, and the impact is more likely in habitats with more productivity. During development, impacts may be large in productive habitats, and only slightly less in stressful habitats. The importance of invertebrates increases in mature vegetation because periodic outbreaks of insects in mature vegetation are a major disturbance that can reset the succession clock. Stressful communities are less likely to be strongly affected by outbreaks due to low population densities.

A sere may fail to develop similarities with adjacent mature vegetation for reasons associated with landscape context, a different pool of available species, stochastic events and stress. During colonization, the chance that vegetation of a stressful site will approach that of the regional vegetation is lower than the chances on a productive site. The productive site will offer fewer barriers for local species to establish and will develop more quickly, reducing the time during which the sere may be deflected. (In extreme environments, the pool of viable species may be so small that convergence to the regional vegetation is the only alternative.) Later during development, probabilities for convergence increase because local species are likely to establish and help to foster similar facilitation, herbivory and inhibition regimes as well as an environment similar to that of the surrounding vegetation. If a site has not begun to converge during the developing stage, then it becomes less likely that it will do so as strong dominance is achieved during the mature phase.

Isolation plays a crucial role in primary succession because nothing happens without immigration. The importance of isolation declines during the sere because processes become increasingly dependent on the site conditions. This factor produces strong density and composition gradients along the edges of newly formed sites, which implies that the difference between no isolation and great isolation can be merely a short distance because the seed rain of most species attenuates over very short distances (cf. Nathan *et al.*, 2001). Successional patterns are as varied as the landscape. However, they are unified by processes such as competition and herbivory, by the effects of environmental factors on rates, by landscape effects and recent history and by the importance of stochastic elements. Seres typically develop greater biodiversity and structural complexity, but it is difficult to predict the ultimate floristic composition of a sere. As we will show in Chapter 8, these characteristics suggest new approaches for restoration of barren landscapes.

8 · *Applications of theory for rehabilitation*

8.1 Theory of rehabilitation ecology

8.1.1 Introduction and definitions

Continual growth of the human population leads to declines in the quality of pastures, rangelands, farms and forestlands through erosion, salinization and desiccation. Further, the natural landscapes that provide many ecosystem services are rapidly being converted to agriculture, industrial and urban sites and even wasteland. The biodiversity and habitability of the planet is now more threatened than ever before. Therefore, it is imperative that degraded land is rehabilitated and that adjoining natural landscapes be protected. However, it is clear that degradation thresholds have been crossed in many habitats and succession alone cannot restore viable and desirable ecosystems without intervention (cf. Rietkerk *et al.*, 1997). In this chapter, we explore rehabilitation ecology, applying lessons of the previous chapters. Processes of primary succession, from dispersal to interactions, are crucial for ecosystem recovery but have not been adequately evaluated for their applicability. Despite advances in the practical aspects of creating new habitats (Gilbert & Anderson, 1998), reclaiming degraded ones (Wali, 1992) and rehabilitating sites to productive uses (Dobson *et al.*, 1997), restoration ecology remains in its formative stages. Rehabilitation offers hope for a better future, particularly when individual projects with limited scope can be integrated with larger conservation programs.

Even when faced with severe economic and time constraints, most practitioners aim beyond mere reclamation to provide valuable resources that include pastures, wildlife habitat and recreational land. They also strive to develop sites that contribute to biodiversity or that provide habitats useful for passive recreation. Few rehabilitation projects have been evaluated longer than their legal mandate and most of these fail to meet performance criteria (McKinstry & Anderson, 1994; Kondolf, 1995). Such failures can result from a lack of appreciation of the dynamics

Table 8.1. *Definitions for concepts used in restoration ecology*

Term	Definition
Reclamation	Actions that stabilize a landscape and increase the utility or economic value of a site. Usually involves amelioration that permits vegetation to establish and become self-sustaining. Rarely uses indigenous ecosystems as a model
Reallocation	Management or development actions that deflect succession of a site from one land use to another, with the goal of increased functionality
Rehabilitation	Actions that seek quickly to repair damaged ecosystem functions, particularly productivity. Indigenous species and ecosystem structure and function are the targets for rehabilitation
Bioremediation	The use of plants and microbes to reduce site toxicity (a special form of rehabilitation)
Restoration *sensu stricto*	Actions that lead to the full recovery of an ecosystem to its pre-disturbance structure and function
Restoration *sensu lato*	Actions that seek to reverse degradation and to direct the trajectory in the general direction of one aspect of an ecosystem that previously existed on the site

After Aronson *et al.* (1993).

of developing systems and the powers of biotic interactions. In this chapter we describe how theory relates to rehabilitation and how to improve results. Because natural succession may be protracted, restoration actions must abet natural processes and direct development along desirable trajectories. The challenge is to develop self-sustaining, useful ecosystems with as many natural elements as possible. We recognize that economic imperatives differ throughout the world, so we explore models from simple reclamation to full restoration. We will describe problems faced by restoration ecologists and suggest how to improve common practices. Important topics such as bioengineering or naturalistic landscape design are beyond the scope of this chapter.

There are several, sometimes contradictory, terms used to describe efforts to accelerate primary succession on barren landscapes (Bradshaw, 1983a,b; Wali, 1992; Hobbs, 1999). Table 8.1 summarizes the main terms employed in restoration ecology (Aronson *et al.*, 1993).

We use the term *rehabilitation* for any action aimed to improve ecosystem structure or function by directing trajectories towards local mature ecosystems. We use *reclamation* when the immediate goals do not include the return to even a quasi-natural system. Rehabilitation may seek only to repair ecosystem functions (rehabilitation in the narrow sense). It may seek to develop an ecosystem generally similar to one previously on the site (*restoration* in the broad sense) or it may seek to fully recreate an ecosystem previously present (restoration in the narrow sense).

Reclamation seeks to establish some functional productivity, for example pasture or playing fields. Reclamation often involves post-industrial, severely degraded or toxic sites. One result is that land is reallocated from its once pristine condition to a new use. Reallocation may occur outside the realm of restoration, as when a forest is converted to a farm. In our context, it occurs when a degraded system is beyond a threshold and a simple, artificial ecosystem is the only feasible endpoint. Rehabilitation seeks to repair a damaged ecosystem. The emphasis is on structure, productivity and stability, not composition. The goals of rehabilitation are modest: to establish a simplified version of one of the ecosystems that may have occupied the site. Restoration has been used in several ways. The goal of a restoration plan is to recreate fully a specific ecosystem. Restoration requires removing alien species, introducing desirable natives, ameliorating any stresses and removing sources of disturbance. Ultimately, the structure, function, diversity and dynamics of a specific pre-disturbance ecosystem are sought. This is both difficult (assuming the target ecosystem is known) and unlikely. A more usual restoration plan seeks to guide the system in the general direction of some presumptive ecosystem. This view recognizes that there are *alternative steady states*. We use restoration in this second, less stringent meaning.

8.1.2 Interdependency between rehabilitation and ecological theory

Rehabilitation is linked to classical succession theory. Clements' (1916) paradigm (see section 3.3 on holism) that primary seres are deterministic and converge to a single equilibrium climax underlies much restoration thinking. This view is congruent with an engineering approach: all actions have unique, predictable consequences. However, we demonstrated in Chapter 7 that primary succession is stochastic, at best only generally directional and often reticulate, regressive or cyclic. Predictability at the species level is poor. Mosaics of plant communities on the landscape are common and vegetation often fails to converge to become a single homogeneous vegetation type. The elements of a sere can change owing to internal or external forces. This reality is unsettling if specific performance standards are mandated within a short time frame.

Natural recovery is too slow and useful land is scarce. Therefore, rehabilitation must return land to productivity to help support growing human populations. Rehabilitated habitats must be marketable and therefore physically stable. They must support construction activities (e.g. roads or buildings) or provide other societal values (e.g. conservation, ecosystem services or education). Studies of local primary succession reveal potential constraints for rehabilitation. Successful rehabilitation requires coordinated maintenance to overcome bottlenecks and to guide a sere toward desired targets. There may be a series of bottlenecks, some of which are not revealed until others are eliminated, so contingencies must be considered.

The perspectives of research and restoration ecologists overlap (Zedler, 2001). However, they differ in important ways. Research ecologists are question-driven, biased towards the study of natural systems, skeptical, patient and usually impractical. As a class, their mission is to test hypotheses. Restoration ecologists work with degraded or entirely barren systems. They may be impatient and must get results within temporal and fiscal constraints. They are apt to conduct operations, not experiments. Sometimes, the results of monitoring or experiments indicate that the desired goals are not being reached, yet positive 'spin' is placed on the results. Such instances do not advance understanding of restoration processes.

Improved results will occur when ecologists and restorationists communicate with each other. Studies of primary succession provide a basic understanding of processes and mechanisms that drive community development. No sere unfolds without modification and the effects of chance. Succession studies rarely address practical problems, but when studies of primary seres do consider ramifications for rehabilitation and discuss them explicitly, they can facilitate restoration. Rehabilitation studies published in refereed journals that include experiments or monitoring of rehabilitation results enlighten our understanding of succession. Experiments in rehabilitation projects, dominated by unique species combinations in novel substrates, can be used to test principles. The result of this communication is a feedback loop that accelerates effective restoration and understanding of succession.

Ecosystem structure and function are clearly related, but the correspondence is inexact. Dobson et al. (1997) implied that either natural succession or rehabilitation could lead to a complete restoration of an original ecosystem, comparable to that of natural processes. We modified

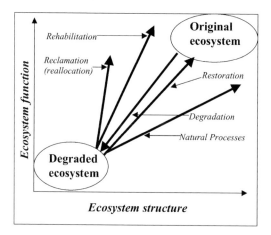

Fig. 8.1. Ecosystem structure (e.g. diversity and complexity) is related to ecosystem function (e.g. biomass accumulation, nutrient cycling). Reclamation produces functioning ecosystems of a different structure than primary succession. Rehabilitation attempts to restore the original system, but it may fail for many reasons. In a fragmented landscape with many alien species, natural processes may also fail to achieve a semblance of the original ecosystem. Modified with permission from Dobson *et al.* (1997), *Science*, **277**, 515–22. © (1997) American Association for the Advancement of Science.

this view (Fig. 8.1) to suggest that rehabilitation may create functionally similar ecosystems resembling the original, but that neither rehabilitation nor natural processes can fully recreate the original ecosystem. Rehabilitation may return a system to functionality more rapidly than do natural processes, but it may be arrested in many ways.

Predictive models can be used to assist with planning species introductions. Ecosystems may develop through predictable sequences of species, although this is less likely than commonly believed. Predictable species outcomes are uncommon, although structural features are more likely to be predictable. Non-equilibrium communities that develop under stochastic regimes are common (Fig. 8.2). Palmer *et al.* (1997) suggested that the degree of predictability should be a guide to the planning of restoration efforts. If species composition and other structural elements are predictable (e.g. low biodiversity), then restoration should focus on key species. These models can also be used to predict when to expect a sustained community to develop. When seres are more stochastic, functional attributes should drive restoration. Decomposition rates and biomass accumulation parameters can be monitored. Variation in the results can be predicted by using probabilistic models.

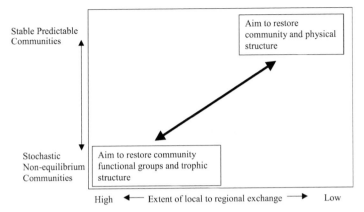

Fig. 8.2. Utility of theory varies with predictability and extent of biotic interaction. From Palmer *et al.* (1997). Reprinted by permission of Blackwell Science, Inc.

In the context of shrinking land devoted to biotic production and expanding urban areas, impervious surfaces and human populations, rehabilitation is imperative. Because productive land is precious, it no longer suffices to just get grass to grow. Where the goals of rehabilitation do not include the production of grazing, farming or forestry lands, or land for urbanization or industrialization, maintenance of biodiversity should be a high priority (Bakker *et al.*, 2000). Rehabilitated sites sustain a biota that can disperse into newly disturbed sites. They can serve as part of a landscape that promotes biodiversity, not one that further degrades the biological integrity of a region.

8.2 Rehabilitation processes

8.2.1 Conceptual framework

Many natural primary seres share features with anthropogenic sites subject to rehabilitation. For example, cinder cones are similar to many mine tailings (Fig. 8.3). Just as the local flora may not contain colonists for extremely stressful naturally produced substrates, it is unlikely that natural dispersal will provide effective colonists (Mueller-Dombois, 2000; Wijdeven & Kuzee, 2000) of unnatural sites. For example, Ninot *et al.* (2001) compared coal mine waste dumps that were reclaimed by several methods with those for which only spontaneous invasion was permitted. They found that the best results occurred next to natural vegetation and when appropriate native species were sown. Therefore, a major component of rehabilitation is to ensure that suitable species are provided naturally or by direct intervention.

A

B

Fig. 8.3. Naturally produced substrates are often mimicked by anthropogenic ones. (A) Tolbachik Volcano cinder cone; (B) a mine tailing in Yorkshire, England. Each substrate was characterized by low fertility, drought and instability.

Two fundamental questions arise. Can the rate of rehabilitation be accelerated compared to a natural sere? Can the product of rehabilitation resemble the product of a natural sere? The answer to the first question is clearly 'yes.' Table 8.2 summarizes solutions to rehabilitation problems that accelerate succession. Fig. 8.4 shows the relationships of terms used

Table 8.2. *Approaches that accelerate rehabilitation*

Category	Problem	Initial treatment	Long-term treatment
Texture	Coarse	Organic matter or fines	Vegetation
	Fine	Organic matter	Vegetation
Structure	Compact	Scarify	Vegetation
	Loose	Compact	Vegetation
Stability	Unstable	Nurse plants or physical barriers	Contour the land or vegetation
Moisture	Wet	Drainage	Drain
	Dry	Irrigation or mulch	Drought-tolerant vegetation
Nutrients	Deficient	Fertilizers	N-fixing species
pH	Low	Lime	Tolerant species
	High	Pyritic waste or organic matter	Weathering
Heavy metals	High	Organic matter or tolerant plants	Bioremediation (using plants to extract heavy metals)
Organic wastes	High	Inert barriers	Microbial decomposition
Salinity	High	Weathering or irrigate	Weathering or salt-tolerant species

Modified with permission from Dobson *et al.* (1997), *Science*, **277**, 515–22. © (1997) American Association for the Advancement of Science.

in restoration ecology (Aronson *et al.*, 1993). Disturbance leads to altered steady states, but chronic disturbances result in degraded systems. Continued disturbance drives a degraded site beyond a threshold from which the site will not recover without direct human intervention. The resultant barren sites may be reclaimed so that reallocation results in some new use. Intense intervention in an attempt to rehabilitate the site may produce simplified ecosystems with either native or alien species. These sites may be further rehabilitated to restore an original alternative state or may form a new ecosystem. Degraded ecosystems may also be brought under improved management and restored to an alternative steady state. In rare cases, intense restoration efforts may re-establish a system with many qualities of the original pristine ecosystem.

The answer to the second question is less clear. The conventional paradigm is to reinitiate succession and accelerate biotic change along a pre-determined and illusory trajectory. Unfortunately, targets suggested by existing mature vegetation may not be reached. Unforeseen

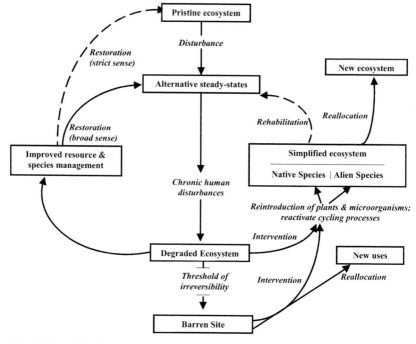

Fig. 8.4. Model of ecosystem degradation and potential management responses. Modified from Aronson *et al.* (1993). Reprinted by permission of Blackwell Science, Inc.

interactions, such as the effects of herbivory or indirect competition (see section 6.4.2 on herbivory) often produce unstable vegetation (MacLeod *et al.*, 1993; Grice & McIntire, 1995). It is not clear how management can push vegetation to a stable equilibrium. Turner *et al.* (1998; see also Dale *et al.*, 1998; Franklin *et al.*, 1998) correctly noted that the predictability of species replacement in primary succession is low and that alternative seres are likely. Thus we should expect restoration projects to explore uncharted territory.

A newly formed site may be incapable of developing without restoration efforts. Fig. 8.5 shows a hypothetical situation in which a mature, undamaged system (State 1) is related to three degraded ones (cf. Hobbs & Norton, 1996). Mild degradation (State 2) can be repaired, perhaps by secondary succession. Recovery from more intense degradation (State 3) may require intervention, but is still possible. The completely degraded site (State 4) will not recover unless rehabilitation is imposed (cf. MacLeod *et al.*, 1993; Grice & McIntire, 1995) because it is below

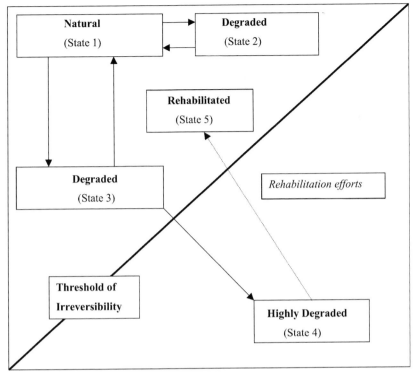

Fig. 8.5. State and transition model of a hypothetical system. There are five states, with transitions from State 1 (mature) to 2 (degraded) and from State 1 to 3 (degraded) mediated by disturbance intensity. State 4 is highly degraded, a result of very intense disturbance on State 3. As a result, State 4 exists below a threshold of irreversibility and is barred from recovering to even State 3. Rehabilitation is required to assist State 4 to re-cross the threshold. The result may be State 5, a rehabilitated form of this community. Modified from Hobbs & Norton (1996), with State 5 added. Reprinted by permission of Blackwell Science, Inc.

the threshold of irreversibility and no natural process can return it to any semblance of a natural ecosystem in a reasonable time. However, informed rehabilitation can force the system back above the threshold (State 5) to a rehabilitated state that differs significantly from the original state.

Rehabilitation creates functional vegetation, but the complete reconstruction of a diverse, working ecosystem from barren surfaces is rare. The goal of rehabilitation should not be to create ecosystems indistinguishable from a putative natural endpoint. Instead, it should be to create an ecosystem with as many functional elements as possible, substantial productivity relative to local conditions, strong biotic interactions and little need for maintenance.

8.2.2 Planning

Reclamation that aims to produce land suitable for industry, residences or amenities has goals that are limited and readily achievable. A purely engineering approach can be effective, but is often costly and rarely forms complex or diverse habitats. Biologically sensitive approaches can enhance even simple reclamation projects, whose goal may be only to provide land for housing, by retaining intrinsic diversity and using structurally diverse elements. Rehabilitation provides a closer approximation to independently functioning ecosystems (Schuller et al., 2000), but practitioners recognize that success is rarely achieved quickly. In contrast, true restoration seeks to create sites that are integrated with the adjacent landscape by including many structural and floristic similarities. If rehabilitation is started after many undesirable species have become established, then recovery must be manipulated to hasten and direct the sere to a desirable goal.

Primary seres develop slowly and ecosystem components develop at different rates. Most biological processes occur on a scale from 1 to 100 yr, whereas many physical processes occur on a scale of centuries. Stress (environmental factors that limit productivity) and grazing limit productivity and can arrest development. Fragmentation may preclude the return of desirable species. Restoration ecologists must balance these constraints with their goals and the ultimate function of a project with a 'naturalistic' feeling. A project may be completed with biodiversity substantially lower than surrounding mature vegetation, but if the system is physically stable and productive, it may gradually diversify.

Realistic goals of restoration ecology must be appropriate to the scope and conditions under which the project is undertaken. Restoration ecology has developed from at least four lineages (Ehrenfeld, 2000) that work synergistically and provide it with 'hybrid vigor': conservation biology (emphasis on endangered species and communities); wetland creation (emphasis on ecosystem functions and services); landscape ecology (emphasis on management) and rehabilitation (emphasis on establishment).

Ecological rehabilitation programs must be based on a process that allows for mid-course modifications and for contingencies resulting from stochastic processes. The planning must be integrated with site-specific information. Goals must be amenable to changes mandated by

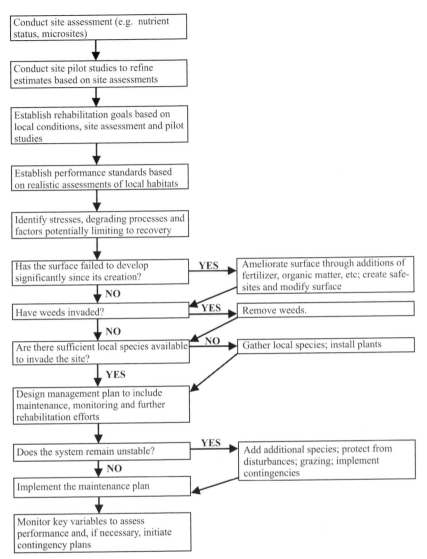

Fig. 8.6. Flow diagram for rehabilitation of a barren site. Modified from Holmes & Richardson (1999), Hobbs (1999) and Tordoff *et al.* (2000). We assume that initially there is no soil or vegetation on the site. Reprinted with permission from Elsevier Science and by permission of Blackwell Science, Inc.

information obtained during implementation and monitoring. Fig. 8.6 combines the work of Hobbs (1999) and Tordoff *et al.* (2000) concerning conceptual rehabilitation planning with that of Holmes & Richardson (1999), who emphasize operational planning. Prior to major planning efforts, the site should be analyzed for problems (e.g. infertility, toxicity), opportunities (e.g. intrinsic heterogeneity) and hints concerning the species capable of growing well (e.g. relict species). Assessment includes habitat conditions and the presence of naturally occurring species (Strykstra et al., 1998). The results of previous efforts under similar circumstances in the region should be considered. The literature may offer guidance concerning which factors may require amelioration and the levels of any such treatments. Pilot studies may be required to determine appropriate levels of resource application. For example, tests might show that the addition of phosphorus would be superfluous whereas excessive nitrogen would lead to strong growth of one species at the expense of many others. The balance of fertility is crucial in determining project success and often cannot be determined without site-specific information.

Project goals must be flexible and established in local terms. The balance between specific goals that almost guarantee failure and general goals that accept any result must be struck. Goals include at least the general nature of the biota and the vegetation structure. They may include functional targets as well. Prach et al. (1999) developed an expert system, based on factors such as geography, substrate, relief, moisture, nutrients, landscape position and size to determine reasonable goals for severely degraded habitats. Their method predicted outcomes on derelict sites to serve as the basis of goal setting. This process produced results that recognized an array of suitable targets. Having even a fuzzy goal permits the manager to detect when the sere has stalled or deflected. In addition, less precise goals offer flexibility later in the project. Alternative contingency plans may be implemented to facilitate the development of a sere in a suitable, but not foreseen, direction.

Goals are formulated as performance standards that permit quantitative assessments of the progress of the project. Performance standards must account for site conditions likely to persist after the project is initiated, relate directly to the goals and be quantifiable. These standards should emphasize structure and function and be less concerned with the particular species present. Functionally equivalent species (cf. Pugnaire & Vallodares, 1999) may be permitted to satisfy performance standards. Project management should allow for modifications of these goals. However, explicit reasons for changes in goals must be set forth a priori. Factors that could limit the success of the project should be identified and plans for determining their full impact should be developed. This will suggest how adverse effects can be ameliorated in an efficient way.

Rehabilitation is not an event. Rather, it is a process started by thoughtful amelioration of the stresses and degrading processes likely to limit successful vegetation development. If the surface has failed to develop after disturbance, stresses must be ameliorated (Fig. 8.6). If there has been a major invasion of undesirable species, these must be controlled, both initially and during maintenance of the project.

It is rare that desirable species have naturally invaded a site, but any such species should be protected. Locally produced species should then be gathered and installed according to commonly accepted installation practices designed to maximize the survival of individuals. If the system remains biologically or physically unstable, corrective actions must be undertaken. Additional species may be installed, mortality compensated, disturbance abetted and grazing reduced.

Without a management plan, the project is unlikely to succeed. This plan determines the nature and frequency of maintenance, specifies long-term monitoring to assess performance and develops ways to meet unfavorable developments (contingencies).

8.3 Implementation

Implementation of a well-planned rehabilitation program does not guarantee success. Many subtle factors can subvert even well-planned projects. The most crucial factor in implementation is the selection of appropriate species in proper combinations, densities and patterns. However, aggressive plant species and herbivores must be controlled, slopes stabilized, hydrology adjusted and soils amended (Figs. 8.6, 8.7) to provide plants with a fair chance of success. All of this involves several initial actions to ameliorate stressful physical and chemical conditions and alleviate infertility (Table 8.2). Only then is the first suite of desired species planted. Substantial maintenance may be required soon after plants are introduced, or it may be delayed until undesirable species require control.

There are six levels of intervention (Lockwood, 1997) that describe the combined manipulations of biological and physical symptoms. Typical rehabilitation programs fall into the first three levels, which require little to extensive physical manipulations and little biological manipulations. Often the plan depends on natural reassembly and rarely are more than a few species introduced. Typical of these interventions are wetlands, small prairies and landfills. However, in situations that resemble primary succession, soil is often lacking, and the project requires direct management of the developing community. Restoration of entire communities (level 4)

Fig. 8.7. The grass *Leymus* on scoria on Mount Tolbachik (Kamchatka, Russia). Sod-forming grasses are important for reclamation on many unstable substrates including volcanic deposits, mine tailings and dunes.

as well as the development of 'industrial communities' for use in bioremediation and water treatment (level 5) requires extensive construction of the physical habitat and the precise timing and sequencing of species introductions. The most extreme form of modification (level 6) requires construction of completely enclosed and isolated ecosystems. The self-contained, artificial environment called Biosphere 2 is the only large-scale example of such a habitat. Implementation can be initiated with suitable short-term amendments and many species can grow well even without soil. However, the long-term success requires soil formation, so some attention must be paid to fostering a soil biota (e.g. decomposing bacteria, mycorrhizal fungi, nematodes; Allen & Allen, 1988; Allen & MacMahon, 1988).

8.3.1 Dispersal

Natural dispersal rarely can supply the requisite biota (Chapter 5) for a diverse species assemblage in a disrupted environment. Failure to introduce the critical species results in undesirable trajectories. 'Engineers' that modify or even create a habitat dominate some systems (Jones *et al.*, 1994; see section 6.2.1 on plant impacts on soils). Some natural engineers transform materials (e.g. alligators, ants, elephants and beavers) and if a sere requires such species, it is essential that their introduction be incorporated at the appropriate stage. A vexing problem is how to decide when a sufficient number of species has been introduced. Many species should be chosen for their functional attributes, but redundant species must be included to help insure against selection errors and stochastic failures.

In urban areas, industrialized sites and agricultural landscapes, the pool of native species suitable for restoration that can naturally reach a site is limited. Further, crucial pollinators may be missing in these sites. Therefore, sowing seeds and planting individuals are common practices, which can be augmented by inexpensive methods such as installing bird perches, promoting *in situ* soil formation and encouraging appropriate pollinators. Resource extraction that leaves patches of vegetation untouched provides native seed sources and reduces rehabilitation costs (van Aarde *et al.*, 1996). Installing strategically placed perches to attract birds in secondary succession can accelerate invasion of woody species (Debussche *et al.*, 1985; McClanahan & Wolfe, 1993). The effectiveness of perches increased in primary succession on Puerto Rican landslides when there was already some existing ground cover (A. Shiels, pers. comm.), suggesting limited usefulness of perches on large, barren landscapes. Even with perches, attention to seed predation and safe-site requirements is required and care must be taken to exclude weeds. Of course, this method can work only if suitable bird-dispersed species are present. Many restoration projects concentrate on dominant species and assume that natural processes will augment the development of an understory, but this view is optimistic. Parrotta (1995) found that failure to include plant species attractive to bird and bat dispersers in the restoration mix reduced the diversity of naturally recruited seedlings in a tropical forest. Even in temperate climates, restoration projects should consider the indirect effects of introduced dominants (see section 6.3.2 on inhibition). If the local seed bank is depleted, then sods containing suitable seeds can be added after the sere is established (Pärtel *et al.*, 1998).

8.3.2 Establishment

The first task to promote good establishment is to stabilize and improve the site. After required physical stabilization, site amelioration can commence. Loosening surfaces and promoting water infiltration and drainage by ripping and furrowing a surface enables the entrapment and germination of seeds. Fertilization can be helpful, particularly with organic matter from the site. Over-fertilization favors aggressive species (Ross, 1999; see section 8.4.3 on infertility). Mulches improve soil moisture (Schuman et al., 1998; Andrés, 1999) and may also promote seed entrapment, seed germination and soil fertility (Tongway & Ludwig, 1996). However, mulches high in carbon (e.g. straw) can lead to immobilization of nutrients by soil microbes (Zink & Allen, 1998). Straw mulch was less effective than simple rock mulch in arid Nevada (U.S.A.) because it promoted drier conditions on mined substrates (Walker & Powell, 2001).

Topsoil is often added as a quick fix (Andrés & Jorba, 2000), because it is more fertile, and may contain mycorrhizae, decomposers or a useful seed bank. Native topsoil will support planted seeds well, but be more hospitable to invading species (Smyth, 1997). However, all of these characteristics decay rapidly (Koch et al., 1996) with storage, and stored soil can accumulate a propagule bank of undesirable species (e.g. rhizomatous perennials such as *Typha* or *Equisetum*). Using topsoil should be considered only after testing because the seed bank often does not represent the vegetation from which it is derived (Van der Valk et al., 1992). Seed banks available for restoration often lack propagules of woody species (Tekle & Bekele, 2000), so even with topsoil application, plans leading to woody vegetation should include planting the desired shrubs or trees. Shrubs can ameliorate stressful conditions in many cases. The addition of small amounts of some topsoil will often provide suitable microbes (Strykstra et al., 1998). Amelioration may also require creation of safe-sites. Soils may require detoxification, aeration and pH adjustment. Many other considerations depend on specific circumstances, but most projects must make a series of decisions like those in Table 8.3. Other site manipulation decisions may be required after the initial establishment stage.

We strongly recommend the use of native species even in such inhospitable cases as steep, dry road cuts (Paschke et al., 2000). Further, species selected should be natives of local provenance (Walker & Powell, 1999b). When nitrogen-fixing species are used to enhance site amelioration, both herbaceous (e.g. *Lupinus*, *Trifolium*) and woody nitrogen-fixers (e.g. *Alnus*, *Myrica*, *Acacia*; Animkwapong & Teklehairanot, 1995) should be incorporated if possible to reduce long-term costs and provide steady levels of nitrogen. Transplanting individuals already infected by their N-fixing symbiont and mycorrhizal fungi can speed the restoration process (Lumini et al., 1994). However, care should be taken that these species do not come to dominate the site. When volunteers occur, criteria should exist to determine their desirability. Species unlikely to persist or inhibit the establishment of desirable species can be tolerated. However,

Table 8.3. *Decisions prior to planting seeds or young plants*

Question	Options	Comments (see examples)
Stabilize substrate	No	Substrate not subject to erosion
	Yes	Grade site; construct physical barriers; biodegradable mats; mulch (steep slopes)
Eliminate existing vegetation	No	No vegetation or desirable vegetation
	Yes	Consists of aggressive undesirable aliens; treat with herbicides, etc.
Create safe-sites	No	Not in plan
	Yes	Germination rate problematic (derelict urban sites); natural invasion part of plan
Add fertilizer	No	Substrate moderately fertile and non-toxic (e.g. wetlands)
	Yes	Substrate infertile, short-term project (e.g. mine tailings rehabilitation)
Add organic matter, mulch	No	Substrate moderately fertile
	Yes	Substrate bare, requires nutrient holding, microbe hosting, etc.
Add topsoil	No	Not available, too costly, may import weeds
	Yes	Soil may improve fertility (landscape projects)
Add soil with seed bank	No	Rarely available; seed viability not known
	Yes	Available from similar habitats, augmenting planted species (e.g. salt marsh, other wetlands; bauxite mine wastes)
Alter pH, salinity	No	Substrate with moderate pH and low salinity
	Yes	Lime acid sites (e.g. coal or iron wastes); leach saline sites
Alter toxicity	No	Site not toxic
	Yes	Toxicity can be physically isolated; use tolerant strains to mine toxic materials; leach system
Introduce microorganisms	No	Not required or already present
	Yes	With legumes or soil inoculations (e.g. mycorrhizae)

invaders are often alien species likely to compromise the success of the project and they should be controlled.

There are at least a few alternative viable outcomes for most seres, so several alternative targets are acceptable (cf. Chambers & Linnerooth, 2001; Young *et al.*, 2001). More species than may survive should be included early in a project; later in the project, maintenance can cull poorly performing species. This approach depends on local conditions to fine-tune the developing vegetation. Competition, herbivory and the interactions between a plant and its immediate environment combine to select for a well-adapted combination of species. Planners should create micro-variation (e.g. safe-sites, mound and pit topography). Functional type diversity also should be an integral part of the plan. This enhances biodiversity, offers some redundancy and provides alternative nutrient cycling paths. Variable disturbance regimes (e.g. mowing frequencies, selective culling) may be viable management tools.

Desirable outcomes should include a time dimension to recognize that the dominant vegetation may change through time. Although it is commonly accepted that desirable species are unlikely to reach the site by natural or directed dispersal (*sensu* Wenny, 2001), it is often overlooked that dispersal limitations apply also to species normally found in mature vegetation. These species must be introduced at appropriate times.

8.3.3 Monitoring

Monitoring is essential for success because early warnings of deflections will spur corrective actions. The details of how to establish a monitoring program are beyond the scope of this chapter, but principles from studies of primary succession can be applied. Monitoring must meet statistical and sampling standards because changes in composition and survival must be assessed rigorously. Permanent plots are ideally suited for monitoring because they reduce sampling variation and ensure that spatial variation does not confound temporal change. They are the most cost effective and reliable way to detect directional change (Austin, 1981; Herben, 1996; Bakker *et al.*, 1996). Photo documentation also can be very valuable (Parikh & Gale, 1998).

Routine parameters to monitor include species cover or abundance and survival. However, these measures are insufficient for complex projects. Structural features can be used to estimate the development of ecosystem functions such as wildlife use, sediment accretion, hydrology, soil organic matter and nutrient retention. Mine restoration may emphasize soil development, loss of toxicity and biomass accumulation, whereas wetland studies may emphasize soil organic matter and biomass. For each measure, the observed values can be compared to the range of acceptable conditions determined from the literature (Hobbs & Norton, 1996) and to performance standards.

Statutory monitoring usually terminates before a mature community has developed (Zedler & Callaway, 1999). External (e.g. hydrology, pollution) or internal (e.g. grazing) changes may alter trajectories. Long-term monitoring is essential because a system that fails to meet performance standards in the short run may still be on a successful trajectory (Mitsch & Wilson, 1996). Monitoring teaches much about rehabilitation, but this knowledge is often buried by a collective unwillingness to report failures. We appreciate the political and economic reasons for this, but we suggest that 'post-mortem' analyses and management responsive to developing conditions would improve future results.

8.3.4 Maintenance

The real work of a rehabilitation program starts once the first plants have been installed. Maintenance may be required for long periods. Left unmanaged, a rehabilitation project rarely follows a desired trajectory and short-term success does not guarantee that the trajectory will be maintained. The project may develop a mosaic of successful and unsuccessful patches. Reducing the competitive effects of less desirable species through planned disturbances (Lesica & Cooper, 1999), augmenting the growth of desirable species, reducing herbivory and supplementing soil fertility are some actions that may be required. Species expected to be dominant later in the sere may not invade so they must be introduced subsequently. Practical maintenance actions should be based on obvious contingencies such as developing drought or infertility and on potential adverse biotic interactions such as excessive herbivory or intense competition.

8.4 Overcoming adverse conditions

Rehabilitation seeks to promote strong establishment and rapid growth. How this can be achieved depends on the types of stress and on the project's goal. An immediate goal may be only to reclaim it from a weedy, unproductive condition and allocate the site to some productive function. Therefore, even traditional reclamation has a role in conservation. However, rehabilitation is also the first stage in more completely restoring function or structure. Here we explore problems created by common stress factors and provide examples of creative approaches to their solution that enhance the role of such projects in regional conservation.

8.4.1 Drought

Drought is a leading cause of rehabilitation failure, but sufficient water is hard to supply. Tree species are particularly difficult to establish (Prach & Pyšek, 1994) because their seedlings are sensitive to drought. In some regions, strongly vegetative, drought-tolerant shrubs can serve as nurse plants by providing shade. However, the benefit of shade can be offset by competition for scarce water (see section 6.3.1 on facilitation by nurse plants). Landscape sculpting can create swales or furrows to trap runoff. Trees are then planted in the most favorable sites, from which they can expand. Desert succession is slow owing to long-lived, persistent species and very slow recruitment rates (Cody, 2000). Therefore, rehabilitation under desert conditions is slow. To accelerate developing a shrub matrix, lower planting densities may enhance individual plant growth rates (Fonteyn & Mahall, 1978). Trickle irrigation enhances the success of desert rehabilitation. Soil compaction not only results in destruction of vegetation, but increases surface evaporation, thus intensifying drought. Compacted soils should be worked to provide greater aeration and safe-sites (Bolling & Walker, 2000).

Drought is also a challenge in the arid parts of the Arctic. Rehabilitating arid Arctic sites is difficult because short growing seasons, low temperatures and infertility combine with drought to slow growth. Bishop & Chapin (1989b) showed that gravel sites regenerated poorly because suitable colonists were absent and drought and infertility restricted establishment and growth of planted species (Bishop & Chapin, 1989a). Rehabilitating dry soils should involve mulches or topsoil additions to conserve water and to improve survival and growth of planted species.

Open pit coal mines extend across much of the arid western U.S.A. Even the dominant shrub *Artemisia* establishes with difficulty on recontoured land. Drought-tolerant weeds establish quickly and out-compete seeded shrubs. Using fresh topsoil, not stored soil, and applying mulch can improve results (Schuman et al., 1998). Because drought limits seedling establishment, it is imperative to manage sites to reduce competition. Management plans also should mimic natural primary succession by using a wide variety of species in the initial planting mixes and sowing at high density (Booth *et al.*, 1999). In this way, the species subsequently will be sorted according to local conditions. Higher sowing density also reduces proportionate seed predation, but may attract herbivores.

In many parts of the world, little rehabilitation has occurred, so it is not surprising that many landscapes remain degraded. Even where landscape factors are favorable, with many available species and few weeds, natural regeneration is inefficient. For example, natural reforestation of coal mines in eastern Oklahoma (U.S.A.) produced forests that appeared natural and served well as wildlife habitat. However, they were floristically impoverished (Johnson *et al.*, 1982) and unlikely to support wildlife populations comparable to those of natural forests. Rehabilitation that emphasizes natural invasions should supplement natural processes with uncommon species (cf. Jochimsen, 1996). Most natural mine recolonization results are far less successful than the Oklahoma case because of a combination of a lack of suitable colonists, alien invasions, high temperatures and drought or short growing seasons. Fully successful rehabilitation must include active introduction of a variety of species and life forms (Aronson & LeFloch, 1996).

8.4.2 Hydric conditions

Periodically and permanently inundated sites pose special challenges to rehabilitation and are discussed in many handbooks (e.g. Hammon, 1997; Zedler, 2001) and journals such as *Wetlands* and *Restoration Ecology*. Challenges include achieving the right hydrological regime, managing nutrients, controlling erosion and reducing herbivory. Creating wetlands is a special case of reallocation. The uncertain response of planted species to the conditions that actually develop, problems due to alien species and herbivores and extended development time are some problems associated with creating fresh water wetlands. These problems routinely lead to constructed wetlands that are smaller than their designed size (McKinstry & Anderson, 1994).

Natural colonization leads to dominance by a few wind-dispersed species (e.g. *Typha, Salix* or *Populus*). Reinartz & Warne (1993) showed that the distance from established wetlands strongly affected diversity in created marshes and that planting produced higher species richness. Created wetlands should include many more species than standards require, use a soil seed bank salvaged from other wetlands (Brown & Bedford, 1997) and be planted with more mature plants to overcome establishment problems. Using diverse species provides safeguards against miscalculations of the hydrological regime and permits heterogeneity to develop. Using wetland soil from similar wetlands may be superior to sowing seeds (VivianSmith & Handel, 1996). However, the soil seed bank must be tested before application to ensure that the included species are desirable (Tu et al., 1998). Newly excavated soil is superior to stored soil, where seed viability declines (Van der Valk et al., 1992).

Bottomland forests may be destroyed by a variety of disturbances including lahars or thermal pulses, as happened at the Savannah River (Nelson et al., 2000). Swamp creation may be needed to replace filled wetlands or to reclaim industrial waste sites. In the design of swamp restoration, careful attention must be paid to the flooding and drought tolerance of the planted species (Dulohery et al., 2000) and to herbivores (Keeland & Conner, 1999; McLeod, 2000). Where annual fluctuations are extreme, seedling establishment may require practices such as digging deep channels along which woody plants can persist during draw-down. Woody vegetation appears to establish best during declining water periods. Competition from herbs is less important in these habitats than is the elevation relative to the water table. In New South Wales (Australia), swamps were restored by planting woody species as tube stock to avoid competition from meadow species (de Jong, 2000). Directly planting rooted cuttings can enhance biodiversity in degraded swamps where natural dispersal is limited.

Dramatic fluctuations in water levels pose major challenges to rehabilitation. Planted seedlings often fail owing to exposure or submersion. Planting seeds or introducing seed banks is often more successful in these situations because seed germination is more likely at low water levels than high, and seedlings can gradually gain tolerance for flooding (Nelson et al., 2000). Rehabilitation is therefore possible if the system is well understood. Lake margin vegetation can establish best when an organic layer is present to provide protection to young invading seedlings. Fens are mesotrophic (more fertile than oligotrophic bogs), diverse and require careful management. Widespread desiccation of European fens results in the spread of weeds, loss of productivity, release of large quantities of greenhouse gasses, fragmentation and loss of biodiversity (Pfadenhauer & Grootjans, 1999). To rehabilitate fens the proper hydrology must be re-established through engineering methods and fen species must be introduced directly by seeding, planting or seed bank transfer. Fens can only be established by removing *Sphagnum*, which creates strongly acidic conditions and inhibits diverse vegetation (Beltman et al., 1996). Constructed fens will develop greater species richness if they are connected to existing remnants to facilitate dispersal. Alternatively, management practices must replace natural dispersal when it is inadequate.

Mature salt marshes are productive, but temperate zone marshes have low diversity. Created tidal marshes often fail to develop past an early stage because the planted species grow fast and are strongly vegetative, a combination that can inhibit further development. Functional development under very good conditions may require several decades if highly diverse systems with complex food webs are the target. When combined with uncertain initial conditions, pollution and dispersal limitations, it is virtually certain that functional equivalence will require much more time than is available (Zedler & Callaway, 1999). Because aeration permits high marsh diversity and rapid structural development, several authors have recommended that dense networks of tidal creeks be a feature of constructed marshes. Functional age of *Spartina* marshes was greater near creeks because aeration, organic matter, nitrogen and productivity were all higher there (Tyler & Zieman, 1999; Zeff, 1999). West & Zedler (2000) reached a similar conclusion because interconnecting tidal creeks enhanced fish dispersal and breeding. Craft *et al.* (1999) compared created *Spartina* marshes 25 yr old to nearby natural marshes. Within ten years, values of plant biomass, benthic fauna density and diversity were all greater than those of natural marshes. However, soil organic matter and nitrogen were much lower in the constructed marshes than in the natural marshes. Studies such as this suggest that monitoring should also include structural features over longer periods of time (cf. Mitsch & Wilson, 1996). Bernhardt & Handke (1992; Bernhardt, 1992) found that newly created salt marshes in Germany were more variable than mature natural marshes due to vagaries in the residual seed bank, stochastic seed dispersal and site variations. These seres may converge to locally common vegetation as competition overwhelms initial chance establishment and habitat heterogeneity. Onaindia *et al.* (2001) found that young recovering salt marshes along the Bay of Biscay (Spain) had about half the species found in adjacent mature salt marshes and were structurally distinct. They suggested that restoration of salt marshes was restricted by lack of dispersal and the slow rate at which soils developed.

Mangroves form an important barrier along tropical coasts, significantly reducing erosion. They provide nurseries for marine species and provide many products for human use, but exploitation has led to large-scale degradation. Proper rehabilitation of mangrove ecosystems requires attention to tidal forces, hydrological patterns, seedling predation and zonation. Ellison (2000) suggests that mangrove seres develop predictably if the hydrology is well established. This may be because there are so few stable mangrove community types in a single region.

Tidal flushing of mangroves is required to prevent hyper-salinity. Biomass production, food webs and nutrient cycling are important functional aspects that may not develop despite an adequate structure. McKee & Faulkner (2000) compared structure and function in two restored mangroves to mature mangroves. They determined that the younger site was on a trajectory to resemble the existing mature site, but the older site was not. The latter was hypersaline, which reduced faunal diversity, biomass and mineral cycling. They recommended that mangrove restoration projects pay very close attention to the hydrological regime.

Coal slurry ponds combine several stresses and are difficult to revegetate. Based on natural regeneration patterns, one solution is to lower the water level periodically to permit emergent vegetation to establish, then to re-flood the site (Middleton, 1995). Fertilizer can accelerate growth rates and biomass accumulation. Lime can be added if acidity is high. Ponds can become biologically more diverse if wetland soil is added to the exposed sediments because the seed banks will contribute a suite of species adapted to emergent wetland habitats.

8.4.3 Infertility and toxicity

Infertility is a common rehabilitation problem. It is relatively easy to address the simple lack of nutrients. However, inappropriate fertilization can doom a rehabilitation project almost before it starts. There is a delicate balance between addressing infertility and adding excess fertilizer that favors strong competitors (Prach, 1994a). Marquez & Allen (1996) demonstrated that legume nurse plants often inhibit growth of plants designed to dominate the community, so legume density must be controlled. Mitchley et al. (1996) demonstrated that heavy fertilization of chalk–marl spoil resulted in dominance by Lolium, whereas light fertilizer applications produced biologically diverse, but unstable, communities. Subsequent field trials permitted appropriate fertilization regimes to be prescribed. An insurance policy is to plant saplings of desired woody species at the initial planting time and perform routine maintenance to reduce competition from the ground cover.

Urban sites usually pose no adverse physical traits other than low fertility. However, owing to their locations, they are often very 'high-profile', and require sensitivity to the local human community. Bradshaw &

Chadwick (1980) summarized methods of rehabilitation of derelict urban sites using ecological methods that were less intrusive on the surroundings. By avoiding topsoil, seeding directly into the substrate, limiting fertilizer and using sparse legumes, Bradshaw could create self-sustaining grasslands and woodlands.

Rehabilitation should not rely on succession models that prescribe strict sequencing. While primary succession often develops from herbs to shrubs to trees, this is more often due to dispersal, not facilitation. Rebele (1992) showed that primary seres accumulated species from their immediate surroundings, but did not necessarily go through well-defined stages. There is always a balance between competition and facilitation and fertile substrates often resist invasion unless species adapted to very high fertility levels are available. Therefore, rehabilitation projects are more likely to succeed when lower fertility precludes strong competition and when most elements of the sere are introduced at the start of the project (cf. Bornkamm, 1984).

Sometimes a pulse of nutrients is essential. Infertile sands derived from lava in Iceland are hard to rehabilitate (Arnalds *et al.*, 1987). Greipsson & Davy (1997) found that fast-release nitrogen was better than slow-release fertilizer to establish *Leymus*. This grass used the nutrient pulse to establish, then persisted. Greipsson & El-Mayas (1999) found that mature vegetation in similar Icelandic habitats could be promoted by aerial seeding of *Festuca* and aerial fertilization for 5 yr. After 10 yr, *Agrostis* and mosses excluded *Festuca*, and after 25 yr, *Salix* became established. Here the grasses served as nurse plants and their negative impacts were less than their facilitative impacts.

Landfills, even when covered with topsoils, can be contaminated by gas migration and toxic seepage, but soil microbial development and plant growth (particularly legumes) can be more rapid than on less contaminated sites (Chan *et al.*, 1997). Simmons (1999) noted that landfills usually are reclaimed only as grasslands, but uses that are more creative are possible. An urban landfill (Seattle, Washington, U.S.A.) was hydro-seeded with standard grasses and legumes. However, after 20 yr, subsidence led to wetland formation and management shifted to promote wildlife and native wetland species. In such cases, habitats may be augmented or created. Trees may be planted (Rawlinson *et al.*, 2000) and can be used to develop woodlands where forests form an integral part of the landscape. In other places, meadows with enhanced richness may be developed. On Staten Island (New York, U.S.A.), none of 22 woody species penetrated the landfill cap (Handel *et al.*, 1997) because caps are anoxic and acidic.

A

B

Fig. 8.8. Reclaimed sites in Liverpool, England. (A) A reclaimed landfill was converted for several uses including demonstration gardens and community festivals. (B) Rehabilitation of a small derelict site using local materials and limited resources.

Surface soils were rich in mycorrhizae, so roots remain. In Liverpool (U.K.), a large landfill was capped and used for passive recreation that greatly improved the local environment (Fig. 8.8). Therefore, landfills can be converted into productive habitats. Because landfills are increasingly common, often in close proximity to urban areas, this resource cannot be ignored.

Infertility on mined lands, especially on tailings and excavations, often combines with other factors. One method to rehabilitate mined lands is to add topsoil, then seed to provide erosion control. Sydnor & Redente (2000) showed that biomass accumulation on oil shale wastes was related to the depth of initial topsoil application. They noted that seeding native species was as effective as introducing seed mixes and that any mix persisted. This study also demonstrated a priority effect (see section 7.1 on trajectories) in that the pioneer species present at high density persisted.

Coal mining in India produces large areas in need of rehabilitation. Here, producing commercially valuable forests is a principal goal. Singh et al. (1999) described how plantations could quickly establish and grow. Fast-growing species with high litterfall recycled nutrients more quickly, so these were preferred for further reclamation. However, we note that such practices may have little long-term viability and that multi-species forests may reduce erosion, provide steady employment levels and produce more even revenue flows.

There are huge challenges in the reclamation of land mined for heavy metals in the arid tropics, where resources are scarce. Piha et al. (1995a,b) presented economical ways to establish vegetation on tropical tin mined lands in Zimbabwe that were very nitrogen-deficient. Greenhouse trials identified a successful mix of native species, including legumes, to be planted. Effective *Rhizobium* strains were developed under high temperatures on low nutrient wastes and mixed species vegetation that included valuable species was established economically. Tailings and wastes from heavy metal mines present difficult challenges for rehabilitation, so mere reclamation has been the normal goal. Surfaces are usually infertile, often toxic, frequently desiccated and commonly weed-infested. At best, rehabilitation results in only a caricature of the target.

Reduction of pollution potential is the primary goal on heavy metal mine wastes, although rehabilitation may be a long-term goal. Dealing with toxins has become a partnership between the engineer and the restoration professional. Bradshaw (1952) was among the first to develop ecotypes adapted to specific heavy metals (e.g. *Agrostis* strains adapted to lead). His studies led to reclamation of very large areas in the U.K. and evolved from only grass-covered sites to vegetation that is more complex. Where resources are limited, low-input mine reclamation methods are still important. Ye et al. (2000) used several ameliorants of lead/zinc mine ponds and tested growth of common tolerant grasses. Using physical barriers of fly ash or coal residues improved the results compared with soil treatment alone. Adding either manure or compost, with a barrier, permitted a grass to cover the spoils. Based on other studies, after-care is likely to be needed for continued effective growth where nutrients can be depleted. In addition, the introduction of mycorrhizae may be required to support species common in mature communities (Reddell *et al.*, 1999; see section 4.4.3 on mycorrhizae). Johnson (1998) studied unreclaimed taconite mine tailings and discovered that adding suitable mycorrhizae with sludge to avoid a large flush of nutrients promoted late seral species, and inhibited weeds. One way to inoculate substrates is to incorporate small amounts of topsoil from mature vegetation (Titus & del Moral, 1998b).

Grant & Loneragan (1999) described the problems of rehabilitating bauxite waste in Western Australia in the 1970s. The rehabilitated landscape had little of the native *Eucalyptus*. Instead, *Acacia*, introduced species of *Eucalyptus* and many weeds dominated it. Although the practice of using exotic species of *Eucalyptus* has ended in Australia, it remains very common in Mediterranean climate parts of the world (e.g. Spain, Italy, Chile). Drought precluded the rapid development of planted species and suppressed development of closed canopies that could restrict invading species (Zobel *et al.*, 1996). Native vertebrate dispersers were not attracted (cf. Foord *et al.*, 1994). Rehabilitation on bauxite wastes can be aided by natural invasion if mixed natives are planted at high diversity with sufficient spacing for natural invasion (Parrotta *et al.*, 1997; Parrotta & Knowles, 1999). This strategy avoided the risk of arrested succession because diversity was high, yet even in this study, the best rehabilitated sites had only 50% as many species as the reference stand. Planting large-seeded forest species would improve the success of the project, as would shrubs attractive to vertebrate dispersers.

Reallocation of mined lands to wetlands is popular because it is relatively simple. Constructed wetlands on Alabama (U.S.A.) mines developed rapidly and their soils quickly developed wetland characteristics that contrasted sharply with the adjacent mine spoil (Sistani *et al.*, 1995). Although the wetland was floristically and structurally simple, it was comparable to nearby natural emergent wetlands. McCabe & Otte (2000) found lead-tolerant *Glyceria* grass invading ponds on lead/zinc-mined land. After adding fertilizer, plants grew well, although they survived without fertilization. This ecotype could be used to form useful wetlands on certain mined lands and excavations. We suggest that where water is available, the creation of wetlands on mined land offers a significant opportunity for habitat creation where weeds and adverse herbivory might be minimal.

Bioremediation is the use of plants and microbes to reduce site toxicity. Specially adapted ecotypes can stabilize a site and begin to detoxify it (Lefèbvre & Simon, 1979). It can be the first step in longer-term rehabilitation. At present, bioremediation emphasizes site amelioration and few projects consider the development of natural vegetation. Reviews by Khan et al. (2000) and Zaalishvili et al. (2000) explored bioremediation. This approach is a cost-effective way to reduce heavy metal concentration and detoxify organic wastes (Macek et al., 2000). Once toxicity is reduced, more traditional rehabilitation approaches can be used (Gunn, 1995; Gunn et al., 1995; Wali, 1999a). The use of appropriate mycorrhizae may improve the efficiency of bioremediation (Luken, 1990; Munshower, 1993; Kamnev & van der Lelie, 2000) and earthworm additions are effective incorporators of organic matter into mine tailings (Vimmerstedt & Finney, 1973). Composted yard wastes, including lawn clippings and leaf litter, have also proven effective in increasing mycorrhizal content of mined land (Noyd et al., 1996).

8.4.4 Salinity

Saline soils cover a large, expanding part of the planet. Saline and alkaline sites can develop from poor irrigation practices, over-grazing and diversion of irrigation water. Dry saline flats (playas) produce dust storms that pollute the surrounding areas. Restoration of playas has been attempted in such locations as Lake Texcoco, the Kony Basin, Mono Lake and the Aral Sea (Gill, 1996). Rehabilitation of playas is crucial, but difficult. Excluding herbivores can allow recovery, but if degradation due to grazing has crossed a threshold, the unpalatable range weed community merely becomes denser (Le Houérou, 2000). Rehabilitation plantings are successful if there is sufficient water, appropriate species are planted and local human populations support the project. Salinity is often too great for direct reclamation, so desalinization must occur. One approach to desalinization in waterlogged soils is to install soil drainage systems that can reclaim saline soils if there are periodic heavy rains (Sharma et al., 2000). Plant-oriented methods can also be economical and appropriate. One method is to plant fast-growing salt-tolerant tree species (e.g. *Tamarix*), then harvest and remove the biomass, along with the accumulated salts (cf. Qadir et al., 1996). However, *Tamarix* has contributed to the salinization of arid lands outside its native range of northern Africa and the Middle East. The efforts in the southwestern U.S.A. (Busch & Smith, 1995; Walker & Smith, 1997) and Australia (Griffin *et al.*, 1989) are to remove, not plant, *Tamarix*. These differences highlight that each reclamation effort must be geared to site-specific conditions.

8.4.5 Extreme pH values

Bogs are difficult to restore because they combine hydrology sensitive to minor surface alterations, species with narrow tolerance ranges, and fragile soils. Rochefort & Bastien (1998) improved their success by planting multiple species of *Sphagnum* under artificial shade to ensure that some would establish (see also Grosvernier *et al.*, 1997). Poulin *et al.* (1999) compared natural bogs to remnants in mined bogs. They concluded that remnants only marginally aided in the recovery of a bog because they desiccate and lose *Sphagnum* species quickly. While these authors recommended retaining remnants in the design of bog mining, the crucial factor is to restore the hydrology. Active introduction of bog species will nearly always be required.

Standen & Owen (1999) transplanted blanket bogs in Durham (U.K.) in an effort to hasten restoration of heaths. Salvaged blanket bog vegetation was transplanted into prepared sites, but the project was only marginally successful. *Sphagnum* declined sharply in the transplants and only *Calluna* was promoted in the peat between transplants.

Northern European heaths are fragmented, grazed and transformed into bracken fern and woodlands (Komulainen *et al.*, 1999). Heath rehabilitation (dominated by *Calluna* spp., *Erica* spp. and *Ulex* spp.) is difficult. Restoring heaths from bare soil is slow and often ineffective, so rehabilitation of degraded sites has been emphasized. A logical approach to the huge task is to focus first on sites with a high probability of successful rehabilitation. Mitchell *et al.* (1997) identified the most readily rehabilitated vegetation using multivariate analyses. Succession trajectories radiated from intact heath vegetation in ways that depended on the identity of the invading species. Changes in species combinations were associated with increased fertility. The invading species differed in the ease with which they could be excluded. Because soils invaded by *Pinus* changed least and because *Pinus* was most easily removed, efforts to rehabilitate heaths concentrated on sites invaded by *Pinus*, with considerable success.

Mined lands can be stabilized, then reclaimed to form productive systems, but rarely are they returned to natural vegetation (Cooke, 1999).

Amelioration of acid spoils requires lime and fertilizer, followed by planting legumes. However, contrary to agronomic recommendations, nitrogen levels should be kept low to reduce competition from herbs if woody vegetation is the goal. Reclamation may require that the acid be physically isolated before reclamation efforts commence. Isolating the rhizosphere with clay or inert layers and deep layers of organic matter are two ways to reduce the impact of extreme pH. Borgegård & Rydin (1989) showed that sewage sludge and manure could ameliorate an acid site sufficiently for pasture grasses, after which it may be possible to direct the trajectory towards more complex vegetation.

Unassisted rehabilitation of surface coal-mine wastes is rarely satisfactory (Russell & Laroi, 1986; Prach, 1987). In Appalachia (U.S.A.), such sites remained dissimilar to natural forests with no evidence of convergence (Holl & Cairns, 1994). They had low diversity and lacked uncommon species. In central Europe, many mined lands are also left to undergo natural recolonization, with very poor results and a poor prognosis for recovery to any semblance of natural vegetation (Wiegleb & Felinks, 2001). We suggest that mine wastes be actively rehabilitated and that rehabilitation emphasize species with limited dispersal and uncommon species unlikely to establish by natural means. Care must be taken to use species capable of growing on the ameliorated substrate.

Some industrial processes have produced sites with extremely high pH. Soda ash deposited in Great Britain during the late nineteenth century filled valleys with basic material (pH > 11). Sites weathered to pH = 8, and now sustain many species adapted to calcareous soils, so the conservation value of old, basic industrial sites should not be overlooked. Owing to low productivity, many interesting species can invade and persist and there are opportunities for introducing rare species and for amending portions of the site to provide habitat heterogeneity.

8.4.6 Low temperatures

Rehabilitation under Arctic conditions can be challenging. Human impacts in the Arctic are widespread (e.g. Canada, Russia, Iceland; Komárková & Wielgolaski, 1999) and transportation corridors permit the northward migration of aggressive alien plants that could transform many mildly disturbed landscapes.

Low-temperature habitats have low mineralization rates; so barren sites will develop more quickly if available nutrients are supplied as topsoil, organic amendments or fertilizers. Densmore (1994) showed that stockpiled topsoil could accelerate natural establishment on gold mine spoils in central Alaska. This soil contained nitrogen that supported a vigorous community of *Salix* and *Alnus*, in stark contrast to untreated spoils. In this study, using topsoil was not problematic because there was no store of weeds. However, even Arctic sites are subject to constraints imposed by intense competition. When the immediate goal is to stabilize a site, care must be exercised that the created community does not arrest further vegetation development. Graminoids introduced to start rehabilitation often persist, eliminate safe-sites and sequester nutrients, thus reducing opportunities for establishment by shrubs species (cf. Densmore, 1992).

Sand and gravel normally support slow rates of recovery (100 to 1,000 years). Species that are designed for use merely as nurse species can persist and restrict, rather than foster, natural succession. In Iceland, *Lupinus* was introduced to help rehabilitate grasslands, but it has resisted invasion and reforestation (Forbes & Jeffries, 1999). Arctic systems have so few available species that a more viable strategy may be to plant rhizomatous native species into sites amended with nitrogen. Such sites will also normally require protection from grazing (e.g. by geese or reindeer).

8.4.7 Unstable substrates

Many types of infertile substrate present difficult rehabilitation challenges because soil movement restricts plant establishment. Instability is often combined with drought or infertility to compound the problems.

Quarries are often left to natural recovery. They may recover in ways that have natural analogs on cliffs (Anderson *et al.*, 1999; Larson *et al.*, 2000). Ursic *et al.* (1997) studied limestone quarries that had been developing over 17–92 yr and found strong landscape effects. The invading flora was closely correlated with the surrounding vegetation. When trees on the quarry floor matured enough to shade the slopes, heterogeneity declined and structure began to resemble local cliffs. Though slow, this process resulted in higher diversity than in rehabilitated mines because the local flora was diverse. In contrast, Cullen *et al.* (1998) and Wheater & Cullen (1997) took an aggressive approach by blasting the limestone quarry faces in an effort to recreate local landforms. They seeded the resultant scree with local native grassland species. Their short-term results indicated that communities similar to the local natural grasslands could be developed. However, they did not exclude rabbits and their overall results were poor. This result suggests that exclusion and trapping of rabbits may be required during the establishment phase. These studies suggest that steep cliffs be left to natural regeneration if there is any natural vegetation in the vicinity, but that gentle slopes and quarry floors be rehabilitated with native species (see Jefferson & Usher, 1989).

Sand dunes combine instability with drought and infertility, so dune rehabilitation is difficult. Dunes often must be stabilized by physical structures and then by planting appropriate species (Nordstrom *et al.*, 2000), such as the grasses *Leymus*, *Ammophila* and *Festuca*. Because seed germination is problematic, initial plantings should emphasize mature plants to hasten stabilization. In many cases, protection from human impacts is essential. For example, fences that direct foot traffic, but do not trap sand, can be used to divert pedestrians. Subsequent rehabilitation may emphasize the control of aliens and attraction of animal dispersers. Urban and industrial settings are isolated from sources of dune colonists; so planning for rehabilitation under such conditions should not depend on natural dispersal (see Andersen, 1995). Proactive planting is required to overcome dispersal problems. Sometimes succession is not desirable. In some European dunes, sod-cutting and mowing of later successional grasses is done to preserve rare, early successional species (Ernst *et al.*, 1996).

8.4.8 Alien plants

Most rehabilitation programs are conducted in a novel biological context. Urban sites are intensely disturbed within a context of ruderal species that rarely occur in natural vegetation (Rebele, 1994). Road edges act as corridors that extend alien invasions into rural landscapes (see section 2.2.5 on humans and transportation). Weeds often invade a site before reclamation can begin, compounding intrinsic site problems. Competition from weeds affects recruitment and community structure on seres developing on anthropogenic surfaces, just as it does on other surfaces (Olafsson *et al.*, 1994). Weed control, including any buried seed bank, is often the first step in rehabilitation. Overcoming the detrimental effects of weeds is difficult, sometimes made worse by planting weeds to stabilize the site or to 'facilitate' natural establishment (Sukopp & Starfinger, 1999).

MacDonald *et al.* (1991) provided a dramatic example of how alien species corrupt natural systems. On La Réunion in the Indian Ocean, weeds have invaded all recent lava flows to form new communities. Natives are unlikely to replace these species because they have been restricted to isolated fragments and they cannot compete successfully. Many less common native species are threatened. Lavas that formed in 2001 have little chance of being invaded by native species.

Phosphorus-deficient soils make rehabilitation of mined lands in New South Wales, Australia, difficult and weed competition makes it worse. Owing to slow growth rates and little root growth, erosion is rapid, leading to variable vegetation cover (Dragovich & Patterson, 1995). It is likely that these mines have crossed the rehabilitation threshold and that they will remain degraded, alien-rich deviations from the target vegetation.

Legislation often mandates performance standards that can only be met by seeding with exotic species. For example, the Surface and Mining Control and Reclamation Act in the U.S.A. in 1977 required a defined and high percentage of ground cover to be established on recontoured mine tailings within several years of the cessation of mining (Holl & Cairns, 1994). However, common rehabilitation practices can inhibit the development of more natural and diverse projects. For example, Parmenter *et al.* (1985) reported that sown exotic grasses replaced volunteer forbs in Utah (U.S.A.) rehabilitation sites. Although sown native shrub species were relatively abundant, they did not approach natural levels after six years and appeared to be retarded by the sown grasses. Similar inhibition of native species colonization by sown grasses occurred on tundra following construction of an oil pipeline in Alaska (Densmore, 1992).

Alien species not only arrest succession, but they often also deflect it either directly through interference or subtly by altering the disturbance regime or nutrient cycling (see section 2.1.3 on plants as agents of disturbance). *Myrica* is an N-fixing native tree of the Canary Islands that is widely distributed on Hawaiian lava (Vitousek *et al.*, 1987). *Myrica* overtops the dominant native tree, *Metrosideros*, arresting succession. By rapidly adding 10-fold more N than normal to early successional habitats, *Myrica* has promoted the establishment of other plants (e.g. the shrub *Rubus*; Aplet *et al.*, 2002), earthworms (Aplet, 1990) and pigs (Aplet *et al.*, 1991) adapted to high soil nutrients. Invasions of other woody N-fixers into low-N ecosystems, such as the tree *Robinia* on Indiana (U.S.A.) dunes (Peloquin & Hiebert, 1999) and *Acacia* shrubs in South African fynbos (Witkowski, 1991) have had similar effects on successional trajectories. Another way aliens alter succession is by accelerating fire cycles. Invasion of grasses into Hawaiian forests growing on lava (Ley & D'Antonio, 1998) and into perennial shrub vegetation in deserts has altered ecosystem dynamics. In the Mojave Desert (Nevada), exotic annual grasses such as *Bromus tectorum* may accelerate fire frequency, reduce biodiversity and eventually eliminate native shrubs (Smith *et al.*, 2000). Rehabilitation efforts in these ecosystems will require both grass removal and, at least in Hawaii, some amelioration of the nitrogen budget because the grass-dominated system has lowered rates of nitrification. In contrast, rehabilitation where N-fixers have invaded may need artificial reductions in N to re-establish native species (Peloquin & Hiebert, 1999).

Rehabilitation of moderately degraded landscapes (secondary succession) provides insight for reclamation. Annuals from Mediterranean Europe now dominate California (U.S.A.) grasslands, having established owing to heavy grazing and fire suppression. Restoration of coastal sage was prevented by competition. Eliason & Allen (1997) suggested that restoration could be accomplished either by planting container seedlings of *Artemisia* (thus avoiding the inhibiting effect of grasses) or by grass removal followed by seeding in the shrub. Exclusion of the grasses will be a very long-term process. Ashton *et al.* (1997) suggested that grasslands arrested by fire and herbivore grazing be planted with legumes. Legumes can improve the soil and reduce grazing to permit forest species to develop.

Note that alien species sometimes may be useful, so evaluations should be undertaken if there are questions. Vanier & Walker (1999) studied *Salsola* on mined land. While it traps seeds of native plants, it also traps its own seeds, creating an intense competitive environment for germinating native seeds. Its net effect on natives is not yet known, but it is probably negative (Allen, 1982; Itoh & Barber, 1983). However, there is room to experiment on the potential values of exotic species in accelerating primary succession. Posada *et al.* (2000) found that exotic shrubs such as *Rosa* protected native forest seedlings from grazing (Columbia), so that restoration was accelerated. *Rosa* is also a nurse shrub in Argentina that protects natives from grazing and provides a favorable microclimate (Elba de Pietri, 1992) (see section 6.3.2 on thickets).

8.4.9 Grazing

Unlike natural primary succession, where grazing can accelerate or arrest succession, grazing can wreck a rehabilitation project. Young seedlings are quite susceptible to herbivory. For example, geese, beaver, nutria (Hester *et al.*, 1994) and deer all can kill young seedlings. Opperman & Merenlender (2000) demonstrated that deer would eat virtually all plantings along a riparian corridor in California. Therefore, protection of plants or control of vertebrates is required for success. In most cases, well-established plants can withstand limited browsing, so protection eventually can cease. However, populations of potentially damaging vertebrates should be monitored, with contingency plans for their control or exclusion. Invertebrates also may arrest succession, but planting a variety of species can alleviate this problem.

Grazing is not always harmful to restoration efforts and may be essential to the establishment of effective mineral cycling in some systems. Parmenter *et al.* (1991) monitored arthropod recolonization on coal strip mines in Wyoming (U.S.A.). Sites 6 yr old had low insect diversity. Biomass processing was important to nutrient cycling and biomass turnover in this system, so the lack of insects retarded development. Rabbits on mine wastes in Yorkshire (U.K.) had several successional benefits (Hambler *et al.*, 1995). Middens that developed from plant and soil accumulation around rabbit excretions increased spatial heterogeneity, soil fertility and species richness. However, grazing on seedlings of less common later successional tree species was expected to slow mine reclamation, and fertilization plus rabbit exclusion hastened succession (Davis *et al.*, 1985). Mathis & Middleton (1999) demonstrated that simulated grazing could improve the diversity of wetlands developing on coal slurry ponds. Cutting *Typha* and *Phragmites* below the water line permitted a large increase in other species. Once a site has been rehabilitated, it may be possible to permit some grazing to enhance biodiversity by reducing cover of the dominant species. The impacts of grazers extend beyond what they consume because herbivory will alter the competitive balance and alter conditions for successful invasion.

8.4.10 Air pollution

Sites damaged by air pollutants can be rehabilitated once pollution is abated. In extreme cases, the soil carries a large burden of heavy metals. Two examples of severely damaged sites are mined land in Sudbury, Canada, and the copper smelter 'Legnica' in Poland. Both spewed pollutants that degraded surrounding areas with copper, lead, zinc, cadmium and sulfur dioxide (Rebele *et al.*, 1993). Only scattered ruderal species such as *Convolvulus*, *Agropyron* and *Calamagrostis* remained, creating major challenges for rehabilitation. Pyšek (1992) tried several tactics to rehabilitate these kinds of sites. Soil was removed and mounds created to produce infertile, barren landscapes without any seed bank. After sowing only with grasses, richness was low and grasses were dominant. Subsequently, wind-dispersed woody plants invaded and rhizomatous grasses expanded to exclude other grasses and some forbs. Although distinct from natural

Table 8.4. *Summary of primary succession concepts with ramifications for rehabilitation*

Concept	Ramification
Alternative steady states	Target of trajectory is 'fuzzy'; functional redundancy is required
Functional redundancy helps to ensure success	Install many species with similar functional traits; exact species composition less important than ecosystem function
Dispersal limitations are severe	Introducing species is necessary
Soil structure, safe-sites and stability determine initial species composition	Sculpting the landscape will restrict potential trajectories
Facilitation by plants not required	Direct amelioration of infertility preferable; provide physical amelioration; mature plants of latter stages can be introduced
There is a balance between facilitation and inhibition	Creating high plant cover early in the sere may arrest further development
Primary seres depend on many variables	Predicting the species composition is unlikely; use a spectrum of species to hedge the bet
Mycorrhizae are important	Inoculation required for longer term seres
Herbivory usually retards succession	Herbivores must be controlled
Disturbance during a sere is common	Maintenance is part of the process

vegetation, once *Betula* became established on mounds and the sites were considered rehabilitated, moderate diversity and substantial cover were recorded. This study demonstrated that even with limited resources, productive sites able to develop further with no further intervention can be established. It is important to note that the system was infertile, precluding strong dominance by rhizomatous grasses.

8.4.11 Overcoming adversity: a summary

When Humpty Dumpty had his notorious fall, the King's unlimited resources could not rehabilitate him. Restorationists have progressed a lot since then, but they still could not put Humpty completely back together again. However, studies of primary succession *have* improved our understanding of recovering barren ecosystems. Further application of this understanding will result in rehabilitation projects approaching their goals more efficiently and more closely (Table 8.4).

Restoration ecologists may specify one result, but other results are equally valid. Mature vegetation results from a series of unique events, landscape factors and historical processes, so goals should be general and cast more in structural terms rather than in specific species compositions. Rehabilitation projects are conducted in landscapes that are increasingly fragmented, impoverished and dominated by alien species. Most, if not all, species desired in a program must be introduced immediately or during subsequent planned stages. The process of sowing seeds reduces stochastic elements associated with seed dispersal and should not be limited to species considered 'pioneers'. Pioneer species are usually those with good dispersal, but are not necessarily good colonizers (Wood & del Moral, 1987). Installing seedlings or mature plants overcomes establishment problems and reduces losses to herbivory and seed predation. The well-considered use of topsoil provides a biological legacy that can accelerate succession. The importance of selecting species for introduction will become increasingly critical because natural recruitment will become progressively unlikely.

Many rehabilitation projects are modeled on natural succession, but species sequences in natural seres often result from dispersal limitations, not ecological requirements (Wood & del Moral, 1987). The beneficial effects of 'nurse plants' are often balanced by their inhibitory potential. Rather than trying to achieve this precarious balance, sites can be ameliorated directly, and more mature plants installed. This combination obviates a need for facilitation by pioneer plants. Facilitation may indeed occur later in the sere when mature plants cast shade, create litter and reduce soil temperatures. Amelioration should be tuned finely to the system so that pioneer species do not continue to dominate. A compromise between the need to limit erosion and the desire to permit the sere to develop is needed. Only moderate fertilization should be undertaken, combined with use of species unlikely to achieve strong dominance. For example, rhizomatous perennials should be introduced well after significant fertilization has ceased, along with more mature woody species.

Sculpting a site is a common practice. Wetlands must be excavated, mine spoils contoured and landfills capped. This process must assure that soil structure will permit establishment of the desired species. Compacted substrates restrict root growth and slow vegetation development. The provision of safe-sites (e.g. pits and swales) will enhance survival and diversity. Sites that are too steep or erosion-prone will develop more slowly and may cause problems to adjacent habitats. In some cases, bioremediation (e.g. salt mining, growing metal-tolerant species) precedes physical sculpting. Species used may not be those that will ultimately occupy the rehabilitated site.

Amelioration usually will create surprises, so many more species should be used and at higher densities than normally specified. In this way, species will establish where they are best adapted and more of the site will be vegetated. This approach may be impractical in low-diversity systems (e.g. mangroves).

Many species common to barren substrates lack mycorrhizae, whereas those of mature vegetation often require them (see section 4.4.2 on soil microbes). However, mycorrhizae should be introduced only when it is certain that they will be beneficial. Where extreme nutrient limitations are likely to persist or recur, mycorrhizae may act as parasites and slow the sere or even permit degradation. The introduction of mycorrhizae should be done for a reason, not just merely because others have done it.

Although herbivores sometimes facilitate a sere (see section 6.4.2 on herbivores), this is rarely true during establishment. Therefore, rehabilitation projects should be protected from vertebrate herbivores that congregate in large numbers. It is very unlikely that insects can destroy a project, but extreme outbreaks (e.g. locust) could occur. Limited herbivory will accelerate mineral cycling and reduce the degree to which nutrients are sequestered in plant biomass.

Natural seres are subject to recurrent disturbances that may redirect their course. Therefore, there is no logic to precluding maintenance, a disturbance planned to direct the trajectory in a desirable way, even if the goal is a natural habitat. Appropriate maintenance permits trajectory correction and improvements upon the original plan.

The underlying theme of rehabilitation may well be 'balance.' The greatest success is achieved when no species becomes strongly dominant. Fertility levels that support moderate biomass increases but do not permit strong dominance by one species are ideal. Some grazing accelerates mineral cycling, can reduce dominance and creates safe-sites. Too much grazing may eliminate useful plants, cause retrogression or cause total failure. Erosion control is essential, but some erosion and surface flux creates important heterogeneity. Mycorrhizae are generally useful, but may be detrimental when nutrients are limited. Balance will only be achieved by careful analysis of the conditions of the site and of the likely interactions between these conditions, the species to be introduced and the species likely to establish without aid.

8.5 Feedback between theory and practice

Primary succession rates are related to gradients of temperature, moisture, growing season and fertility and to the degree of spatial isolation. Rehabilitation strives to accelerate development. Rehabilitation emphasizes rates, sometimes at the expense of trajectory. Here we consider both speed and accuracy. A large-scale restoration project is like a cross-country race. Speed is essential, but the route matters. The system must be monitored throughout and maintenance applied at critical junctions to coax the sere along the desired trajectory.

8.5.1 Increasing restoration rates

The rate of restoration is measured by how quickly the site becomes fully occupied or by how long it takes to achieve a desirable state. Maximizing the rate of site occupancy may delay reaching the project goal if the first species (e.g. pioneers, nurse plants, stabilizers) form a dense cover and inhibit subsequent species. Here are several practices that can be applied to accelerate the development of desirable stable endpoints.

Surface compaction occurs because of the disturbance (e.g. roads) and because of initial treatments (e.g. grading overburden). Surfaces should be loosened to provide greater aeration and heterogeneity to trap seeds and wind-blown organic matter. Additional forms of heterogeneity (e.g. large rocks, logs) can be installed to provide shade and more opportunities for natural invasion of woody species and to attract seed dispersers. Surfaces prepared in this way will support more surviving seedlings, which will grow more rapidly than in untreated surfaces. We note that existing favorable heterogeneity should not be eliminated during site preparation. Natural depressions will accumulate water, nutrients and seeds and could serve as establishment points from which desirable species spread.

Fertility is always important and managing the nutrient budget is imperative. Soil organic matter is usually very low. Projects may start with a cover crop of legumes that are plowed back into the soil. Mulches that increase soil carbon, nitrogen and phosphorus levels should be used when possible. Carbon is a substrate for microbes and therefore essential in establishing a functional ecosystem. Mulches can double biomass accumulation, increase soil water retention and enhance the cation exchange capacity of most soils (cf. Fierro et al., 1999).

Many substrates may be fertile except for nitrogen and will sustain excellent growth provided this element is provided. Directly planting into barren substrates can be effective if care has been taken to address deficiencies (Chu & Bradshaw, 1996). Rich organic wastes rather than topsoil can be used in reclamation projects where biomass accumulation is more important than a particular species composition (Coyne *et al.*, 1998). The key is to use materials with high carbon : nitrogen ratios because the aim is to build the soil, not to grow crops. Topsoil can be applied when it is available, for example when surface mining creates a large pool of topsoil to be returned to the tailings or excavations. Topsoil contains nutrients, organic matter, a seed bank (cf. Parikh & Gale, 1998) and fungi (cf. Andrés *et al.*, 1996). It is best when it is used immediately and when it does not contain alien species. Wetland rehabilitation benefits greatly from using topsoil from mature wetlands that is similar to topsoils from the target ecosystem. Small amounts of topsoil can be added to barren substrates to provide an inoculum of valuable mycorrhizae.

Major problems befall rehabilitation programs when the herbaceous cover inhibits trees and shrubs. It is sometimes possible to plant clusters of young woody species before seeding the herbs. This permits plants to become well established and reduces maintenance costs.

Wali (1992) proposed a checklist to determine a rehabilitation strategy. A crucial step is to select species based on knowledge of the ameliorated conditions. However, most barren sites retain heterogeneity and exact conditions after site preparation are rarely known in detail. We recommend that several more species than required should be sown simultaneously and allowed to sort themselves. Higher sowing densities should be used to ensure that cover is adequate. This strategy may be expanded to planted materials if the initial cost can be recaptured in lower maintenance. Many native plant nurseries have been established during the past 20 yr in response to stringent rehabilitation regulations in many countries. The increased supply of native nursery stock should help reduce initial costs of installing larger individuals and decrease the time required for an established project. Reduced project time limits the duration of risk and hastens the time of full utility of the project.

Residual species and aliens may slow rates. Weed control is beyond our scope, but we note that Bradshaw (1983a,b) reports on many successful rehabilitation projects that start by using pre-emergent herbicides to control weed growth and allow shrubs to become established.

8.5.2 Improving the aim

We noted previously that stochastic processes implicit in succession combine with uncertain rehabilitation practice to make it unlikely that natural systems can be replicated. It is probable, however, that we can do better. Successional patterns are complex and rehabilitation cannot be accomplished merely by following standard recipes. Instead, much site-specific data must be combined with general principles and filtered through economic constraints. Ultimately a plan is developed that describes how a goal will be achieved.

One key to improved aim is to understand the nature of the target. It is insufficient merely to establish a particular habitat without consideration of the species that will occupy the habitat. However, a target based on functional types or growth forms is the most readily achievable. Biodiversity targets are more restrictive and should take into account the usually poor dispersal of plants and provide for their introduction. Often, a specific community is the stated target, although such targets are often based on limited information. Bakker *et al.* (2001) reports an exception. Definitive target communities have been developed for The Netherlands that are models for restoration. However, if a particular plant community is the goal, continued introductions and management of the developing sere remain important, owing to dispersal limitations. Communities not previously recognized can be valid results of rehabilitation because they represent communities that may have once occurred or which are variants of recognized communities. The possibility of reaching such targets should be acknowledged during planning. Targeting specific species may be valid if these species are 'engineers' or indicators of the desired resulting habitat conditions. Targeting endangered species, or other politically sensitive species, is rarely the goal of rehabilitation. It represents the narrowest, most difficult target to achieve. Therefore, planners should recognize that the hierarchy of targets ranges from very narrow (endangered species), constrained (particular species), narrow (communities), moderate (biodiversity), to wide (growth-forms, dominants or functions).

A second key to improving the aim is to realize that maintenance is essential. For example, an airplane leaves Seattle for Copenhagen. During its trajectory, pilot and computer make hundreds of mid-course corrections based on in-flight navigation readings. Without these the plane might arrive in Rome. Maintenance, based on monitoring, is analogous, although the arrival point is far less crucial. Maintenance can involve planned disturbances, late planting and reducing the density of some native species, as well as fertilizing, irrigating and weeding.

A third key is to realize that monitoring is an informational tool, not an onerous requirement. Monitoring must include both species and the environment because the environment is also changing and these changes impact the subsequent community trajectory. Developing conditions may suggest new opportunities for enhancement as well as signal potential problems.

A final key is to realize that the natural environment does not necessarily provide a template for either the trajectory or the target. The existing natural vegetation developed in a different landscape context and experienced a different history. Stages of succession are abstractions based on descriptions of what often happens, not on what must happen. Many rehabilitation programs attempt to mimic natural seres, but fail when facilitation becomes inhibition. We have learned that 'priority can dictate trajectory', so introducing desired species should occur as soon as feasible. Tactics that differ from their natural analogs may be required to achieve similar goals.

We should continue to aim for diverse, fully functional ecosystems. However, we should not insist on particular species combinations unless that is a specific conservation requirement. Rather, goals should be expressed in terms of functions and structures, with any specific species goals limited to ecological functional types that represent these structures and functions.

8.5.3 Enlarging the target

A lesson from successional trajectories (Chapter 7) is that pathways often diverge to form alternative stable states. Which trajectory will be followed may be due to factors unique to the site that are not completely predictable. Further, Zedler (2001) comments that predictability can be further compromised by each specific restoration action that can change the trajectory of ecosystem development. Although terrestrial primary succession may be more deterministic than aquatic or island systems, it appears to be less deterministic than is suggested by traditional succession models (Young et al., 2001). Such qualitative models are not yet designed to predict short-term, site-specific or small-scale patterns, and quantitative models often fail to predict the several alternative communities that develop from homogeneous initial conditions (cf. Childress et al., 1998). To the extent that species compositions are merely surrogates for specific structural and functional targets, goals expressed in terms of the species should be fuzzy. Any of several alternative stable states is satisfactory. Priority effects imply that some species will persist for several generations due to strong competition. This suggests that strong competitors should be handicapped by being installed later in the process. Just as a landscape filters the available species to permit a particular vegetation to develop, the restoration ecologist should use more species than in the target (Keddy, 1999) and draw species from a wide context.

Many, if not most, degraded systems have reached a threshold of irreversibility. In most cases, reallocation is the only viable alternative, so many alternatives can be a target. However, reaching the vicinity of the target requires continued maintenance because degraded systems are particularly susceptible to further invasion by undesirable species.

Another way to 'enlarge the target size' is to consider the targeted area in as broad a spatial context as possible. Lessons from landscape ecology are highly relevant to restoration and suggest the importance of interdisciplinary and hierarchical approaches (Urban et al., 1987). Watersheds provide excellent examples of how any given restoration effort must address the integrated impacts of hydrological, biological, political and conservation uses (Malanson, 1993; Dynesius & Nilsson, 1994; Higgins et al., 1996).

8.5.4 Summary of feedback between theory and practice

The lessons of primary succession must be applied to rehabilitation to produce more effective land use. There is still no general theory of community assembly (see section 3.8 on ecosystem assembly), although many themes can be applied to habitats dominated by humans. Just as many site-specific responses occur in primary seres, the restoration ecologist knows that the site-specific details will determine success or failure. Our view is that using the principles of primary succession can stack the deck in favor of the restorationist.

We emphasize, as have such ardent advocates of ecological restoration as A. D. Bradshaw, M. J. Chadwick, R. P. Gemmell, O. L. Gilbert, R. J. Hobbs, M. K. Wali and J. B. Zedler, that experiments and analysis of the site will prevent many errors. Knowledge of the specific habitat will guide the selection and rejection of candidate species. Properly communicated, the results of such experiments will reduce future errors and increase the efficiency of rehabilitation.

We know from the body of descriptive and experimental studies of primary seres some valid general principles for establishing and developing vegetation on barren substrates. Sites should be stabilized before rehabilitation starts. Using grasses to stabilize a site may work, but there will be a major cost when more mature vegetation fails to establish in the dense sward. Fertilization should be used carefully to reduce the risk of establishing invasion proof vegetation. Site heterogeneity should be retained or created to offer a variety of opportunities for establishment. Intense herbivory can destroy a project, so grazing must be addressed in any plan. Maintenance based on monitoring is essential to overcome unforeseen problems and to adjust trajectories.

8.6 Politics

It is said that John F. Kennedy once sighed wistfully that he only wanted a one-handed science advisor. As ecologists, we are quite addicted to couching our statements and advice in multiple alternatives ('. . . on the other hand . . . '). There is rarely patience or time for a thorough, careful analysis of the site before a project must commence. Rarely are there resources for proper investigations to guide the management plan. When time and money are constraints (as they always are), advice must be crystal-clear, based on the best available evidence and general principles. The plan's implementation must be practical, economical and understandable. Where multiple goals can be achieved, a project has a greater chance of success because there will be more people with a stake in that success. For example, White & Bayley (1999) reported on marshes established with wastewater that resulted in multiple benefits (new marshes, pollution abatement, wildlife habitat, job creation).

A good project must have clear goals and support by the social, economic or political infrastructure. For many reasons, the project may not develop as designed, so monitoring tied to contingency plans is essential to provide warnings for early action. Unfortunately, most rehabilitation projects require at least ten years to reach some form of stability. This strains the infrastructure. We propose that in many cases fruitful partnerships can be forged between the private sector, government agencies, educational groups and volunteer organizations to produce effective long-term programs.

9 · *Future directions*

9.1 Paradigm shifts

During the twentieth century, the study of succession has experienced a series of transformations. The dominant view between 1900 and 1950 described succession as progressive and deterministic. In this view, rugged pioneers changed the raw earth to facilitate the invasion of more advanced life forms, a process that reached its climax when only the best-adapted, largest and most persistent species survived. Stable communities were believed to develop reliably and to display emergent qualities not discernible among the individual components. Observations of chronosequences on the landscape were used extensively because of the assumption that the present repeats the past. A complex, rigorous and internally consistent philosophy was built on what we have come to see as a generally erroneous assumption.

A concurrent view, which gained favor only after mid-century, held a less optimistic and more reductionist view of succession. According to this alternative, succession was an individualistic, population-based process in which inhibitory and facilitative processes were equally likely to dominate. This shift in perspectives has gone from a belief in strict determinism to one in which stochastic behavior borders on chaos. More recently, modeling has suggested that some prediction of succession may be possible. Patterns have emerged, if not at the species level, then in terms of functional attributes. The search for assembly rules continues but prediction of entire successional trajectories remains elusive.

Traditional reliance solely on observation and logic has given way during the past 30 years to experimentation. These experimental studies began to demonstrate that the analogy between space and time was often flawed. Simultaneously, permanent plot studies became more common and were used to supplement chronosequence approaches. The combination of experiments with permanent plot studies led to the view of succession outlined in this book. Succession is not always progressive and it is not deterministic. Pioneers in primary succession normally require some assistance by physical processes but, once established, they can persist indefinitely or they may promote several alternative trajectories leading to different stable states. Selective removal, transplant and addition experiments demonstrated that facilitation is rarely obligate. We now view the effects of any species on its neighbors as a mixture of positive (facilitative) and negative (inhibitory) impacts that differ during the life history of an individual and with respect to the different species with which it interacts. The roles of facilitation and inhibition vary during a sere and also depend upon the degree of stress.

Primary succession is stochastic and contingent upon landscape and historical processes. Two identical devastating disturbances can produce very different recovery trajectories if the surrounding landscapes differ. Given identical impacts and contexts, the resulting seres may still develop in unique ways owing to random events. Primary succession is substantially less predictable than is secondary succession because it is entirely dependent upon early, low-probability events to commence and because small differences early in the process can ramify over time. Nearly identical sites may develop in different ways to form a network of seres. Animals may promote or retard a sere, and may be responsible for stable vegetation that is not dominated by the largest or longest-lived plants. During the past half-century, scenarios that feature multiple paths to several destinations have replaced the concept of predictable seres.

We expect that successional theory will develop along three interlaced trajectories: modeling, experimentation and the search for generalities. Modeling will become a more important tool as stochastic models advance and as more is understood about how to describe contingencies and landscape effects in a mathematically rigorous way. Models will have application to the design and execution of better rehabilitation projects and allow better predictions of the consequences of environmental change. Modeling efforts will be aided by the results of a growing number of permanent plot studies that can help to validate models.

Experimental studies, even of long-term processes like primary succession, will continue to add to the understanding of successional mechanisms. Experiments in the 1950s demonstrated that succession could proceed without facilitation. Subsequent experiments have helped us to understand the balance between facilitation and competition. Experiments can also determine what community properties (e.g. productivity, nitrogen use efficiency) are best correlated with facilitation and inhibition, to what extent primary succession is driven by key species and to what extent succession is a product of the effects of spatial patterning and the aggregation of species. With few exceptions, the role of animals as herbivores and seed predators during succession is poorly understood. Experimental treatments have revealed that animals have widespread effects, but no patterns have yet emerged.

The search for generalities about succession took various guises during the past century, from Clements' comparison of succession to the development of an organism, through neo-Clementsian trends (Tansley, 1935; Odum, 1969), to Connell & Slatyer's synthesis of models (tolerance, inhibition and facilitation; 1977). The search for assembly rules since the 1970s (Diamond, 1975) represents an optimistic attempt to develop a more predictive and deterministic theory of ecology. It represents a return to an earlier paradigm dominated by deterministic processes, but it is more mathematically sophisticated. Although we believe that assembly rules at the species level are weak at best, a search for general patterns should continue for practical reasons (see below). It is more likely that rules will be found with respect to functional types and will echo rules developed for life forms. We cannot extract from the literature, nor do we foresee any trends that would suggest that there are rules that will permit accurate prediction of entire seres.

It would be comforting if all of this effort were designed only to achieve a purely theoretical understanding of primary succession. Unfortunately, there are pragmatic reasons to develop this field further. Devastated landscapes must be returned to productive use more quickly and effectively than is now the norm. Restoration requires better models of succession and the mechanisms that govern species change. Models will require better understanding of any rules or, failing that, better understanding of the range of potential trajectories, of species interactions and of stable points. Understanding the mechanisms that drive succession is essential in the design of rehabilitation programs. It will require directed experiments by academic and applied ecologists. There must also be a shift from mere reclamation to full rehabilitation and restoration. Fully rehabilitated systems are more stable, productive and efficient and they play a large role in local conservation of biodiversity. Global climatic change, already evident, has begun to reshuffle the biota, often in surprising ways. A full understanding of primary succession is essential to the design of any mitigation project that may develop. We will need to know how to ensure the survival of transplanted species that thrive at any of several stages of succession and how to maintain rehabilitated systems in a mosaic of stages. We will need to understand when concepts and models appropriate for primary succession should be replaced by those appropriate for a mature system. We hope that human attitudes toward the natural world can soon undergo another paradigm shift. We need to learn to put greater value on the maintenance of healthy ecosystems and attend more promptly to the repair of damaged ones.

9.2 Development of standard protocols

To further our understanding of primary succession and test the general applicability of conclusions reached from site-specific data, we need comparable data collected across many seres. Without such data we are often comparing apples and oranges and successional theory is unlikely to advance. Some types of simple measurement (e.g. species densities and cover classes) or analyses (total Kjeldahl nitrogen and Olsen phosphorus) have been informally standardized. Others remain too variable for easy comparison (e.g. species diversity or depth of soil sampling). Huge differences in site factors (compare sand dunes and lava flows) make some measurements such as soil depth difficult to contrast, but other variables such as soil organic matter or total system biomass can be more readily compared. A set of easily repeated protocols that could be used by researchers throughout the world would greatly assist the integration and comparison of data from many sites.

We next briefly address the role of protocols and some pitfalls in three common approaches to the study of primary succession. For more on the influence of scales on the choice of sampling technique, see section 1.3 on methods.

9.2.1 Permanent plots

There is no better approach to studying successional change than to make direct observations at the same site over many years, yet permanent plots have many pitfalls. A lack of standard protocols for the observational or sampling methods can greatly diminish the value of such measurements. Another challenge with permanent plots is the potential damage to the plots by the investigators. Although physical damage (e.g. trampling, erosion, compaction) has long been recognized, more subtle influences can occur from repeated visits to a site. For example, the mere presence of a researcher can affect insect herbivores on leaves, even if the leaves are not touched. Further problems include relocating permanent plots and issues of long-term data storage and management. Highly accurate global positioning systems, iron stakes that can be detected by metal detectors, good site maps and awareness of the existence of the plots by appropriate land managers all help in plot relocation. Advances in data archiving on websites are making huge, cumulative data sets more accessible to more researchers and planners. Permanent plots remain an increasingly important tool for the study of long-term change, even in our increasingly impatient culture.

9.2.2 Removal experiments

Impatient researchers have other tools within their reach, of course, than sitting around waiting for century-long changes to occur. Removal experiments (and their corollaries, exclusion and addition experiments) have a growing following (e.g. http://GCTE.org/focus4.htm) because they offer at least a glimpse of how species interact and the possible unexpected consequences of removing dominant species (Zavaleta *et al.*, 2001). Yet the placement of removal experiments in a successional context is still uncommon. At best, long-term removal experiments can clarify the impacts of such interactions on successional trajectories. These types of experiment are popular with those who are concerned with the processes that drive succession, but again, pitfalls such as investigator influence, poor design or lack of protocols can hinder data interpretation. Protocols are less relevant when the experiment is short-term and only one set of investigators is involved. One common dilemma is whether to conduct a one-time removal (exclusion, addition; pulse experiment) or to maintain the artificial condition through repeated efforts (press experiment). A more fundamental issue is how to convert short-term experimental results into long-term effects on successional rates or trajectories. This conversion is often done with modeling. Despite the artificiality of removals and models, these tools can clarify successional patterns and aid in rehabilitation planning.

9.2.3 Chronosequence studies

Chronosequence studies are popular because they give immediate results for minimal effort and greatly extend the potential time span of a study. One visit to several sites that are assumed to represent different developmental stages of a sere can provide enough data for an interpretation of the entire trajectory and an estimate of the rate of succession. Standard protocols are less critical than with permanent plots or long-term experiments because of the limited number of repeat measurements involved. However, the broader comparability of the study is compromised if unusual methods are used. Pitfalls to chronosequence studies have been amply addressed (Pickett, 1989). Two problems can occur. First, the sites that are chosen may not be temporally related in a linear fashion. Verification of site age by factors independent of the vegetation is critical. The second and subtler problem is when the sites have different vegetation histories due to landscape effects, stochastic factors or other effects. This was perhaps most dramatically shown by Fastie (1995), who demonstrated that the well-researched Glacier Bay chronosequence in Alaska consisted of two separate successional trajectories, depending on the initial presence or absence of the N-fixer *Alnus*. All three standard approaches to successional studies (permanent plots, experiments and chronosequence observations) have problems, only some of which would be resolved with better protocols. Modeling approaches can help, but do not substitute for repetitive, standardized data collection from multiple sites. Unfortunately, comparisons among seres are still very rare. There is much to do, in both natural and anthropogenic settings.

9.3 Questions for the future

9.3.1 The end of succession

In this book, we have addressed many questions and offered at least provisional answers. We know that primary succession starts, often with a literal bang, but when does it end (see section 1.4 on questions that still remain)? Because succession is a continuous process of change responding to a dynamic background of disturbance across the landscape, we do not mean to imply that the end of primary succession is a closure to a Clementsian climax but only a relative decrease in the type and rate of change of biotic and abiotic variables. Determining when this type of change occurs is critical because the forces that drive succession on barren landscapes are qualitatively different from those that drive succession on fertile landscapes and management practices must change appropriately. Primary succession is unpredictable, but at some point before the relative stability of a mature stage, the system becomes more predictable. Stochastic elements of dispersal and establishment become less important and environmental factors become more closely tied to species composition. The physiological properties and population dynamics of individual species become increasingly important. Does primary succession transition into a phase that resembles changes during secondary succession, in which biotic interactions and minor disturbances drive further change, or are multiple, shorter secondary seres embedded in the longer primary sere (Mueller-Dombois & Ellenberg, 1974)? The answer is dependent on what factors are of interest, the goals of the analysis and temporal and spatial scales that are therefore chosen to examine those factors.

Primary succession may end with a whimper, but it would be useful to have a robust way to mark its demise. We suspect that a general answer will not involve inflections in the curve of various processes over time. Rather, when a decelerating process approaches zero, primary succession might be viewed as functionally equivalent to secondary succession. What might this function be? It is unlikely that any single function can be applied to all systems. To take a plant-centered aboveground perspective, one might argue that primary succession ends when plant cover has reached a relatively stable maximum. In most systems, this means that an open, probably stressful, habitat has been converted into one where a response to shade is a crucial element. The system might or might not have converged to resemble the physical structure and biodiversity of surrounding systems that have merely undergone secondary succession (see section 7.1.1 on convergence). Once a primary sere enters a realm of stable biomass, low surface light intensities and relatively stable soil nutrients, it can be treated in a similar way to later stages of secondary succession, even if species composition continues to change. It then becomes useful to apply the concepts of secondary succession to any analysis or management of the system under study.

Species-based functions are poor candidates for determining when primary succession ends because richness, diversity and turnover change in contradictory ways through succession in different systems. However, if species replacements continue, but represent no significant change in functional attributes, then perhaps primary succession has ceased. For example, replacement of *Alnus* shrubs by *Betula* trees in Kamchatka is surely part of primary succession, but replacement of one *Betula* species by another may not be (Grishin *et al.*, 1996). Once further species turnover ceases to change the diversity of growth forms or of functional groups, primary succession may have ceased.

Ecosystem attributes make alternative candidates (Odum, 1969) to assess the transition from primary to secondary succession, but variations in the importance of ecosystem attributes to successional change vary across the wide range of primary seres, making generalizations risky (see section 3.8 on ecosystem assembly). Different ecosystem attributes change at different rates and reach their asymptotes at very different ages.

Once biomass accumulation approaches zero, it is likely that many factors that characterize primary succession will no longer be recognized. However, a system may have changed into one controlled by the factors of secondary succession long before biomass stabilizes.

Another approach is to look at changing environmental factors. Severe disturbances may become less influential as the vegetation moderates the effects of wind, rain, nutrient depletion and erosion. The rate of change of nutrients and soil organic matter may slow compared to early stages of primary succession. Mean light intensity at the surface changes dramatically in most seres and can serve as an index of canopy closure, competition and other functions. Variations in soil moisture and temperature become dampened during primary succession and may also signal a transition to secondary succession. Ideally, then, several variables can serve as indicators of a shift from forces unique to primary succession to environmental factors common to both primary and secondary succession.

Although plant structure and physical environment may begin to stabilize at a point where primary succession becomes less distinguishable from secondary succession, soil properties will continue to develop. Primary succession can also have a very long-term effect on other abiotic and biotic parameters. For example, asymptotes for the accumulation of biomass or nitrogen may be at much lower levels for primary than for secondary seres. The signature of the severe disturbance that initiated primary succession on site stability, fertility, productivity and other parameters may be detectable for many centuries. This signature may include floristic as well as ecosystem traits. As stated in earlier chapters, one's perspective (focusing on the forest or the trees) is paramount. A focus on pedogenesis or disturbance regimes will lead to a different idea of when forces unique to primary succession end than a focus on species attributes and population-level processes. The value lies, then, in knowing when to adjust management strategies rather than in determining an arbitrary end to primary succession.

9.3.2 Trajectories

We have described several types of trajectory that primary seres can follow, but it remains to be determined what factors are likely to produce particular trajectories. Progressive seres are more likely in less stressful and less isolated habitats. In these cases, there are fewer constraints on development, immigration can continue and species dominate that are similar to those of mature vegetation. Retrogressive seres are more likely to occur in stressful or isolated habitats, particularly in response to soil changes. Leaching reduces fertility of sand dunes, which leads to the elimination of shrubs and persistence of rhizomatous grasses. Soil erosion commonly lowers fertility and can spark a retrogressive response in the vegetation. When salinity builds near the soil surface in an arid environment, shrub regeneration is prevented and the sere regresses to more salt-tolerant species. Further examples may be found in alpine environments affected by local cooling trends and soil erosion or isolated volcanic surfaces where leaching rates are high and immigration limited. Both progressive and retrogressive seres may be viewed as directional, which implies that there is a stable terminus for each. Non-directional trajectories include arrested, reticulate and cyclic seres. Succession is often arrested by a dominant life form such as a thicket or a chronic disturbance. Reticulate patterns lack clear causal relationships between stages and the initial stage may develop into one of several alternative trajectories. Cyclic succession alternates between progressive and retrogressive phases but has neither directionality nor stable endpoint. Most seres appear to be progressive and the special circumstances that cause exceptions can now be anticipated.

The rate at which a sere develops is related to both climatic and soil stresses. In the most extreme climates, such as the dry deserts of Antarctica, succession may not occur. Only a few species can colonize a site and their impact remains negligible so the habitat is not ameliorated sufficiently to support invasion by other species. Microsites in these habitats seem to have random accumulations of lichen and algae, and there is no hint of progressive development. Succession in slightly more benign habitats also may not occur because species representing the local mature vegetation invade and dominate any disturbed habitats in a process of self-replacement. Seres in such situations may simply accumulate species (in no particular order) and develop biomass rather than show directional species turnover through time. Eventual dominants may be among the colonists, but some pioneers must be lost during the process.

9.3.3 Predictions

An important question is, to what extent is the progress of a sere predictable? If primary seres converge upon local mature vegetation, then predictability is high. However, many primary seres fail to converge floristically. Primary seres are affected at least by landscapes, contingent factors and chance, each of which is unlikely to be similar to those factors that led to the formation of the current mature vegetation. Because primary seres are subject to stochastic processes at several stages, convergence remains only a possibility.

Extrapolations based on the assumption of convergence have many pitfalls and provide a poor basis for rehabilitation planning. However, it may be possible to predict seres by models that account for the local conditions of a site. Any such model must be stochastic, developing multiple alternatives. In effect, we see such a model as being a rigorous, quantified type of expert system. Major inputs would include colonization probabilities for all species in the vicinity; establishment probabilities under the given conditions; the probability that sequence effects will alter trajectories; rates of habitat amelioration; interactions among the establishing species; alterations of colonization probabilities through time as new dispersal agents are attracted, facilitation occurs, inhibition develops and herbivory becomes more common; and subsequent environmental changes in such factors as light, soil moisture and fertility. Clearly, such a data-intense model for primary succession does not yet exist. A goal for future research is to determine how to minimize the number of parameters, while retaining reasonable predictability. We believe that predictability will improve when we learn which factors can be excluded without losing vital information (Thompson *et al.*, 2001).

Predictability will also increase as we perform 'post-mortem' studies of failed rehabilitation attempts. By determining the causes of failure in well-designed and executed studies, we may develop a better understanding of what can be restored and what is beyond the current technology. Unfortunately, predictability of trajectories will inevitably increase as the accelerating trend towards a globally homogeneous biota accelerates. As the available species pools become limited and similar, the number of alternative trajectories will become increasingly restricted.

9.4 Missing data and poorly studied habitats

After a century, what remains to be studied? Enough primary seres have been examined that we know that it is not likely that a grand synthesis will emerge. We know what is likely to occur when a system starts to recover. We know which factors will hasten and which will slow the pace of succession. Yet we cannot with certainty predict the trajectory of any complex sere. If civil engineers had the same success rate as restoration ecologists, our cities would be in shambles! But perhaps restoring ecosystems is more analogous to brain reconstruction than bridge construction.

The largest gains in our understanding are to be found by linking such processes as biomass accumulation, nutrient turnover and biotic interactions to the control of species turnover. Such linkages, of course, are crucial in other areas of ecology, but are highlighted in dealing with the dynamics of primary seres where the influences of environment and individual organisms can be isolated and examined with minimum background noise or biological legacies to confound the researcher. Predictive models that link stochastic factors that are dominant early in primary succession with the role of ecosystem processes and biotic interactions will greatly improve our ability to manage developing ecosystems.

Studies of exotic, remote or unique primary seres will continue, and they will contribute to an understanding of the range of possible trajectories and species interactions in primary succession. Studies of both common and accessible seres should also continue in order to examine between-sere variability and test the generality of successional theories. Urban habitats, for example, are readily available, provide opportune habitats for replication and can offer immediate lessons for rehabilitation (Bradshaw, 1987). Comparison of seres along environmental gradients provides a mostly untapped natural experiment to test the role of soil fertility, temperature, elevation, precipitation, compaction, urban pollution or other environmental variables on successional processes.

Basic information on plant and animal demography is required for any study (McCook, 1994) and is often obtained hastily or from old and perhaps inappropriate sources, if not overlooked entirely. Yet without understanding life cycles, how can one compare species interactions or design experiments that account for life cycle phases that particularly influence succession? Traditional monitoring, for as long as possible, helps clarify the fundamental biology of the organisms and has the added benefit of providing a baseline for understanding cycles in population behaviors. Long-term measurements have always been the best way to study succession and fortunately are becoming more common (see section 9.2 on protocols).

Soil processes are intimately involved with species turnover, yet little is known about the mechanisms or effects of soil organisms. The importance of decomposers, mycorrhizae, N-fixers and other soil organisms in regulating species turnover remains virtually unexplored (Thompson *et al.*, 2001; but see Hüttl & Bradshaw, 2001). Whereas long-term leaching and soil erosion clearly retard or reverse succession, little is known about how nutrient pulses or other changes affect the trajectories and pace of succession. Measurements of nutrient inputs are common, but quantifying losses, particularly through deep soil leaching, is difficult and hinders the development of full nutrient budgets. Experimental removals of N-fixers are now being tried in a few primary seres, but what are the impacts of selectively removing (or adding) mycorrhizae or decomposers? As with all successional studies, conclusions will depend on the spatial and temporal scales at which the question is addressed. Studies at all possible scales (and modeling to supplement at larger and longer scales) are needed.

An exciting area of research (that represents a renewed focus on Clements' concept of reaction) addresses how plants alter soils (Binkley & Giardina, 1998). The successional implication of what is actually a two-way interaction between plants and soils is particularly relevant to primary succession when nutrients are concentrated. Successional insights will also be gained from studying the influence of spatial patterns of soil resources and plants on species change. We have noted the importance of system engineers, or species with particularly important impacts on the environment, but perhaps most species alter their environment in subtle yet important ways.

Species interactions remain a focus for much experimental work in primary succession. But we still understand little of the successional implications of competitive, facilitative and other interactions. Do they alter trajectories or just rates? Species interactions also occur in a spatial context where the density and relative size of species are important influences. Aside from some nearest-neighbor studies, however, little attention has been paid to the actual distribution of individuals. Are they single, clustered, regularly spaced? In North American deserts such patterning has been linked with successional processes over long time scales (McAuliffe, 1988).

Successional impacts of plant–animal interactions such as herbivory and predation are poorly known, particularly in primary succession. We have a broad understanding of the effects of animals on plant species composition, but clearly designed experiments are rare, especially in the context of primary succession. A special case of our ignorance involves the role of alien herbivores on native species. Alien plants have well-understood effects on many seres, but there are few analogous studies for animals. It may be that introduced animals (e.g. rabbits in Australia, pigs in the Great Smoky Mountains, U.S.A.) have such devastating effects that there seems little reason to study them in primary succession. However, there may be more subtle and pernicious effects such as losses of native pollinators or disruption of nutrient cycles.

We suspect that the necessity to find solutions to rehabilitation challenges will drive ecologists and managers to fill in many of the gaps we have noted. Practitioners responsible for guiding the trajectories of a project will need to understand basic demography, soil processes and species interactions in order to prescribe proper maintenance and protection. But will site-specific solutions apply to the management of other anthropogenic disturbances or more pristine areas? Without more comparative studies and the lessons from the trial and errors of managers, it is too soon to tell.

9.5 Conclusions

The twentieth century has been characterized by a succession of ideas about succession. As with most scientific progress, there has been a steady interplay of theoretical and methodological advances, complemented by accumulation and comparison of data sets. We believe we are entering a pivotal period where research questions and societal needs are merging and increasing attention is being focused on the dynamics of community coalescence (Thompson *et al.*, 2001). New tools allow the manipulation of large data sets, a prerequisite for cross-site comparisons and ecosystem modeling. Progress in understanding the role of microbes and the interplay of plants and soils is helping to identify functional groups and their successional impacts. More effort will be needed to address animal impacts on plants and soils and the successional consequences of invasive species that may not follow established patterns of assembly (Drake *et al.*, 1996). Yet the twin motivations of curiosity (about pattern and cause) and societal need (for improving rehabilitation efforts) will hopefully provide a strong impetus to address the unresolved issues of succession (Young *et al.*, 2001).

We now have a broad understanding of the outlines of primary succession. We will fill in the outline in the future with more data and generalizations. Models that accurately reflect the complexity of reality can best test these generalizations, but successful prediction of successional outcomes is still beyond reach. One reason to pursue the completion of the outline is that successional studies can provide useful links between the mechanistic world of physiology and population ecology to the broader perspectives of ecosystem and landscape ecologists. Also, the inherent dichotomies between primary and secondary succession, holist and reductionist perspectives, autogenic and allogenic processes, competition and facilitation provide a useful counterpoint that prevents overextending lessons from particular sites. We urge the broadest possible replication and comparison among successional studies to advance most quickly our understanding of primary succession.

If the growing urgency to rehabilitate valuable land does accelerate the feedback between data collection and conceptual advances in primary succession, it will be an odd benefit of human destruction – the constructive focus on how to solve the problems we have generated. The Robinson Crusoe paradox (Giampietro, 1999) suggests that human ingenuity is best expressed in times of tension. Unfortunately Jevon's paradox suggests that as we build solutions (e.g. roads) we also enable further resource abuse (e.g. traffic), thereby offsetting some of the advantages of our passion to solve urgent problems. We certainly have a growing number of problems that a better understanding of primary succession could help address. Will humans rise to the challenge in time? We are hopeful that timely solutions will be found.

Glossary

abiotic	Pertaining to non-biological factors such as wind, temperature or erosion
aerobic	Occurring in the presence of air
alien species	Non-native organism (see exotic species)
allelopathy	A form of inhibition based on the release of chemicals
allogenic	Driven by forces exterior to the system, such as erosion
alpha-diversity	Changes in the number and distribution of species within a community
alternative steady states	Stable vegetation that results from a common origin
alvar	Flat, open area with shallow soil over calcareous bedrock
amelioration	Physical processes that reduce environmental stresses
anaerobic	Occurring in the absence of air
anoxic	Lacking oxygen; particularly in water-saturated soils
anthropogenic	Caused by humans
assembly rules	Predictions concerning mechanisms of community organization
autogenic	Driven by forces within the system, such as competition
beta-diversity	Changes in the number and distribution of species along environmental (and temporal) gradients
biodiversity	Number and distribution of species; a measure of biocomplexity
bioremediation	Reclamation based on the use of plants to reduce toxicity

bog	A wetland with low pH, usually saturated soil and dominance by mosses
browsing	Herbivory by vertebrates on leaves and stems of woody species
carousel model	Model describing replacement of individuals that is stochastic, not deterministic
cation exchange capacity	The ability of a soil to attract and hold such cations as potassium and ammonium
chronosequence	A series of communities arrayed on the landscape presumed to represent a successional sequence (a space-for-time substitution)
climax vegetation	Vegetation that has reached a stable state
colonization	The process of arrival and establishment in a new habitat
commensalism	One-way facilitation with no effect in the reverse direction
community assembly	The process by which species accumulate during succession
community similarity	A measure of the degree to which two communities resemble each other
competition	The negative influence of one species on another due to sharing of limited resources
continuum	The concept that species change in a continuous rather than discrete pattern across the landscape
convergent seres	Two seres that become increasingly similar as they mature
cryptogamic crust	A biotic crust on the soil surface composed of mosses, lichens, algae and liverworts
cybernetics	The study of organisms as complex systems
degradation	Any process that reduces the biodiversity, productivity or other desirable trait of an ecosystem
denudation	Any act that removes the biota from a site (see nudation)

derelict sites	Habitats that have been severely degraded, usually in an urban setting
desalinization	Reduction of salt levels at a site using physical or biotic processes
disclimax	Stable vegetation that is maintained by chronic disturbances such as fire or grazing and does not reach climax
disharmony	The biogeographic concept that isolated biota are represented by taxa in proportions that are significantly different from the norm
dispersal	The process by which an organism or its reproductive units are transferred from their place of origin to another location
ballistic	Occurring by self-propelled mechanisms
diffusion	Gradual spread across the landscape
diplochorous	Having two or more common forms of dispersal
jump	An initial long-distance transfer
lottery	Low-probability dispersal across long distances
myrmecochorous	Dispersed by ants
nucleation	Spread from an initial establishment point
scatter-hoarding	Cacheing of seeds by seed predators throughout the habitat
disturbance	A relatively discrete event in time and space that alters habitat structure and often involves a loss of biomass
disturbance regime	The composite influence of all disturbances at a particular site
divergent seres	Two seres that become increasingly distinct as they mature
donor control	Dependence of any vegetation on external sites for recruitment of seeds
dune slack	The protected, relatively stable and often moist habitat inland from the leading coastal dune
ecosystem services	Seen from a human perspective, such products as clean water and wildlife habitat
ecotypes	Populations of a species that differ morphologically or physiologically

edaphic	Pertaining to soils
eutrophication	The process by which an aquatic system becomes more fertile; usually a negative result ensues
exotic species	Species not native to the location; often a weed (see alien species)
expert system	A method of trajectory prediction that involves collating questionnaires from many knowledgeable people
extinction	The loss of a species from the system under study
exudates	Chemicals released from roots that may alter the soil microbiota
facilitation	The positive influence of one species on another in a successional context
mutual	Describes cases where each partner improves conditions for the other
obligate	Describes a linkage between two species in which the facilitator is required by the dependent species
facilitation model	A model of succession that describes how species alter the environment in ways that permit other species to invade the site
fen	An oligotrophic, acidic habitat dominated by herbaceous species, not mosses; frequently saturated
forb	Any herbaceous species excluding grasses
fragmentation	The biogeographic process of dividing a landscape, as through urbanization
functional group	Species that share physiological, morphological or behavioral traits
gamma-diversity	Changes in the number and distribution of species across landscapes
gap dynamics	Replacement of individuals in small disturbances within a largely intact matrix
glacial moraines	The debris deposited by the retreat of a glacier
global warming	The gradual increase in the Earth's temperature due to human activities
gradient analysis	Statistical method to relate species patterns to environmental patterns

grazing	Herbivory on grasses and other herbs
growth forms	Fundamental units of plant morphology, such as trees or annuals
phanerophtye	Woody plant taller than 50 cm
chamaephyte	Perennial plant with buds near the ground
hemicryptophyte	Perennial herb that dies back to the surface
geophyte	Perennial herb that dies back annually to buried perennating organs
therophyte	Annual herb
holism	An approach that emphasizes connectivity and views the whole as greater than the sum of the parts
hydrarch sere	Succession in wet habitats
inhibition	Any mechanism by which one species reduces the success of another (see inhibition model) in a successional context
allelopathy	Inhibition by the deposition of growth inhibitors
competition	Inhibition by the pre-emptive accumulation of resources required at least by the inhibited species
contramensalism	Later inhibition of a facilitator by the initial beneficiary
hierarchical	With increasing competitiveness during succession
non-hierarchical	Without successional pattern; effects depend on species life histories
symmetric	In which either member of an interaction can inhibit the other, depending on circumstances
inhibition model	A model of succession that describes how species preclude other species from invading a site
initial floristic composition	Egler's hypothesis that trajectories are determined from the species inhabiting a site immediately after the disturbance

intermediate disturbance	The hypothesis that maximum diversity is obtained at an intermediate level of disturbance
keystone species	Species that are crucial to the development or maintenance of a system
lahar	A slurry of mud and debris normally created by rapidly melting ice during a volcanic eruption
life history characteristics	The species-specific patterns of arrival, growth and longevity
local control	Population dynamics are based on recruitment from within the community
mangrove	Subtropical and tropical woodland that occupies shallow tidal ecosystems
marsh	Any wetland dominated by herbaceous species
mature ecosystems	A well-developed ecosystem in which the rate of biomass accumulation and of species turnover are slow
microsite	Small-scale habitat (see safe-site)
mine tailings	The wastes remaining after extraction of minerals or fossil fuels
monitoring	The record of the progress of biological and physical features during a rehabilitation project
mosaic	A patchwork of vegetation that results from small disturbances, differential succession rates or other factors
mutualism	A biotic interaction among different species that is beneficial to both
mycorrhizae	Fungi that form mutualistic interactions with higher plants
nascent focus	The location of the initial establishment of a plant that serves as the center of population expansion
neogeoaeolian biota	Aerial fallout of insects, spiders, pollen, etc., that often constitutes the initial biological input to a newly formed surface

nucleation	The invasion process in which isolated founders spread by seed or by vegetative means (see nascent foci)
nudation	The Clementsian term for the removal of existing communities by disturbance prior to the initiation of succession
nurse plant	An established individual that alters its immediate surroundings in ways that favor the establishment of another plant
organic matter	That portion of the soil derived from organisms
patch dynamics	The concept that the vegetation on a landscape is composed of groups of organisms at different stages of succession and subject to different disturbance regimes
pedogenesis	The formation of soil
performance standards	That part of a rehabilitation project that specifies the parameters (e.g. survival, composition) that will define success
permanent plots	Marked sites that are repeatedly sampled to determine the course of succession through direct observations
pioneer	A plant that colonizes a disturbed area, thereby initiating succession
playas	Exposed desert flats, often saline or alkaline, in which succession is retarded by the absence of safe-sites
podzol	A soil profile with extensive leaching of minerals to the lower B horizon
pollination vectors	Abiotic or biotic factors that transfer pollen; wind, insects and birds are the most common vectors
predation	(a) The capture and consumption of one animal by another; (b) a form of herbivory in which consumption results in the death of the target (e.g. seed predation, herbivory on an annual plant).
primary productivity	Production of plant biomass

primary succession	Ecosystem development on barren surfaces where severe disturbances have removed most biological activity; species change following removal of all plants and soil
priority effects	The consequences of arrival order that condition subsequent compositional changes
process model	A successional model that specifically addresses mechanisms or processes
propagule	Any reproductive unit that is adapted to dispersal
pumice	A silica-rich volcanic rock usually ejected during explosive eruptions
pyroclastic flow	Volcanic material ejected at extreme temperatures and which moves rapidly downslope
reaction	Clements' term for the effect of a species on its environment
reallocation	The conscious transformation of a landscape to a condition or use distinct from its original one
reclamation	The conversion of wasteland to some productive use by conscious intervention
recontoured land	Mine tailings and overburden that have been shaped for purposes of rehabilitation
reductionism	An approach that emphasizes that a system is no more that the sum of its parts
refugia	Isolated patches that escaped the disturbance that initiates a succession (see relict species)
regeneration niches	Microsites that satisfy germination and establishment requirements
rehabilitation	Any manipulation of a sere to enhance its rate or to deflect its trajectory towards a specified goal; includes reclamation and restoration as two extremes of intervention
relay model	The Clementsian model of succession that emphasizes transitions among discrete stages
relict species	A species surviving in a refuge within a large, newly created landscape

resource ratio	In Tilman's model, species changes driven by shifts in ratios of limiting resources; specifically light and nutrients
restoration	Returning the land to its former biological status
return interval	The time between disturbance events at a given site
rhizosphere	The soil around a root influenced by root activities such as exudates
riparian	Pertaining to growth along a stream corridor
rock outcrop	Bare rock surface undergoing primary succession
ruderal	A weedy plant that colonizes recent disturbances
safe-site	A microsite where seeds have an enhanced chance to lodge, germinate and establish
salinization	The process by which soil becomes increasingly saline
salt flat	A saline playa
salt marsh	A coastal wetland characterized by tidal fluctuations, steep gradients and anoxic soils
scoria	A dense form of tephra that dominates many explosive volcanoes
secondary succession	Species change on habitats where soils remain relatively intact
seed bank	Dormant seeds found in the soil; often useful in rehabilitation
seed rain	The input of plant propagules onto a denuded site
sere	A successional sequence (series)
species turnover	A measure of succession, describing the sequential replacement of species through time
stability	A community characteristic expressing (a) the lack of change or (b) resistance to disturbance
stabilization	The process by which vegetation of a sere ceases to change dramatically; other attributes may continue to develop
stochastic	Referring to any random process

strategy	Describes a syndrome of traits adapted to particular conditions (e.g. Grime's 'ruderal' strategy describes attributes normally successful in productive, chronically disturbed habitats)
stress	Any factor that limits the rate of productivity (e.g. infertility, drought, cold, heat, toxicity)
succession rate	The rate of species replacement
superorganism	The Clementsian concept that succession was analogous to the development of an organism
swamp	A wetland dominated by woody species
talus	A collection of rocks fallen from a slope
target	The goal of a restoration project including the desired species composition
tephra	Any volcanic ejecta that is expelled into the air before falling to Earth (see pumice and scoria)
tepui	Table-form sandstone mountain 400–2000 m above forests in northern South America, with many endemic species
threshold of irreversibility	The level of degradation below which an ecosystem is unlikely to recover without direct intervention
tolerance	The ability of a plant to persist despite adverse environmental conditions
tolerance model	A hypothesis that suggests that species replacements result from invasion (or persistence) of species more able to tolerate adverse environmental conditions
toposequence	A series of communities arrayed on the landscape in response to physical factors, not time
topsoil	The organically rich surface layer that often contains seeds and other regenerating organs

trajectory	The temporal path traveled by vegetation from its initiation to stability
accelerated	The rate of species replacement occurs more rapidly than normal due to management intervention
allogenic	The trajectory is controlled by external factors such as seed dispersal
arrested	The development of a sere is delayed in response to such factors as grazing dominance by one species such as a thicket-former
autogenic	The trajectory is controlled by internal factors such as competition
convergent	(a) A sere develops increasing similarity to a local mature community; (b) two seres become increasingly similar to each other
cyclic	No stable point is reached; stages recur in repeatable patterns
deflected	A sere does not follow an expected trajectory owing to management intervention or to the introduction of exotic species
deterministic	(a) Referring to models that always produce the same result; (b) referring to the concept that a trajectory can be confidently predicted
divergent	Two seres develop decreasing similarity through time
parallel	Two seres neither converge nor diverge; each retains its relationship to its microsite
reticulate (network)	A set of trajectories coalesce and diverge during succession
retrogressive	A sere develops reduced stature, biomass or diversity at odds with the prediction of progressive development; often due to erosion, mineral leaching or chronic disturbance; identical to regressive
stochastic	Referring to probabilistic transitions between seres; or to certain succession models

under-saturated vegetation	Immature vegetation with fewer species than a system at equilibrium
wetland	Any habitat in which soils are saturated or inundated for a significant time
xerarch sere	Succession that starts in a dry (xeric) habitat and gets progressively wetter
zonation	A pattern of vegetation in response to a steep environmental gradient that can be recognized by the rapid replacement of the dominant species

Illustration credits

Ned Fetcher: 2.14

Roger del Moral: 2.3A; 2.15; 5.2A, B; 5.3; 5.4; 5.11; 5.12; 5.14; 7.6;
  8.3A, B; 8.7; 8.8A, B

Stan Smith: back cover

Lawrence Walker: front cover; 1.2A, B, C, D; 2.3B; 2.4A, B, C; 2.7;
  2.10; 2.11; 2.13; 2.16; 4.15; 5.5A, B; 5.7A, B; 6.3A

References

Aarssen, L. W. & Epp, G. A. (1990). Neighbor manipulations in natural vegetation: A review. *Journal of Vegetation Science*, **1**, 13–30.

Abbott, I. (1989). The influence of fauna on soil structure. In *Animals in Primary Succession*, ed. J. Majer, pp. 39–50. Cambridge: Cambridge University Press.

Abramovitz, J. N. (2001). Averting unnatural disasters. In *State of the World 2001, The Worldwatch Institute*, ed. L. R. Brown, C. Flavin, H. French *et al.*, pp. 123–42. New York: Norton.

Adachi, N., Terashima, I. & Takahashi, M. (1996a). Nitrogen translocation via rhizome systems in monoclonal stands of *Reynoutria japonica* in an oligotrophic desert on Mt. Fuji: Field experiments. *Ecological Research*, **11**, 175–86.

Adachi, N., Terashima, I. & Takahashi, M. (1996b). Central die-back of monoclonal stands of *Reynoutria japonica* in an early stage of primary succession on Mount Fuji. *Annals of Botany*, **77**, 477–86.

Adams, P. & Sidle, R. (1987). Soil conditions in three recent landslides in southeast Alaska. *Forest Ecology and Management*, **18**, 93–102.

Aellen, M. (1981). Recent fluctuations in glaciers. In *Switzerland and Her Glaciers*, ed. P. Kasser & W. Haeberli, pp. 70–89. Berne: Kümmerly & Frey.

Aikio, S., Vare, H. & Strommer, R. (2000). Soil microbial activity and biomass in the primary succession of a dry heath forest. *Soil Biology and Biochemistry*, **32**, 1091–100.

Allen, E. B. (1982). Water and nutrient competition between *Salsola kali* and two native grass species (*Agropyron smithii* and *Bouteloua gracilis*). *Ecology*, **63**, 732–41.

Allen, E. B. & Allen, M. F. (1980). Natural re-establishment of vesicular-arbuscular mycorrhizae following stripmine reclamation in Wyoming. *Journal of Applied Ecology*, **17**, 139–47.

Allen, E. B. & Allen, M. F. (1988). Facilitation of succession by the monomycotrophic colonizer *Salsola kali* (Chenopodiaceae) on a harsh site: Effects of mycorrhizal fungi. *American Journal of Botany*, **75**, 257–66.

Allen, E. B. & Allen, M. F. (1990). The mediation of competition by mycorrhizae in successional and patchy environments. In *Perspectives on Plant Competition*, ed. J. G. Grace & D. Tilman, pp. 367–89. New York: Academic Press.

Allen, E. B., Chambers, J. C., Conner, K. F., Allen, M. F. & Brown, R. W. (1987). Natural re-establishment of mycorrhizae in disturbed alpine ecosystems. *Arctic and Alpine Research*, **19**, 11–20.

Allen, M. F. (1991). *The Ecology of Mycorrhizae*. Cambridge: Cambridge University Press.

Allen, M. F., Allen, E. B., Zink, T. A., Harney, S., Yoshida, L. C., Sigüenza, C., Edwards, F., Hinkson, C., Rillig, M., Bainbridge, D., Doljanin, C. & MacAller, R. (1999). Soil microorganisms. In *Ecosystems of Disturbed Ground, Ecosystems of the World 16*, ed. L. R. Walker, pp. 521–44. Amsterdam: Elsevier.

Allen, M. F., Crisafulli, C., Friese, C. F. & Jenkins, S. (1992). Reformation of mycorrhizal symbioses on Mount St. Helens, 1980–1990: Interactions of rodents and mycorrhizal fungi. *Mycological Research*, **96**, 447–53.

Allen, M. F. & MacMahon, J. A. (1988). Direct VA mycorrhizal inoculation of colonizing plants by pocket gophers (*Thomomys talpoides*) on Mount St. Helens. *Mycologia*, **80**, 754–6.

Allen, R. B., Bellingham, P. J. & Wiser, S. K. (1999). Immediate damage by an earthquake to a temperate montane forest. *Ecology*, **80**, 708–14.

Allison, F. E. (1973). *Soil Organic Matter and its Role in Crop Production*. New York: Elsevier.

Alpert, P. & Mooney, H. A. (1996). Resource heterogeneity generated by shrubs and topography on coastal sand dunes. *Vegetatio*, **122**, 83–93.

Amman, G. D. (1977). The role of the mountain pine beetle in lodgepole pine ecosystems: Impact on succession. In *The Role of Arthropods in Forest Ecosystems*, ed. W. J. Mattson, pp. 3–18. New York: Springer.

Amundson, R. & Jenny, H. (1997). On a state factor model of ecosystems. *BioScience*, **47**, 536–43.

Andersen, D. C. & MacMahon, J. A. (1985). Plant succession following the Mount St. Helens volcanic eruption: Facilitation by a burrowing rodent, *Thomomys talpoides*. *American Midland Naturalist*, **114**, 62–9.

Andersen, E. (1999). Seed dispersal by monkeys and the fate of dispersed seeds in a Peruvian rain forest. *Biotropica*, **31**, 145–58.

Andersen, U. V. (1993). Dispersal strategies of Danish seashore plants. *Ecography*, **16**, 289–98.

Andersen, U. V. (1995). Succession and soil development in man-made coastal ecosystems in the Baltic Sea. *Nordic Journal of Botany*, **15**, 91–104.

Anderson, D. J. (1986). Ecological succession. In *Community Ecology: Pattern and Process*, ed. J. Kikkawa & D. J. Anderson, pp. 269–85. London: Blackwell.

Anderson, R. C., Fralish, J. S. & Baskin, J. M. (1999). *Savannas, Barrens, and Rock Outcrop Plant Communities of North America*. Cambridge: Cambridge University Press.

Anderson, W. B. & Polis, G. A. (1999). Nutrient fluxes from water to land: seabirds affect plant nutrient status on Gulf of California islands. *Oecologia*, **118**, 324–32.

Andrés, P. (1999). Ecological risks of the use of sewage sludge as fertilizer in soil restoration: Effects on soil microarthropod populations. *Land Degradation and Development*, **10**, 67–77.

Andrés, P. & Jorba, M. (2000). Mitigation strategies in some motorway embankments (Catalonia, Spain). *Restoration Ecology*, **8**, 268–74.

Andrés, P., Zapater, V. & Pamplona, M. (1996). Stabilization of motorway slopes with herbaceous cover, Catalonia, Spain. *Restoration Ecology*, **4**, 51–60.

Animkwapong, G. & Teklehairanot, Z. (1995). Reclamation of degraded cocoa lands using *Albizia zygia*. *Land Degradation and Rehabilitation*, **6**, 109–23.

Antor, R. J. (1994). Arthropod fallout on high alpine snow patches of the Central Pyrenees, northeastern Spain. *Arctic and Alpine Research*, **26**, 72–6.

Aplet, G. H. (1990). Alteration of earthworm community biomass by the alien *Myrica faya* in Hawai'i. *Oecologia*, **82**, 414–16.

Aplet, G. H., Anderson, S. J. & Stone, C. P. (1991). Association between feral pig disturbance and the composition of some alien plant assemblages in Hawaii Volcanoes National Park. *Vegetatio*, **95**, 55–62.

Aplet, G. H., Hughes, R. F. & Vitousek, P. M. (1998). Ecosystem development on Hawaiian lava flows, biomass and species composition. *Journal of Vegetation Science*, **9**, 17–26.

Aplet, G. H., Loh, R. L., Tunison, T. & Vitousek, P. M. (2002). Experimental restoration of a dense faya tree stand. In *Studies on the Ecology and Management of Faya Tree: Hawaii Volcanoes National Park, 1986–1995*, ed. T. Tunison. Technical Report, Cooperative Ecological Study Unit, University of Hawaii at Manoa. (In press.)

Aplet, G. H. & Vitousek, P. M. (1994). An age-altitude matrix analysis of Hawaiian rain-forest succession. *Journal of Ecology*, **82**, 137–47.

Arianoutsou, M. (1998). Aspects of demography in post-fire Mediterranean plant communities of Greece. In *Landscape Disturbance on Biodiversity in Mediterranean-type Ecosystems*, ed. P. W. Rundel, G. Montenegro & F. M. Jaksic, pp. 273–95. Berlin: Springer.

Armesto, J. J., Pickett, S. T. A. & McDonnell, M. J. (1991). Spatial heterogeneity during succession: A cyclic model of invasion and exclusion. In *Ecological Heterogeneity*, ed. J. Kolasa & S. T. A. Pickett, pp. 256–69. New York: Springer.

Arnalds, O., Aradottir, A. L. & Thorsteinsson, I. (1987). The nature and restoration of denuded areas in Iceland. *Arctic and Alpine Research*, **19**, 518–25.

Aronson, J. & LeFloc'h, E. (1996). Vital landscape attributes: Missing tools for restoration ecology. *Restoration Ecology*, **4**, 377–87.

Aronson, J., Floret, C., LeFloc'h, E., Ovalle, C. & Pontanier, R. (1993). Restoration and rehabilitation of degraded ecosystems in arid and semiarid regions. I. A view from the South. *Restoration Ecology*, **1**, 8–17.

Ash, H. J., Gemmell, R. P. & Bradshaw, A. D. (1994). The introduction of native plant species on industrial waste heaps: A test of immigration and other factors affecting primary succession. *Journal of Applied Ecology*, **31**, 74–84.

Ashmole, N. P. & Ashmole, M. J. (1987). Arthropod communities supported by biological fallout on recent lava flows in the Canary Islands. *Entomologia Scandinavica*, Suppl., **32**, 67–88.

Ashmole, N. P., Ormomi, P., Ashmole, M. J. & Martin, J. L. (1992). Primary faunal succession in volcanic terrain: lava and cave studies on the Canary Islands. *Biological Journal of the Linnean Society*, **46**, 207–34.

Ashton, P. M., Samarasighe, S. J., Gunatilleke, I. A. & Gunatilleke, C. V. S. (1997). Role of legumes in release of successionally arrested grasslands in the Central Hills of Sri Lanka. *Restoration Ecology*, **5**, 36–43.

Auerbach, N. A., Walker, M. D. & Walker, D. A. (1997). Effects of roadside disturbance on substrate and vegetation properties in arctic tundra. *Ecological Applications*, **7**, 218–35.

Augspurger, C. K. (1986). Morphology and dispersal potential of wind dispersed diaspores of neotropical trees. *American Journal of Botany*, **73**, 353–63.

Austin, M. P. (1981). Permanent quadrats: An interface for theory and practice. *Vegetatio*, **46**, 6–10.

Austin, M. P. (1985). Continuum concept, ordination methods, and niche theory. *Annual Review of Ecology and Systematics*, **16**, 39–61.

Austin, M. P. & Smith, T. M. (1989). A new model for the continuum concept. *Vegetatio*, **83**, 35–47.

Avis, A. M. & Lubke, R. A. (1996). Dynamics and succession of coastal dune vegetation in the Eastern Cape, South Africa. *Landscape and Urban Planning*, **34**, 237–53.

Ayyad, M. A. (1973). Vegetation and environment of the western Mediterranean coastal land of Egypt. *Journal of Ecology*, **4**, 18–26.

Bach, C. E. (1994). Effects of a specialist herbivore (*Altica subplicata*) on *Salix cordata* and sand dune succession. *Ecological Monographs*, **64**, 423–45.

Bach, C. E. (2001). Long-term effects of insect herbivory and sand accretion on plant succession on sand dunes. *Ecology*, **82**, 1401–16.

Baker, W. L. & Walford, G. M. (1995). Multiple stable states and models of riparian vegetation succession on the Animas River, Colorado. *Annals of the Association of American Geographers*, **85**, 320–38.

Bakken, L. (1985). Separation and purification of bacteria from soil. *Applied Environmental Microbiology*, **49**, 1482–7.

Bakker, J. P., Olff, H., Willems, J. H. & Zobel, M. (1996). Why do we need permanent plots in the study of long-term vegetation dynamics? *Journal of Vegetation Science*, **7**, 147–55.

Bakker, J. P., Grootjans, A. P., Hermy, M. & Poschlod, P. (2000). How to define targets for ecological restoration? Introduction. *Applied Vegetation Science*, **3**, 3–6.

Balakrishnan, N. & Mueller-Dombois, D. (1983). Nutrient studies in relation to habitat types and canopy dieback in the montane rain forest ecosystem, Island of Hawai'i. *Pacific Science*, **37**, 339–59.

Barbour, M. G., de Jong, T. M. & Pavlik, P. M. (1985). Marine beach and dune plant communities. In *Physiological Ecology of North American Plant Communities*, ed. B. F. Chabot & H. A. Mooney, pp. 296–322. New York: Chapman & Hall.

Bardgett, R. D. (2000). Patterns of below-ground primary succession at Glacier Bay, southeast Alaska. *Bulletin of the British Ecological Society*, **31**, 40–2.

Bardgett, R. D., Mawdsley, J. L., Edwards, S., Hobbs, P. J., Rodwell, J. S. & Davies, W. J. (1999). Plant species and nitrogen effects on soil biological properties of temperate upland grasslands. *Functional Ecology*, **13**, 650–60.

Barrow, C. J. (1991). *Land Degradation*. Cambridge: Cambridge University Press.

Barrow, C. J. (1999). How humans respond to natural or anthropogenic disturbance. In *Ecosystems of Disturbed Ground, Ecosystems of the World 16*, ed. L. R. Walker, pp. 659–71. Amsterdam: Elsevier.

Basher, L. R. (1986). Pedogenesis and erosion history in a high rainfall mountainous drainage basin – Cropp River, New Zealand. Ph.D. dissertation, Lincoln College, University of Canterbury, Christchurch, New Zealand.

Baskin, C. C. & Baskin, J. M. (1998). *Seeds: Ecology, Biogeography and Evolution of Dormancy and Germination*. New York: Academic Press.

Baskin, J. M. & Baskin, C. C. (1988). Endemism in rock outcrop plant communities of unglaciated eastern United States: An evaluation of the roles of the edaphic, genetic and light factors. *Journal of Biogeography*, **15**, 829–40.

Bazzaz, F. A. 1979. The physiological ecology of plant succession. *Annual Review of Ecology and Systematics*, **10**, 351–71.

Beatty, S. W. & Stone, E. L. (1986). The variety of soil microsites created by tree falls. *Canadian Journal of Forest Research*, **16**, 539–48.

Becher, H. H. (1985). Compaction of arable soils due to reclamation or off-road military traffic. *Reclamation and Revegetation Research*, **4**, 155–64.

Begon, M., Harper, J. L. & Townsend, C. R. (1990). *Ecology: Individuals, Populations and Communities*. 2nd edition. Boston: Blackwell.

Bekker, R. M., Lammerts, E. J., Schutter, A. & Grootjans, A. P. (1999). Vegetation development in dune slacks: The role of persistent seed banks. *Journal of Vegetation Science*, **10**, 745–54.

Bellingham, P. J., Walker, L. R. & Wardle, D. A. (2001). Differential facilitation by a nitrogen-fixing shrub during primary succession influences relative performance of canopy tree species. *Journal of Ecology*, **89**, 861–75.

Belnap, J. (1995). Surface disturbances: Their role in accelerating desertification. *Environmental Monitoring and Assessment*, **37**, 39–57.

Belnap, J. & Gillette, D. A. (1998). Vulnerability of desert biological soil crusts to wind erosion: The influences of crust development, soil texture and disturbance. *Journal of Arid Environments*, **39**, 133–42.

Belnap, J. & Harper, K. T. (1995). Influence of cryptobiotic crusts on elemental content of tissue of two desert seed plants. *Arid Soil Research and Rehabilitation*, **9**, 107–15.

Belnap, J., Harper, K. T. & Warren, S. D. (1994). Surface disturbance of cryptobiotic soil crusts: Nitrogenase activity, chlorophyll content, and chlorophyll degradation. *Arid Soil Research and Rehabilitation*, **8**, 1–8.

Belnap, J., Rosentreter, R., Kaltenecker, J., Williams, J, Leonard, S. & Eldridge, D. (contributors) (2000). *Role of Microbiotic Soil Crusts in Rangeland Health*. Bureau of Land Management National Training Center. Course Number 1730–41. Lake Mead National Recreation Area, Nevada, U.S.A.

Belousov, A. & Belousova, M. (1996). Large-scale landslides on active volcanoes in the 20th century: examples from the Kuril-Kamchatka region (Russia). In *Landslides*, ed. K. Senneset, pp. 953–7. Rotterdam: Balkema.

Beltman, B., van den Broek, T., van Maanen, K. & Vaneveld, K. (1996). Measures to develop a rich fen wetland landscape with a full range of successional stages. *Ecological Engineering*, **7**, 299–313.

Belyea, L. R. & Lancaster, J. (1999). Assembly rules within a contingent ecology. *Oikos*, **86**, 402–16.

Benedetti-Cecchi, L. (2000). Predicting direct and indirect interactions during succession in a mid-littoral rocky shore assemblage. *Ecological Monographs*, **70**, 45–72.

Benedetti-Cecchi, L. & Cinelli, F. (1996). Patterns of disturbance and recovery in littoral rock pools: Nonhierarchical competition and spatial variability in secondary succession. *Marine Ecology Progress Series*, **135**, 145–61.

Bénito-Espinal, F. P. & Bénito-Espinal, E. (eds) (1991). *L'Ouragan Hugo: Genese, Incidences Géographiques et Écologiques sur la Guadeloupe*. Co-editors: Parc National de la Guadeloupe, Délégation Régionale a l'Action Culturelle, and Agence Guadeloupéenne de l'Environnement du Tourisme et des Loisirs. Martinique: Imprimerie Désormeaux, Fort-de-France.

Berendse, F., Lammerts, E. J. & Olff, H. (1998). Soil organic matter accumulation and its implications for nitrogen mineralization and plant species composition during succession in coastal dune slacks. *Plant Ecology*, **137**, 71–8.

Berkowitz, A. R., Canham, C. D. & Kelly, V. R. (1995). Competition vs. facilitation of tree seedling growth and survival in early successional communities. *Ecology*, **76**, 1156–68.

Bernhardt, K. G. (1992). Colonizing strategies on sandy sites at the tidal coast. *Flora*, **187**, 271–81.

Bernhardt, K. G. & Handke, P. (1992). Successional dynamics on newly created saline marsh soils. *Oecologia*, **11**, 139–52.

Bertness, M. D. & Callaway, R. M. (1994). Positive interactions in communities. *Trends in Ecology and Evolution*, **9**, 191–3.

Berz, G. (1988). List of major natural disasters, 1960–1987. *Natural Hazards*, **1**, 97–9.

Beyer, L., Bölter, M. & Seppelt, R. D. (2000). Nutrient and thermal regime, microbial biomass, and vegetation of Antarctic soils in the Windmill Islands region of East Antarctica (Wilkes Land). *Arctic, Antarctic and Alpine Research*, **32**, 30–9.

Bhiry, N. & Filion, L. (1996). Holocene plant succession in a dune-swale environment of southern Quebec: A macrofossil analysis. *Ecoscience*, **3**, 330–42.

Binet, P. (1981). Short-term dynamics of minerals in arid ecosystems. In *Arid Land Ecosystems: Structure, Functioning and Management*, ed. D. W. Goodall, R. A. Perry & K. M. W. Howes, pp. 325–56. Cambridge: Cambridge University Press.

Binggeli, P., Eakin, M., Macfayden, A., Power, J. & McConnell, J. (1992). Impact of the alien sea buckthorn (*Hippophaë rhamnoides* L.) on sand dune ecosystems in Ireland. In *Coastal Dunes: Geomorphology, Ecology and Management for Conservation*, ed. R. W. G. Carter, T. G. F. Curtis & M. J. Sheehy-Skeffington, pp. 325–37. Rotterdam: Balkema.

Binkley, D. & Giardina, C. (1998). Why do tree species affect soils? The warp and woof of tree-soil interactions. *Biogeochemistry*, **42**, 89–106.

Binkley, D., Suarez, F., Stottlemeyer, R. & Caldwell, B. (1997). Ecosystem development on terraces along the Kugurorok River, northwest Alaska. *Ecoscience*, **4**, 311–18.

Birkeland, P. W. (1984). *Soils and Geomorphology*. Oxford: Oxford University Press.

Birkeland, P. W., Burke, R. M. & Benedict, J. B. (1989). Pedogenic gradients for iron and aluminium accumulation and phosphorus depletion in arctic and alpine soils as a function of time and climate. *Quaternary Research*, **32**, 193–204.

Birkemeier, W. A., Bichner, E. W., Scarborough, B. L., McConathy, M. A. & Eiser, W. C. (1991). Nearshore profile response caused by Hurricane Hugo. *Journal of Coastal Research*, Special Issue No. **8**, 113–28.

Birks, H. J. B. (1980a). Modern pollen assemblages and vegetational history of the moraines of the Klutlan Glacier and its surroundings, Yukon Territory, Canada. *Quaternary Research*, **14**, 101–29.

Birks, H. J. B. (1980b). The present flora and vegetation of the moraines of the Klutlan Glacier, Yukon Territory, Canada. *Quaternary Research*, **14**, 60–86.

Bishop, J. G. & Schemske, D. W. (1998). Variation in flowering phenology and its consequences for lupines colonizing Mount St. Helens. *Ecology*, **79**, 534–46.

Bishop, S. C. & Chapin, F. S. III. (1989a). Establishment of *Salix alaxensis* on a gravel pad in Arctic Alaska. *Journal of Applied Ecology*, **26**, 575–84.

Bishop, S. C. & Chapin, F. S. III. (1989b). Patterns of natural revegetation on abandoned gravel pads in arctic Alaska. *Journal of Applied Ecology*, **26**, 1073–81.

Bliss, L. C. (2000). Arctic tundra and polar desert biome. In *North American Terrestrial Vegetation*, 2nd edition, ed. M. G. Barbour & W. D. Billings, pp. 1–41. Cambridge: Cambridge University Press.

Bliss, L. C. & Gold, W. G. (1994). The patterning of plant communities and edaphic factors along a high arctic coastline: Implications for succession. *Canadian Journal of Botany*, **72**, 1095–107.

Bliss, L. C. & Gold, W. G. (1999). Vascular plant reproduction, establishment, and growth and the effects of cryptogamic crusts within a polar desert ecosystem. *Canadian Journal of Botany*, **77**, 623–36.

Blizzard, A. W. (1931). Plant sociology and vegetational change on High Hill, Long Island, New York. *Ecology*, **12**, 208–31.

Blundon, D. J. & Dale, M. R. T. (1990). Dinitrogen fixation (acetylene reduction) in primary succession near Mt. Robson, British Columbia. *Arctic and Alpine Research*, **22**, 255–63.

Blundon, D. J., MacIsaac, D. A. & Dale, M. R. T. (1993). Nucleation during primary succession in the Canadian Rockies. *Canadian Journal of Botany*, **71**, 1093–6.

Boag, B. & Yeates, G. W. (1998). Soil nematode biodiversity in terrestrial ecosystems. *Biodiversity and Conservation*, **7**, 617–30.

Boerner, R. E. J. (1985). Alternate pathways of succession on the Lake Erie Islands. *Vegetatio*, **63**, 35–44.

Boerner, R. E. J. DeMars, B. G. & Leicht, P. N. (1996). Spatial patterns of mycorrhizal infectiveness of soils along a successional chronosequence. *Mycorrhizae*, **6**, 79–90.

Bolin, R. & Stanford, L. (1999). Constructing vulnerability in the first world: The Northridge earthquake in southern California, 1994. In *The Angry Earth*, ed. A. Oliver-Smith & S. M. Hoffman, pp. 89–112. New York: Routledge.

Bolling, J. D. & Walker, L. R. (2000). Plant and soil recovery along a series of abandoned desert roads. *Journal of Arid Environments*, **46**, 1–24.

Bolling, J. D. & Walker, L. R. (2002). Fertile island development around perennial shrubs across a Mojave Desert chronosequence. *Western North American Naturalist*, **61**, 88–100.

Bond, W. J. & van Wilgen, B. W. (1996). *Fire and Plants*. London: Chapman & Hall.

Booth, D. T., Gores, J. K., Schuman, G. E. & Olson, R. A. (1999). Shrub densities on pre-1985 reclaimed mines in Wyoming. *Restoration Ecology*, **7**, 24–32.

Borgegård, S.-O. & Rydin, H. (1989). Utilization of waste products and inorganic fertilizer in the restoration of iron mine tailings. *Journal of Applied Ecology*, **26**, 1083–88.

Bormann, B. T. & Sidle, R. C. (1990). Changes in productivity and distribution of nutrients in a chronosequence at Glacier Bay National Park, Alaska. *Journal of Ecology*, **78**, 561–678.

Bormann, F. H. & Likens, G. E. (1979). *Patterns and Process in a Forested Ecosystem: Disturbance, Development, and the Steady State Based on the Hubbard Brook Ecosystem Study*. New York: Springer.

Bornkamm, R. (1984). Experimentell-ökologishce Untersuchungen zur Sukzession von ruderalen Pflanzengesellschaften auf unterschiedlichen Boden. II. Quantität und Qualität der Phytomasse. *Flora*, **175**, 45–74.

Bornkamm, R. (1985). Vegetation changes in herbaceous communities. In *The Population Structure of Vegetation*, ed. J. White, pp. 89–109. Dordrecht, Holland: Junk.

Bornkamm, R., Lee, J. A. & Seaward, M. R. D. (eds) (1982). *Urban Ecology*. Oxford: Oxford University Press.

Bowers, J. E., Webb, R. H. & Pierson, E. A. (1997). Succession of desert plants on flow terraces, Grand Canyon, Arizona, USA. *Journal of Arid Environments*, **36**, 67–86.

Braatne, J. H. & Bliss, L. C. (1999). Comparative physiological ecology of lupines colonizing early successional habitats on Mount St. Helens. *Ecology*, **80**, 891–907.

Braatne, J. H. & Chapin, D. M. (1986). Comparative water relations of four subalpine plants at Mount St. Helens. In *Mount St. Helens: Five Years Later*, ed. S. A. C. Keller, pp. 163–72. Cheney, Washington: Eastern Washington University Press.

Bradshaw, A. D. (1952). Populations of *Agrostis tenuis* resistant to lead and zinc poisoning. *Nature*, **169**, 1098.

Bradshaw, A. D. (1983a). The restoration of mined land. In *Conservation in Perspective*, ed. A. Warren & F. B. Goldsmith, pp. 177–99. London: Wiley.

Bradshaw, A. D. (1983b). The reconstruction of ecosystems. *Journal of Applied Ecology*, **20**, 1–17.

Bradshaw, A. D. (1987). Restoration: An acid test for ecology. In *Restoration Ecology: A Synthetic Approach to Ecological Research*, ed. W. R. Jordan III, M. E. Gilpin & J. D. Aber, pp. 23–9. Cambridge: Cambridge University Press.

Bradshaw, A. D. & Chadwick, M. J. (1980). *The Restoration of Land: The Ecology and Reclamation of Derelict and Degraded Land*. Oxford: Blackwell.

Bradshaw, A. D., Humphreys, M. O. & Johnson, M. S. (1978). The value of heavy metal tolerance in the revegetation of metalliferous mine wastes. In *Environmental Management of Mineral Wastes*, ed. G. T. Goodman & M. J. Chadwick, pp. 311–23. Alphen aan den Rijn, Netherlands: Sijthoff and Nordhoff.

Brady, N. C. & Weil, R. R. (1998). *The Nature and Properties of Soils*, 12th edition. Englewood Cliffs, New Jersey: Prentice Hall.

Bramble, W. C. & Ashley, R. H. (1955). Natural revegetation of spoil banks in central Pennsylvania. *Ecology*, **36**, 417–23.

Braun-Blanquet, J. (1932). *Plant Sociology: The Study of Plant Communities*. New York: Hafner. (English translation.)

Braun-Blanquet, J. (1964). *Pflanzensociologie: Grundzüge der Vegetationskunde*. Vienna: Springer-Verlag.

Breitburg, D. L. (1985). Development of a subtidal epibenthic community: Factors affecting species composition and the mechanisms of succession. *Oecologia*, **65**, 173–84.

Bretz, J. H. (1932). *The Grand Coulee*. New York: Special Publication of the American Geographical Society.

Brittingham, S. & Walker, L. R. (2000). Facilitation of *Yucca brevifolia* recruitment by Mojave Desert shrubs. *Western North American Naturalist*, **60**, 374–83.

Brock, T. D. (1973). Primary colonization of Surtsey, with special reference to the blue-green algae. *Oikos*, **24**, 239–43.

Brokaw, N. V. L. & Walker, L. R. (1991). Summary of the effects of Caribbean hurricanes on vegetation. *Biotropica*, **23**, 442–7.

Brooker, R. W. & Callaghan, T. V. (1998). The balance between positive and negative plant interactions and its relationship to environmental gradients: A model. *Oikos*, **81**, 196–207.

Brookes, P. C., Landman, A., Pruden, G. & Jenkinson, D. S. (1985). Chloroform fumigation and the release of soil nitrogen; a rapid direct extraction method to measure microbial biomass nitrogen in soil. *Soil Biology and Biochemistry*, **17**, 837–42.

Brown, J. H. (1995). Organisms and species as complex adaptive systems: Linking the biology of populations with the physics of ecosystems. In *Linking Species and Ecosystems*, ed. C. G. Jones & J. H. Lawton, pp. 16–24. New York: Chapman & Hall.

Brown, L. R. (1999). Feeding nine billion. In *State of the World 1999: The Millennium Edition*, ed. L. R. Brown, C. Flavin, L. Starke & H. French, pp. 115–32. Washington, D. C.: Worldwatch Institute.

Brown, S. C. & Bedford, B. L. (1997). Restoration of wetland vegetation with transplanted wetland soil: An experimental study. *Wetlands*, **17**, 424–7.

Brown, V. K. & Gange, A. C. (1992). Secondary plant succession: How is it modified by insect herbivory? *Vegetatio*, **101**, 3–13.

Brown, W. H., Merrill, E. D. & Yates, H. S. (1917). The revegetation of Volcano Island, Luzon, Philippine Islands, since the eruption of Taal Volcano in 1911. *The Philippine Journal of Science*, **12**, 177–243.

Bruno, J. F. (2000). Facilitation of cobble beach plant communities through habitat modification by *Spartina alterniflora*. *Ecology*, **81**, 1179–92.

Bruno, J. F. & Kennedy, G. W. (2000). Patch-size dependent habitat modification and facilitation on New England cobble beaches by *Spartina alterniflora*. *Oecologia*, **122**, 98–108.

Bryant, J. P. & Chapin, F. S. III. (1986). Browsing-woody plant interactions during a boreal forest plant succession. In *Forest Ecosystems in the Alaska Taiga, a Synthesis of Structure and Function*, ed. K. Van Cleve, F. S. Chapin III, P. W. Flanagan, L. A. Viereck & C. T. Dyrness, pp. 213–25. New York: Springer.

Bucknam, R. C., Hemphill-Haley, E. & Leopold, E. B. (1992). Abrupt uplift within the last 1700 years at Southern Puget Sound, Washington. *Science*, **258**, 1611–14.

Bullock, J. M. & Clarke, R. T. (2000). Long distance seed dispersal by wind measuring and modeling the tail of the curve. *Oecologia*, **124**, 506–21.

Bunting, M. J. & Warner, B. G. (1998). Hydroseral development in southern Ontario, patterns and controls. *Journal of Biogeography*, **25**, 3–18.

Buol, S. W. (1994). Soils. In *Changes in Land Use and Land Cover: A Global Perspective*, ed. W. B. Meyer & B. L. Turner II, pp. 211–29. Cambridge: Cambridge University Press.

Burbanck, M. P. & Phillips, D. L. (1983). Evidence of plant succession on granite outcrops of the Georgia Piedmont. *American Midland Naturalist*, **109**, 94–104.

Burleigh, S. H. & Dawson, J. O. (1994). Occurrence of *Myrica*-nodulating *Frankia* in Hawaiian volcanic soils. *Plant and Soil*, **164**, 283–9.

Burrows, C. J. (1990). *Processes of Vegetation Change*. London: Unwin Hyman.

Burton, P. J. (1982). The effect of temperature and light on *Meterosideros polymorpha* seed germination. *Pacific Science*, **36**, 229–40.

Burton, P. J. & Mueller-Dombois, D. (1984). Response of *Metrosideros polymorpha* seedlings to experimental canopy opening. *Ecology*, **65**, 779–91.

Busch, D. E. & Smith, S. D. (1995). Mechanisms associated with decline of woody species in riparian ecosystems of the southwestern U.S. *Ecological Monographs*, **65**, 347–70.

Bush, M. B., Whittaker, R. J. & Partomihardjo, T. (1992). Multiple Krakatau pathways, divergence of types in lowland forests. *GeoJournal*, **28**, 185–99.

Cahill, J. F. Jr., Castelli, J. P. & Casper, B. B. (2001). The herbivory uncertainty principle: visiting plants can alter herbivory. *Ecology*, **82**, 307–12.

Cain, S. A. (1959). Henry Allan Gleason – eminent ecologist. *Bulletin of the Ecological Society of America*, **40**, 105–10.

Calderón, F. J. (1993). The role of mycorrhizae in the nutrient absorptive strategy of important landslide colonizers. M.S. thesis, University of Puerto Rico, Río Piedras.

Caldwell, M. M., Dawson, T. E. & Richards, J. H. (1998). Hydraulic lift: Consequences of water efflux from the roots of plants. *Oecologia*, **113**, 151–61.

Callaway, R. M. (1992). Effect of shrubs on recruitment of *Quercus douglasii* and *Quercus lobata* in California. *Ecology*, **73**, 2118–28.

Callaway, R. M. (1995). Positive interactions among plants. *Botanical Reviews*, **61**, 306–49.

Callaway, R. M., Nadkarni, N. M. & Mahall, B. E. (1991). Facilitation and interference of *Quercus douglasii* on understory productivity in central California. *Ecology*, **72**, 1484–99.

Callaway, R. M. & Walker, L. R. (1997). Competition and facilitation: A synthetic approach to interactions in plant communities. *Ecology*, **78**, 1958–65.

Camill, P. (1999). Peat accumulation and succession following permafrost thaw in boreal peatlands of Manitoba, Canada. *Ecoscience*, **6**, 592–602.

Camp, R. J. & Knight, R. L. (1997). Cliff bird and plant communities in Joshua Tree National Park, California, USA. *Natural Areas Journal*, **17**, 110–17.

Campbell, B. M., Lynam, T. & Hatton, J. C. (1990). Small-scale patterning in the recruitment of forest species during succession in tropical dry forest, Mozambique. *Vegetatio*, **87**, 51–7.

Campbell, J. E. & Gibson, D. J. (2001). The effect of seeds of exotic species transported via horse dung on vegetation along trail corridors. *Plant Ecology*, **157**, 23–35.

Carlquist, S. (1974). *Island Biology*. New York: Columbia University Press.

Carlquist, S. (1994). The first arrivals. *Natural History*, **91**, 20–30.

Carlsson, R. B. & Callaghan, T. V. (1991). Positive plant interactions in tundra vegetation and the importance of shelter. *Journal of Ecology*, **79**, 973–84.

Carpenter, F. L. (1976). Plant-pollinator interactions in Hawaii: Pollination energetics of *Metrosideros collina* (Myrtaceae). *Ecology*, **57**, 1125–44.

Carpenter, S. E., Trappe, J. M. & Ammirati, J. Jr. (1987). Observations of fungal succession in the Mount St. Helens devastation zone, 1980–1983. *Canadian Journal of Botany*, **65**, 716–28.

Carr, G. D., Powell, E. A. & Kyhos, D. W. (1986). Self-incompatibility in the Hawaiian Madiinae (Compositae): An exception to Baker's rule. *Evolution*, **40**, 430–4.

Carreiro, M. M., Sinsabaugh, R. L., Repert, D. A. & Parkhurst, D. F. (2000). Microbial enzyme shifts explain litter decay responses to simulated nitrogen deposition. *Ecology*, **81**, 2359–65.

Carter, J. W., Carpenter, A. L., Foster, M. S. & Jessee, W. N. (1985). Benthic succession on an artificial reef designed to support a kelp-reef community. *Bulletin of Marine Science*, **37**, 86–113.

Castillo, S., Popma, J. & Moreno-Casasola, P. (1991). Coastal sand dune vegetation of Tabasco and Campeche, Mexico. *Journal of Vegetation Science*, **2**, 73–88.

Cázares, E. (1992). Mycorrhizal fungi and their relationship to plant succession in subalpine habitats. Ph.D. dissertation, Oregon State University, Corvallis, Oregon.

Chadwick, H. W. & Dalke, P. D. (1965). Plant succession on dune sands in Fremont County, Idaho. *Ecology*, **46**, 765–80.

Chadwick, O. A., Derry, L. A., Vitousek, P. M., Huebert, B. J. & Hedin, L. O. (1999). Changing sources of nutrients during four million years of ecosystem development. *Nature*, **397**, 491–99.

Challies, C. N. (1975). Feral pigs (*Sus scrofa*) on Auckland Island: Status and effects on vegetation and nesting sea birds. *New Zealand Journal of Zoology*, **2**, 479–90.

Chambers, J. C. & Linnerooth, A. R. (2001). Restoring riparian meadows currently dominated by *Artemisia* using alternative state concepts – the establishment component. *Applied Vegetation Science*, **4**, 157–66.

Chan, Y. S. G., Chu, L. M. & Wong, M. H. (1997). Influence of landfill factors on plants and soil fauna – an ecological perspective. *Environmental Pollution*, **97**, 39–44.

Chapin, D. M. (1995). Physiological and morphological attributes of two colonizing plant species on Mount St. Helens. *American Midland Naturalist*, **133**, 76–87.

Chapin, D. M. & Bliss, L. C. (1989). Seedling growth, physiology, and survivorship in a subalpine, volcanic environment. *Ecology*, **70**, 1325–34.

Chapin, D. M., Bliss, L. C. & Bledsoe, C. S. (1991). Environmental regulations of nitrogen fixation in a high arctic lowland ecosystem. *Canadian Journal of Botany*, **69**, 2744–55.

Chapin, F. S. III(1993). Physiological controls over plant establishment in primary succession. In *Primary Succession on Land*, ed. J. Miles & D. H. Walton, pp. 161–78. Oxford: Blackwell.

Chapin, F. S. III (1995). New cog in the nitrogen cycle. *Nature*, **377**, 199–200.

Chapin, F. S. III, Vitousek, P. M. & Van Cleve, K. (1986). The nature of nutrient limitation in plant communities. *The American Naturalist*, **127**, 48–58.

Chapin, F. S. III & Walker, L. R. (1993). Direct and indirect effects of calcium sulfate and nitrogen on growth and succession of trees on the Tanana River floodplain, interior Alaska. *Canadian Journal of Forest Research*, **23**, 995–1000.

Chapin, F. S. III, Walker, L. R., Fastie, C. L. & Sharman, L. C. (1994). Mechanisms of primary succession following deglaciation at Glacier Bay, Alaska. *Ecological Monographs*, **64**, 149–75.

Chapman, C. A. & Onderdonk, D. A. (1998). Forests without primates: Primate/plant codependency. *American Journal of Primatology*, **45**, 127–41.

Charley, J. L. & West, N. E. (1975). Plant-induced soil chemical patterns in some shrub-dominated semi-desert ecosystems of Utah. *Journal of Ecology*, **63**, 945–64.

Chen, R. G. & Twilley, R. R. (1998). A gap dynamic model of mangrove forest development along gradients of soil salinity and nutrient resources. *Journal of Ecology*, **86**, 37–51.

Cheng, L. & Birch, M. C. (1987). Insect flotsam: an unstudied marine resource. *Ecological Entomology*, **3**, 87–97.

Cherrett, J. M. (1989). Key concepts: the results of a survey of our members' opinions. In *Ecological Concepts: The Contribution of Ecology to an Understanding of the Natural World*, ed. M. Cherrett, pp. 1–16. Oxford: Blackwell.

Chestnut, T. J., Zarin, D. J., McDowell, W. H. & Keller, M. (1999). A nitrogen budget for late-successional hillslope tabonuco forest, Puerto Rico. *Biogeochemistry*, **46**, 85–108.

Chiba, N. & Hirose, T. (1993). Nitrogen acquisition and use in three perennials in the early stage of primary succession. *Functional Ecology*, **7**, 287–92.

Childress, W. M., Crisafulli, C. M. & Rykiel, E. J. (1998). Comparison of Markovian matrix models of a primary successional plant community. *Ecological Modelling*, **107**, 92–102.

Choi, Y. D. & Wali, M. K. (1995). The role of *Panicum virgatum* (switch grass) in the revegetation of iron-mine tailings in northern New York. *Restoration Ecology*, **3**, 123–32.

Christensen, N. L. & Peet, R. K. (1981). Convergence during secondary forest succession. *Journal of Ecology*, **72**, 25–36.

Chu, L. M. & Bradshaw, A. D. (1996). The value of pulverized refuel fines (PRF) as a substitute for topsoil in land reclamation. 1. Field studies. *Journal of Applied Ecology*, **33**, 851–7.

Cicolani, B. (1992). Macrochelid mites (Acari: Mesostigmata) occurring in animal droppings in the pasture ecosystem in central Italy. *Agriculture, Ecosystems and Environment*, **40**, 47–60.

Clark, J. S. (1998). Why trees migrate so fast: Confronting theory with dispersal biology and the paleorecord. *The American Naturalist*, **152**, 204–24.

Clark, J. S., Fastie, C., Hurtt, G., Jackson, S. T., Johnson, C., King, G. A., Lewis, M., Lynch, J., Pacala, S., Prentice, C., Schupp, E. W., Webb, T. III & Wyckoff, P. (1998). Reid's paradox of rapid plant migration. *BioScience*, **48**, 13–24.

Clark, J. S., Silman, M., Kern, R., Macklin, E. & Hille Ris Lambers, J. (1999). Seed dispersal near and far: Patterns across temperate and tropical forests. *Ecology*, **80**, 1475–95.

Clarkson, B. D. (1990). A review of vegetation development following recent (<450 years) volcanic disturbance in North Island, New Zealand. *New Zealand Journal of Ecology*, **14**, 59–71.

Clarkson, B. R. & Clarkson, B. D. (1983). Mt. Tarawera: 2. Rates of change in the vegetation and flora of the high domes. *New Zealand Journal of Botany*, **6**, 107–19.

Clarkson, B. R. & Clarkson, B. D. (1995). Recent vegetation changes on Mount Tarawera, Rotorua, New Zealand. *New Zealand Journal of Botany*, **33**, 339–54.

Clein, J. S. & Schimel, J. P. (1995). Nitrogen turnover and availability during succession from alder to poplar in Alaskan taiga forests. *Soil Biology and Biochemistry*, **27**, 743–52.

Clements, F. E. (1916). *Plant Succession: An Analysis of the Development of Vegetation*. Washington, D. C.: Carnegie Institution of Washington Publication 242.

Clements, F. E. (1928). *Plant Succession and Indicators*. New York: H. W. Wilson.

Clements, F. E. (1936). Nature and structure of the climax. *Journal of Ecology*, **24**, 252–84.

Coaldrake, J. E. (1962). The coastal sand dunes of southern Queensland. *Proceedings of the Royal Society of Queensland*, **72**, 101–16.

Cody, M. L. (1981). Citation classic. *Current Contents*, **12**, 14.

Cody, M. (2000). Slow-motion population dynamics in Mojave Desert perennial plants. *Journal of Vegetation Science*, **11**, 351–8.

Cogbill, C. V. (1996). Black growth and fiddlebutts: The nature of old-growth red spruce. In *Eastern Old-Growth Forests: Prospects for Rediscovery and Recovery*, ed. M. B. Davis, pp. 113–25. Washington, D. C.: Island Press.

Cohen, J. E. (1995). *How Many People Can the Earth Support?* New York: Norton.

Coleman, D. C. & Crossley, D. A. Jr. (1996). *Fundamentals of Soil Ecology*. New York: Academic Press.

Colinvaux, P. A. (1973). *Introduction to Ecology*. New York: Wiley.

Collins, B. S. & Quinn, J. A. (1982). Displacement of *Andropogon scoparius* on the New Jersey piedmont by the successional shrub *Myrica pensylvanica*. *American Journal of Botany*, **69**, 680–9.

Collins, S. L., Mitchell, G. S. & Klahr, S. C. (1989). Vegetation-environment relationships in a rock outcrop community in Southern Oklahoma. *American Midland Naturalist*, **122**, 339–8.

Compton, S. G., Ross, S. J. & Thornton, I. W. B. (1994). Pollinator limitation of fig tree reproduction on the island of Anak Krakatau (Indonesia). *Biotropica*, **26**, 180–6.

Conn, C. E. & Day, F. P. (1996). Response of root and cotton strip decay to nitrogen amendment along a barrier island dune chronosequence. *Canadian Journal of Botany*, **74**, 276–84.

Connell, J. H. (1978). Diversity in tropical rain forests and coral reefs. *Science*, **199**, 1302–10.

Connell, J. H. (1983). On the prevalence and relative importance on interspecific competition: Evidence from field experiments. *The American Naturalist*, **122**, 661–96.

Connell, J. H. (1990). Apparent versus "real" competition in plants. In *Perspectives in Plant Competition*, ed. J. B. Grace & G. D. Tilman, pp. 9–26. New York: Academic Press.

Connell, J. H. & Keough, M. J. (1985). Disturbance and patch dynamics of subtidal marine animals on hard substrata. In *The Ecology of Natural Disturbance and Patch Dynamics*, ed. S. T. A. Pickett & P. S. White, pp. 125–52. New York: Academic Press.

Connell, J. H., Noble, I. R. & Slatyer, R. O. (1987). On the mechanisms producing successional change. *Oikos*, **50**, 136–7.

Connell, J. H. & Slatyer, R. O. (1977). Mechanisms of succession in natural communities and their roles in community stability and organization. *The American Naturalist*, **111**, 1119–44.

Cook, J. E. (1996). Implications of modern succession theory for habitat typing: A review. *Forest Science*, **42**, 67–75.

Cooke, J. A. (1999). Mining. In *Ecosystems of Disturbed Ground*, ed. L. R. Walker, pp. 365–84. Amsterdam: Elsevier.

Coomes, D. A. & Grubb, P. J. (2000). Impacts of root competition in forests and woodlands: a theoretical framework and review of experiments. *Ecological Monographs*, **70**, 171–207.

Cooper, R. & Rudolph, E. D. (1953). The role of lichens in soil formation and plant succession. *Ecology*, **34**, 805–7.

Cooper, W. S. (1913). The climax forest of Isle Royale, Lake Superior and its development. *Botanical Gazette*, **55**, 1–44, 115–40, 189–235.

Cooper, W. S. (1916). Plant successions in the Mount Robson region, British Columbia. *The Plant World*, **19**, 211–38.

Cooper, W. S. (1919). Ecology of the strand vegetation of the Pacific coast of North America. *Carnegie Institution of Washington Yearbook*, **18**, 96–9.

Cooper, W. S. (1923). The recent ecological history of Glacier Bay, Alaska: II. The present vegetation cycle. *Ecology*, **4**, 223–46.

Cooper, W. S. (1931). A third expedition to Glacier Bay, Alaska. *Ecology*, **12**, 61–95.

Cooper, W. S. (1939). A fourth expedition to Glacier Bay, Alaska. *Ecology*, **20**, 130–55.

Corkidi, L. & Rincón, E. (1997). Arbuscular mycorrhizae in a tropical sand dune ecosystem of the Gulf of Mexico, 2: Effects of arbuscular mycorrhizal fungi on the growth of species distributed in different early successional stages. *Mycorrhiza*, **7**, 17–23.

Cowles, H. C. (1899). The ecological relations of the vegetation on the sand dunes of Lake Michigan. *Botanical Gazette*, **27**, 95–117, 167–202, 281–308, 361–91.

Cowles, H. C. (1901). The physiographic ecology of Chicago and vicinity: A study of the origin, development, and classification of plant societies. *Botanical Gazette*, **31**, 73–108, 145–82.

Cox, P. A. & Elmquist, T. (2000). Pollinator extinction in the Pacific Islands. *Conservation Biology*, **14**, 1237–39.

Coxson, D. S. (1987). Net photosynthetic response patterns of the basidiomycete lichen *Cora pavonia* (Web.) E. Fries from the tropical volcano La Soufrière (Guadeloupe). *Oecologia*, **73**, 454–8.

Coyne, M. S., Zhai, Q., Mackown, C. T. & Barnhisel, R. I. (1998). Gross nitrogen transformation rates in soil at a surface coal mine site reclaimed for prime farmland use. *Soil Biology and Biochemistry*, **30**, 1099–106.

Craft, C., Reader, J., Sacco, J. N. & Broome, S. W. (1999). Twenty-five years of ecosystem development of constructed *Spartina alterniflora* (Loisel) marshes. *Ecological Applications*, **9**, 1405–19.

Crampton, C. (1987). Soils, vegetation, and permafrost across an active meander of Indian River, Central Yukon, Canada. *Catena*, **14**, 157–63.

Crawford, R. L., Sugg, P. M. & Edwards, E. S. (1995). Spider arrival and primary establishment on terrain depopulated by volcanic eruption at Mount St. Helens, Washington. *American Midland Naturalist*, **133**, 60–75.

Crawley, M. J. (1983). *Herbivory*. Oxford: Blackwell.

Crawley, M. J. (1987). What makes a community invasible? In *Colonization, Succession and Stability*, ed. A. J. Gray, M. J. Crawley & P. J. Edward, pp. 429–53. Oxford: Blackwell.

Crawley, M. J. (1997). Biodiversity. In *Plant Ecology*, 2nd edition, ed. M. J. Crawley, pp. 595–632. London: Blackwell.

Crews, T. (1999). The presence of nitrogen fixing legumes in terrestrial communities: Evolutionary vs. ecological considerations. *Biogeochemistry*, **46**, 233–46.

Crews, T., Kitayama, K., Fownes, J., Riley, R., Herbert, D., Mueller-Dombois, D. & Vitousek, P. (1995). Changes in soil phosphorus fractions and ecosystem dynamics across a long chronosequence in Hawaii. *Ecology*, **76**, 1407–24.

Crocker, R. L. (1952). Soil genesis and the pedogenic factors. *Quarterly Review of Biology*, **27**, 139–68.

Crocker, R. L. & Dickson, B. A. (1957). Soil development on the recessional moraines of the Herbert and Mendenhall Glaciers, south-eastern Alaska. *Journal of Ecology*, **45**, 169–85.

Crocker, R. L. & Major, J. (1955). Soil development in relation to vegetation and surface age at Glacier Bay, Alaska. *Journal of Ecology*, **43**, 427–48.

Cullen, W. R., Wheater, C. P. & Dunleavy, P. J. (1998). Establishment of species-rich vegetation on reclaimed limestone quarry faces in Derbyshire, U.K. *Biological Conservation*, **84**, 25–33.

Curtis, J. T. (1959). *The Vegetation of Wisconsin*. Madison: University of Wisconsin Press.

Dahlskog, S. (1982). Successions in a Lapland mountain delta. *Meddelanden fran Växtbiologiska Institutionen*, **3**, 54–62.

Daily, G. (ed.) (1997). *Nature's Services*. Washington, D. C.: Island Press.

Dale, V. H., Lugo, A.E., MacMahon, J.A. & Pickett, S.T.A. (1998). Management implications of large, infrequent disturbances. *Ecosystems*, **1**, 546–557.

Dancer, W. S., Handley, J. F. & Bradshaw, A. D. (1977). Nitrogen accumulation in kaolin mining wastes in Cornwall. I. Natural Communities. *Plant and Soil*, **48**, 153–67.

D'Antonio, C. M., Dudley, T. L. & Mack, M. (1999). Disturbance and biological invasions: direct effects and feedbacks. In *Ecosystems of Disturbed Ground*, *Ecosystems of the World 16*, ed. L. R. Walker, pp. 413–52. Amsterdam: Elsevier.

Danin, A. (1991). Plant adaptations in desert dunes. *Journal of Arid Environments*, **21**, 193–212.

Da Silva, J. M. C., Uhl, C. & Murray, G. (1996). Plant succession, landscape management, and the ecology of frugivorous birds in abandoned Amazonian pasture. *Conservation Biology*, **10**, 491–503.

Davey, M. C. & Rothery, P. (1993). Primary colonization by microalgae in relation to spatial variation in edaphic factors on Antarctic fell-field soils. *Journal of Ecology*, **81**, 335–44.

Davidson, D. W. (1993). The effects of herbivory and granivory on terrestrial plant succession. *Oikos*, **68**, 23–35.

Davis, B. N. K., Lakhani, K. H., Brown, M. C. & Park, D. G. (1985). Early seral communities in a limestone quarry: An experimental study of treatment effects on cover and richness of vegetation. *Journal of Applied Ecology*, **22**, 473–90.

Davis, M. A., Grime, J. P. & Thompson, K. (2000). Fluctuating resources in plant communities: a general theory of invisibility. *Journal of Ecology*, **88**, 528–34.

Davis, W. M. (1899). The geographical cycle. *Geographical Journal*, **14**, 481–504.

Dawson, J. O., Christensen, T. W. & Timmons, R. G. (1983). Nodulation by *Frankia* of *Alnus glutinosa* seeded in soil from different topographic positions on an Illinois spoil bank. *The Actinomycetes*, **17**, 50–60.

Day, T. & Wright, R. (1989). Positive plant spatial associations with *Eriogonum ovalifolium* in primary succession on cinder cones: Seed trapping nurse plants. *Vegetatio*, **80**, 37–45.

Dean, T. A. & Hurd, L. E. (1980). Development in an estuarine fouling community: The influence of early colonists on later arrivals. *Oecologia*, **46**, 295–301.

Debussche, M., Lepart, J. & Molina, J. (1985). La dissémination des plantes à fruits charnus par les oiseaux: Rôle de la structure de la végétation et impact sur la succession en région méditerranéenne. *Acta Oecologica*, **6**, 65–80.

Degens, B. P. & Harris, J. A. (1997). Development of a physiological approach to measuring the catabolic diversity of soil microbial communities. *Soil Biology and Biochemistry*, **29**, 1309–20.

de Jong, N. H. (2000). Woody plant restoraton and natural regeneration in wet meadows at Coomonderry Swamp on the south coast of New South Wales. *Marine and Freshwater Research*, **51**, 81–9.

de Jong, T. J., Klinkhamer, P. G. L. & de Heiden, J. L. H. (1995). The effect of water and mycorrhizal infection on the distribution of *Carlina vulgaris* on sand dunes. *Ecography*, **18**, 384–9.

De Kovel, C. G. F., Van Mierlo, A. J. E. M., Wilms, Y. J. O. & Berendse, F. (2000). Carbon and nitrogen in soil and vegetation at sites differing in successional age. *Plant Ecology*, **149**, 43–50.

Delgadillo, C. & Cárdenas, A. (1995). Observations on moss succession on Parícutin Volcano, Mexico. *The Bryologist*, **98**, 606–8.

del Moral, R. (1983). Initial recovery of subalpine vegetation on Mount St. Helens. *American Midland Naturalist*, **109**, 72–80.

del Moral, R. (1984). The impact of the Olympic marmot on subalpine vegetation structure. *American Journal of Botany*, **71**, 1228–36.

del Moral, R. (1993a). Understanding dynamics of early succession on Mount St. Helens. *Journal of Vegetation Science*, **4**, 223–34.

del Moral, R. (1993b). Mechanisms of primary succession on volcanoes: A view from Mount St. Helens. In *Primary Succession on Land*, ed. J. Miles & D. H. Walton, pp. 79–100. Oxford: Blackwell.

del Moral, R. (1998). Early succession on lahars spawned by Mount St. Helens. *American Journal of Botany*, **85**, 820–8.

del Moral, R. (1999a). Plant succession on pumice at Mount St. Helens, Washington. *American Midland Naturalist*, **141**, 101–14.

del Moral, R. (1999b). Predictability of primary successional wetlands on pumice, Mount St. Helens. *Madroño*, **46**, 177–86.

del Moral, R. (2000a). Succession and species turnover on Mount St. Helens, Washington. *Acta Phytogeographica Suecica*, **85**, 53–62.

del Moral, R. (2000b). Local species turnover on Mount St. Helens. In *Proc. 41st Symposium of the IUVS*, ed. P. White, pp. 195–7. Uppsala: Opulus Press.

del Moral, R. & Bliss, L. C. (1993). Mechanisms of primary succession: Insights resulting from the eruption of Mount St. Helens. *Advances in Ecological Research*, **24**, 1–66.

del Moral, R. & Clampitt, C. A. (1985). Growth of native plant species on recent volcanic substrates from Mount St. Helens. *American Midland Naturalist*, **114**, 374–83.

del Moral, R. & Grishin, S. Yu. (1999). Volcanic disturbances and ecosystem recovery. In *Ecosystems of Disturbed Ground, Ecosystems of the World 16*, ed. L. R. Walker, pp. 137–60. Amsterdam: Elsevier.

del Moral, R. & Jones, C. C. (2002). Early spatial development of vegetation on pumice at Mount St. Helens. *Plant Ecology*, **161**, 9–22.

del Moral, R. & Muller, C. H. (1969). Fog-drip: A mechanism of toxin transport from *Eucalyptus globulus*. *Bulletin of the Torrey Botanical Club*, **96**, 467–75.

del Moral, R. & Muller, C. H. (1970). The allelopathic effects of *Eucalyptus camaldulensis*. *American Midland Naturalist*, **83**, 254–82.

del Moral, R. & Standley, L. A. (1979). Pollination of angiosperms in contrasting coniferous forests. *American Journal of Botany*, **66**, 26–35.

del Moral, R., Titus, J. H. & Cook, A. M. (1995). Early primary succession on Mount St. Helens, Washington, USA. *Journal of Vegetation Science*, **6**, 107–20.

del Moral, R., Willis, R. J. & Ashton, D. H. (1978). Suppression of coastal heath vegetation by *Eucalyptus baxteri*. *Australian Journal of Botany*, **16**, 102–19.

del Moral, R. & Wood, D. M. (1993a). Early primary succession on a barren volcanic plain at Mount St. Helens, Washington. *American Journal of Botany*, **80**, 981–92.

del Moral, R. & Wood, D. M. (1993b). Early primary succession on the volcano Mount St. Helens. *Journal of Vegetation Science*, **4**, 223–34.

Demarais, S., Tazik, D. J., Guertin, P. J. & Jorgensen, E. E. (1999). Disturbance associated with military exercises. In *Ecosystems of Disturbed Ground, Ecosystems of the World 16*, ed. L. R. Walker, pp. 385–96. Amsterdam: Elsevier.

Densmore, R. V. (1992). Succession on an Alaskan tundra disturbance with and without assisted revegetation with grass. *Arctic and Alpine Research*, **24**, 238–43.

Densmore, R. V. (1994). Succession on regraded placer mine spoil in Alaska, USA, in relation to initial site characteristics. *Arctic and Alpine Research*, **26**, 354–63.

De Scally, F. A. & Gardner, J. S. (1994). Characteristics and mitigation of the snow avalanche hazard in Kaghan Valley, Pakistan Himalaya. *Natural Hazards*, **9**, 197–213.

Diamond, J. M. (1975). The assembly of species communities. In *Ecology and Evolution of Communities*, ed. M. L. Cody & J. M. Diamond, pp. 342–444. Cambridge, Massachusetts (USA): Harvard University Press.

Díaz, S., Cabido, M. & Casanoves, F. (1998). Plant functional traits and environmental filters at a regional scale. *Journal of Vegetation Science*, **9**, 113–22.

Díaz, S., Cabido, M. & Casanoves, F. (1999). Functional implications of trait-environmental linkages in plant communities. In *Ecological Assembly Rules: Perspectives, Advances, Retreats*, ed. E. Weiher & P. Keddy, pp. 338–62. Cambridge: Cambridge University Press.

Dickson, B. A. & Crocker, R. L. (1953). A chronosequence of soils and vegetation near Mt. Shasta, California. I. Definition of the ecosystem investigated and features of the plant succession. *Journal of Soil Science*, **4**, 123–41.

Dickson, L. G. (2000). Constraints to nitrogen fixation by cryptogamic crusts in a polar desert ecosystem, Devon Island, N.W.T., Canada. *Arctic, Antarctic and Alpine Research*, **32**, 40–5.

Dlugosch, K. & del Moral, R. (1999). Vegetational heterogeneity along elevational gradients. *Northwest Science*, **43**, 12–18.

Dobson, A. P., Bradshaw, A. D. & Baker, A. J. M. (1997). Hopes for the future: Restoration ecology and conservation biology. *Science*, **277**, 515–22.

Docters van Leeuwen, W. M. (1936). Krakatau 1883-1983. *Annales du Jardin Botanique de Buitenzorg*, **46/47**, 1–506.

Doing, H. (1985). Coastal fore dune zonation and succession in various parts of the world. *Vegetatio*, **61**, 65–75.

Donnegan, J. A. & Rebertus, A. H. (1999). Rates and mechanisms of subalpine forest succession along an environmental gradient. *Ecology*, **80**, 1370–84.

Douglas, D. A. (1987). Growth of *Salix setchelliana* on a Kluane River point bar, Yukon Territory, Canada. *Arctic and Alpine Research*, **19**, 35–44.

Dragovich, D. & Patterson, J. (1995). Conditions of rehabilitated coal-mines in the Hunter Valley, Australia. *Land Degradation and Rehabilitation*, **6**, 29–39.

Drake, D. (1990). Communities as assembled structures: Do rules govern pattern? *Trends in Ecology and Evolution*, **5**, 159–64.

Drake, D. R. (1998). Relationships among the seed rain, seed bank and vegetation of a Hawaiian forest. *Journal of Vegetation Science*, **9**, 103–12.

Drake, D. R. & Mueller-Dombois, D. (1993). Population development of rain forest trees on a chronosequence of Hawaiian lava flows. *Ecology*, **74**, 1012–19.

Drake, J. A. (1991). Community-assembly mechanics and the structure of an experimental species ensemble. *The American Naturalist*, **137**, 1–26.

Drake, J. A., Huxel, G. R. & Hewitt, C. L. (1996). Microcosms as models for generating and testing community theory. *Ecology*, **77**, 670–7.

Drury, W. H. (1956). Bog flats and physiographic processes in the upper Kuskokwim River Region, Alaska. *Contributions from the Gray Herbarium*, **178**, 1–130.

Drury, W. H. & Nisbet, I. C. T. (1973). Succession. *Journal of the Arnold Arboretum*, **54**, 331–68.

Dublin, H. T., Sinclair, R. E. & McGlade, J. (1990). Elephants and fire as causes of multiple stable states in the Serengeti-Mara woodlands. *Journal of Animal Ecology*, **59**, 1147–64.

Dulohery, C. J., Kolka, R. K. & McKevlin, M. R. (2000). Effects of a willow overstory on planted seedlings in a bottomland restoration. *Ecological Engineering*, **15**(supplement), S57–S66.

Duncan, R. P. (1993). Flood disturbance and the coexistence of species in a lowland podocarp forest, south Westland, New Zealand. *Journal of Ecology*, **81**, 403–16.

Duncan, R. S. & Duncan, V. E. (2000). Forest succession and distance from forest edge in an Afro-tropical grassland. *Biotropica*, **32**, 33–41.

Dynesius, M. & Nilsson, C. (1994). Fragmentation and flow regulation of river systems in the northern third of the world. *Science*, **266**, 753–62.

Dyrness, C. T. & Van Cleve, K. (1993). Control of surface soil chemistry in early-successional floodplain soils along the Tanana River, interior Alaska. *Canadian Journal of Forest Research*, **23**, 979–94.

Early, M. & Goff, M. L. (1986). Arthropod succession patterns in exposed carrion on the island of O'ahu, Hawaiian Islands, USA. *Journal of Medical Entomology*, **23**, 520–31.

Ebersole, J. J. (1987). Short-term vegetation recovery at an Alaskan arctic coastal plain site. *Arctic and Alpine Research*, **19**, 442–50.

Eccles, N. S., Esler, K. J. & Cowling, R. M. (1999). Spatial pattern analysis in Namaqualand desert plant communities: Evidence for general positive interactions. *Plant Ecology*, **142**, 71–85.

Edwards, J. S. (1988). Life in the allobiosphere. *Trends in Ecology and Evolution*, **3**, 111–14.

Edwards, J. S. & Sugg, P. (1993). Anthropod fallout as a resource in the recolonization of Mount St. Helens. *Ecology*, **74**, 954–8.

Edwards, P. J. & Gillman, M. P. (1987). Herbivores and plant succession. In *Colonization, Succession and Stability*, ed. A. J. Gray, J. J. Crawley & P. J. Edwards, pp. 482–501. Symposium of the British Ecological Society, vol. 26. Oxford: Blackwell.

Eggler, W. A. (1941). Primary succession of volcanic deposits in southern Idaho. *Ecological Monographs*, **3**, 277–98.

Eggler, W. A. (1963). Plant life of Parícutin volcano, Mexico, eight years after activity ceased. *American Midland Naturalist*, **69**, 38–68.

Eggler, W. A. (1971). Quantitative studies of vegetation on sixteen young lava flows on the island of Hawaii. *Tropical Ecology*, **12**, 66–100.

Egler, F. E. (1951). A commentary on American plant ecology based on the textbooks of 1947–1949. *Ecology*, **32**, 673–95.

Egler, F. E. (1954). Vegetation science concepts I. Initial floristic composition, a factor in old-field vegetation development. *Vegetatio*, **4**, 412–17.

Ehrenfeld, J. G. (1990). Dynamics and processes of barrier island vegetation. *Review of Aquatic Science*, **2**, 437–80.

Ehrenfeld, J. G. (2000). Defining the limits of restoration: The need for realistic goals. *Restoration Ecology*, **8**, 2–9.

Ehrenfeld, J. G. & Scott, N. (2001). Invasive species and the soil: effects on organisms and ecosystem processes. *Ecological Applications*, **11**, 1259–60.

Elba de Pietri, D. (1992). Alien shrubs in a national park: Can they help in the recovery of natural degraded forest? *Biological Conservation*, **62**, 127–30.

El-Baz, F. (1992). Preliminary observations of environmental damage due to the Gulf War. *Natural Resources Forum*, **16**, 71–5.

Elgersma, A. M. (1998). Primary forest succession on poor sandy soils as related to site factors. *Biodiversity and Conservation*, **7**, 193–206.

Eliason, S. A. & Allen, E. B. (1997). Exotic grass competition in suppressing native shrubland reestablishment. *Restoration Ecology*, **5**, 245–55.

Ellenberg, H. (1956). *Aufgaben und Methoden der Vegetationskunde*. Stuttgart: Eugen Ulmer.

Ellison, A. M. (2000). Mangrove restoration: Do we know enough? *Restoration Ecology*, **8**, 219–29.

Emmer, I. M. & Sevink, J. (1994). Temporal and vertical changes in the humus form profile during a primary succession of *Pinus sylvestris*. *Plant and Soil*, **167**, 281–95.

Engstrom, D. R. (ed.) (1995). *Proceedings of the 3rd Glacier Bay Science Symposium, 1993*. Anchorage, Alaska: National Park Service.

Engstrom, D. R., Fritz, S. C., Almendinger, J. E. & Juggins, S. (2000). Chemical and biological trends during lake evolution in recently deglaciated terrain. *Nature*, **408**, 161–6.

Eriksson, O. (2000). Seed dispersal and colonization ability of plants – Assessment and implications for conservation. *Folia Geobotanica*, **35**, 115–23.

Eriksson, O. & Ehrlen, J. (1992). Seed and microsite limitation of recruitment in plant populations. *Oecologia*, **91**, 360–4.

Eriksson, O. & Eriksson, A. (1998). Effects of arrival order and seed size on germination of grassland plants: Are there assembly rules during recruitment? *Ecological Research*, **13**, 229–39.

Ernst, A. (1908). *The New Flora of the Volcanic Island of Krakatau*. Translation by A. C. Seward. Cambridge: Cambridge University Press.

Ernst, W. H. O., Slings, O. L. & Nelissen, H. J. M. (1996). Pedogenesis in coastal wet dune slacks after sod-cutting in relation to revegetation. *Plant and Soil*, **180**, 219–30.

Ernst, W. H. O., Van Duin, W. E. & Oolbekking, G. T. (1984). Vesicular-arbuscular mycorrhiza in dune vegetation. *Acta Botanica Neerlandica*, **33**, 151–60.

Escudero, A. (1996). Community patterns on exposed cliffs in a Mediterranean calcareous mountain. *Vegetatio*, **125**, 99–110.

Facelli, J. M. & D'Anela, E. (1990). Directionality, convergence and the rate of change during early succession in the Inland Pampa, Argentina. *Journal of Vegetation Science*, **1**, 255–60.

Facelli, J. M. & Pickett, S. T. A. (1990). Markovian chains and the role of history in succession. *Trends in Ecology and Evolution*, **5**, 27–30.

Fagan, W. F. & Bishop, J. G. (2000). Trophic interactions during primary succession: Herbivores slow a plant reinvasion at Mount St. Helens. *The American Naturalist*, **155**, 238–51.

Farrell, T. M. (1991). Models and mechanisms of succession: An example from a rocky intertidal community. *Ecological Monographs*, **61**, 95–113.

Fastie, C. L. (1995). Causes and ecosystem consequences of multiple pathways on primary succession at Glacier Bay, Alaska. *Ecology*, **76**, 1899–916.

Fenner, M. (1985). *Seed Ecology*. London: Chapman & Hall.

Ferreira, S. M. & van Aarde, R. J. (1999). Habitat associations and competition in *Mastomys-Saccostomus-Aethomys* assemblages on coastal dune forests. *African Journal of Ecology*, **37**, 121–36.

Fierro, A., Angers, A. D. & Beauchamp, C. J. (1999). Restoration of ecosystem function in an abandoned sandpit, plant and soil responses to paper de-inking sludge. *Journal of Applied Ecology*, **36**, 244–53.

Finegan, B. (1984). Forest succession. *Nature*, **313**, 109–14.

Finkl, C. W. & Pilkey, O. H. (ed.) (1991). Impacts of Hurricane Hugo: September 10–22, 1989. *Journal of Coastal Research*, Special Issue No. 8.

Fisher, S. G., Gray, L. J., Grimm, N. B. & Busch, D. E. (1982). Temporal succession in a desert stream ecosystem following flash flooding. *Ecological Monographs*, **52**, 93–110.

Fitter, A. H. & Parsons, W. F. J. (1987). Changes in phosphorus and nitrogen availability on recessional moraines of the Athabasca Glacier, Alberta. *Canadian Journal of Botany*, **65**, 210–13.

Fletcher, W. W. & Kirkwood, R. C. (1979). The bracken fern (*Pteridium aquilinum* L. (Kuhn)); its biology and control. In *The Experimental Biology of Ferns*, ed. A. F. Dyer, pp. 591–635. New York: Academic Press.

Fonteyn, P. J. & Mahall, B. E. (1978). Competition among desert perennials. *Nature*, **275**, 544–5.

Fonda, R. W. (1974). Forest succession in relation to river terrace development in Olympic National Park, Washington. *Ecology*, **55**, 927–42.

Foord, S. H., Vanaarde, R. J., & Ferreira, S. M. (1994). Seed dispersal by vervet monkeys in rehabilitating coastal dune forests at Richards Bay. *South African Journal of Wildlife Research*, **24**, 56–9.

Forbes, B. C. (1995). Tundra disturbance studies, III: Short-term effects of aeolian sand and dust, Yamal Region, Northwest Siberia. *Environmental Conservation*, **22**, 335–44.

Forbes, B. C. & Jeffries, R. L. (1999). Revegetation of disturbed arctic sites: Constraints and applications. *Biological Conservation*, **88**, 15–24.

Forbes, S. A. (1887). The lake as a microcosm. *Bulletin Science Association of Peoria, Illinois*, **1887**, 77–87. (Not seen but cited in McIntosh (1985).)

Forster, S. M. & Nicolson, T. H. (1981). Microbial aggregation of sand in a maritime dune succession. *Soil Biology and Biochemistry*, **13**, 20–208.

Fort, K. P. & Richards, J. H. (1998). Does seed dispersal limit initiation of primary succession in desert playas? *American Journal of Botany*, **85**, 1722–31.

Foster, B. L. & Tilman, D. (2000). Dynamic and static views of succession: Testing the descriptive power of the chronosequence approach. *Plant Ecology*, **146**, 1–10.

Fox, B. J. (1982). Fire and mammalian secondary succession in an Australian coastal heath. *Ecology*, **63**, 1332–41.

Fox, B. J. (1990). Changes in the structure of mammal communities over successional time scales. *Oikos*, **59**, 321–9.

Fox, R. L., de la Pena, R. S., Gavenda, R. T., Habte, M., Hue, N. V., Okawa, H., Jones, R. C., Plucknett, D. L., Silva, J. A. and Soltanpour, P. (1991). Amelioration, revegetation, and subsequent soil formation in denuded bauxite materials. *Allertonia*, **6**, 128–84.

Foy, C. D. (1984). Physiological effects of hydrogen, aluminium and manganese toxicities in acid soil. In *Soil Acidity and Liming*, 2nd edition, ed. F. Adams, pp. 57–97. Madison, Wisconsin: American Society of Agronomy. (Not seen but cited in Tyler (1996).)

Francis, P. (1993). *Volcanoes: A Planetary Perspective*. Oxford: Oxford University Press.

Franco, A. C. & Nobel, P. S. (1988). Interactions between seedling of *Agave deserti* and nurse plant *Hilaria rigida*. *Ecology*, **69**, 1731–40.

Franco, A. C. & Nobel, P. S. (1989). Effect of nurse plants on the microhabitat and growth of cacti. *Journal of Ecology*, **77**, 870–86.

Frankl, R. & Schmeidl, H. (2000). Vegetation change in a South German raised bog: Ecosystem engineering by plant species, vegetation switch or ecosystem level feedback mechanisms. *Flora*, **195**, 267–96.

Franklin, J. F., MacMahon, J. A. Swanson, F. J. & Sedell, J. R. (1985). Ecosystem responses to catastrophic disturbances: Lessons from Mount St. Helens. *National Geographic Research*, **1**, 198–216.

Franklin, J. F., Frenzen, P. M. & Swanson, F. J. (1998). Re-creation of ecosystems at Mount St. Helens: Contrasts in artificial and natural approaches. In *Rehabilitating Damaged Ecosystems*, 2nd edition, ed. J. Cairns, pp. 287–334. Boca Raton, Florida: CRC Press.

Franklin, J. F. & MacMahon, J. A. (2000). Messages from a mountain. *Science*, **288**, 1183–5.

Frelich, L. E. & Lorimer, C. G. (1991). Natural disturbance regimes in hemlock-hardwood forest of the Upper Great Lakes region. *Ecological Monographs*, **61**,145–64.

Frenot, Y., Gloaguen, J. C., Cannavacciuolo, M. & Bellido, A. (1998). Primary succession on glacier forelands in the subantarctic Kerguelen Islands. *Journal of Vegetation Science*, **9**, 75–84.

Frenzen, P. M., Krasney, M. E. & Rigney, L. P. (1988). Thirty-three years of plant succession on the Kautz Creek mudflow, Mount Rainier National Park, Washington. *Canadian Journal of Botany*, **66**, 130–7.

Fridriksson, S. (1987). Plant colonization of a volcanic island, Surtsey, Iceland. *Arctic and Alpine Research*, **19**, 425–31.

Fridriksson, S. (1992). Vascular plants on Surtsey (1981–1990). *Reykiavik, Surtsey Research Progress Report*, **10**, 17–30.

Fridriksson, S. & Magnusson, B. (1992). Development of the ecosystem on Surtsey with reference to Anak Krakatau. *GeoJournal*, **28**, 287–91.

Friedman, J. M., Osterkamp, W. R. & Lewis, W. M. Jr. (1996a). Channel narrowing and vegetation development following a Great Plains flood. *Ecology*, **77**, 2167–81.

Friedman, J. M., Osterkamp, W. R. & Lewis, W. M. Jr. (1996b). The role of vegetation and bed-level fluctuations in the process of channel narrowing. *Geomorphology*, **14**, 341–51.

Fröborg, H. & Eriksson, O. (1997). Local colonization and extinction of field layer plants in a deciduous forest and their dependence upon life history features. *Journal of Vegetation Science*, **8**, 395–400.

Frostegård, A. & Bååth, E. (1996). The use of phospholipid fatty acid analysis to estimate bacterial and fungal biomass in soil. *Biology and Fertility of Soils*, **22**, 59–65.

Fuller, R. N. (1999). The role of refugia in primary succession on Mount St. Helens, Washington. M.S. thesis, University of Washington, Seattle.

Gadgil, R. L. (1971). The nutritional role of *Lupinus arboreus* in coastal sand dune forestry. I. The potential influence of undamaged lupin plants on nitrogen uptake by *Pinus radiata*. *Plant and Soil*, **34**, 357–67.

Gagné, J.-M. & Houle, G. (2001). Facilitation of *Leymus mollis* by *Honckenya peploides* on coastal dunes on subarctic Quebec, Canada. *Canadian Journal of Botany*, **79**, 1327–31.

Gallagher, E. D., Jumars, P. A. & Trueblood, D. D. (1983). Facilitation of soft-bottom benthic succession by tube builders. *Ecology*, **64**, 1200–16.

Gallagher, W. (1993). *The Power of Place: How Our Surroundings Shape Our Thoughts, Emotions, and Actions.* New York: Poseidon.

Gallé, L. (1991). Structure and succession of ant assemblages in a north European sand dune area. *Holarctic Ecology*, **14**, 31–7.

Game, M., Carrel, J. E. & Hotrabhavandra, T. (1982). Patch dynamics of plant succession on abandoned surface coal mines: A case history approach. *Journal of Ecology*, **70**, 707–20.

García-Fayos, P., García-Ventoso, B. & Cerdà, A. (2000). Limitations to plant establishment on eroded slopes in southeastern Spain. *Journal of Vegetation Science*, **11**, 77–86.

García-Mora, M. R., Gallego-Fernández, J. B. & García-Novo, F. (1999). Plant functional types in coastal foredunes in relation to environmental stress and disturbance. *Journal of Vegetation Science*, **10**, 27–34.

Gardner, G. (1997). Preserving global cropland. In *State of the World 1997*, Worldwatch Institute, ed. L. R. Brown, C. Flavin, H. French *et al.*, pp. 42–59. New York: Norton.

Gardner, L. R., Michener, W. K., Blood, E. R., Williams, T. M., Lipscomb, D. J. & Jefferson, W. H. (1991). Ecological impact of Hurricane Hugo – salinization of a coastal forest. *Journal of Coastal Research*, Special Issue No. **8**, 301–18.

Garland, J. L. & Mills, A. L. (1991). Classification and characterization of heterotrophic microbial communities on the basis of patterns of community-level sole-carbon-source utilization. *Applied Environmental Microbiology*, **57**, 2351–9.

Garner, W. & Steinberger, Y. (1989). A proposed mechanism for the formation of 'fertile islands' in the desert ecosystem. *Journal of Arid Environments*, **16**, 257–62.

Garwood, N., Janos, P. & Brokaw, N. (1979). Earthquake-caused landslides: A major disturbance in tropical forests. *Science*, **205**, 997–9.

Gaynor, M. & Wallace, A. (1998). Population interactions in primary succession: An example of contramensalism involving rock-colonizing bryophytes. *Lindbergia*, **23**, 81–5.

Gemma, J. N. & Koske, R. E. (1990). Mycorrhizae in recent volcanic substrates in Hawaii. *American Journal of Botany*, **77**, 1193–200.

Gerlach, A. (1993). Biogeochemistry of nitrogen in a coastal dune succession on Spiekeroog (Germany) and the impact of climate. *Phytocoenologia*, **23**, 115–27.

Gerlach, A., Albers, E. A. & Broedlin, W. (1994). Development of the nitrogen cycle in the soils of a coastal dune succession. *Acta Botanica Neerlandica*, **43**, 189–203.

Gerrish, G., Mueller-Dombois, D. & Bridges, K. W. (1988). Nutrient limitation and *Metrosideros* forest dieback in Hawai'i. *Ecology*, **69**, 723–7.

Giampietro, M. (1999). Economic growth, human disturbance to ecological systems, and sustainability. In *Ecosystems of Disturbed Ground*, ed. L. R. Walker, pp. 723–46. Amsterdam: Elsevier.

Gibb, J. A. (1994). Plant succession on the braided bed of the Orongorongo River, Wellington, New Zealand, 1973–1990. *New Zealand Journal of Ecology*, **18**, 29–40.

Gibson, D. J., Ely, J. S. & Looney, P. B. (1997). A Markovian approach to modeling succession on a coastal barrier island following beach nourishment. *Journal of Coastal Research*, **13**, 831–41.

Gibson, D. J., Ely, J. S., Looney, P. B. & Gibson, P. T. (1995). Effects of inundation from the storm surge of Hurricane Andrew upon primary succession on dredge spoil. *Journal of Coastal Research*, **21**, 208–16.

Gilbert, O. L. (1989). *The Ecology of Urban Habitats*. London: Chapman & Hall.

Gilbert, O. L. & Anderson, P. (1998). *Habitat Creation and Repair*. Oxford: Oxford University Press.

Gill, D. (1972). The point bar environment in the Mackenzie River Delta. *Canadian Journal of Earth Sciences*, **9**, 1382–93.

Gill, D. (1973). Floristics of a plant succession sequence in the Mackenzie Delta, Northwest Territories. *Polarforschung*, **43**, 55–65.

Gill, T. E. (1996). Eolian sediments generated by anthropogenic disturbance of playas: Human impacts on the geomorphic system and geomorphic impacts on the human system. *Geomorphology*, **17**, 207–28.

Gitay, H. & Noble, I. R. (1997). What are functional types and how should we seek them? In *Plant Functional Types*, ed. T. M. Smith, H. H. Shugart & F. I. Woodward, pp. 3–19. Cambridge: Cambridge University Press.

Gitay, H. & Wilson, J. B. (1995). Post-fire changes in community structure of tall tussock grasslands: A test of alternative models of succession. *Journal of Ecology*, **83**, 775–82.

Gleason, H. A. (1917). The structure and development of the plant association. *Bulletin of the Torrey Botanical Club*, **44**, 463–81.

Gleason, H. A. (1926). The individualistic concept of the plant association. *Bulletin of the Torrey Botanical Club*, **53**, 7–26.

Gleason, H. A. (1939). The individualistic concept of the plant association. *American Midland Naturalist*, **21**, 92–110.

Gleason, H. A. (1953). Autobiogeographical letter. *Bulletin of the Ecological Society of America*, **34**, 40–2.

Gleeson, S. K. & Tilman, D. (1990). Allocation and the transient dynamics of succession on poor soils. *Ecology*, **71**, 1144–55.

Glenn-Lewin, D. C. (1980). The individualistic nature of plant community development. *Vegetatio*, **43**, 141–6.

Glenn-Lewin, D. C. & van der Maarel, E. (1992). Pattern and process of vegetation dynamics. In *Plant Succession: Theory and Prediction*, ed. D. C. Glenn-Lewin, R. K. Peet & T. T. Veblen, pp. 11–59. London: Chapman & Hall.

Glenn-Lewin, D. C., Peet, R. K. & Veblen, T. T. (eds) (1992). *Plant Succession: Theory and Prediction*. London: Chapman & Hall.

Gliessman, S. R. (1976). Allelopathy in a broad spectrum of environments as illustrated by bracken. *Botanical Journal of the Linnean Society*, **73**, 95–104.

Goddard, J. & Lago, P. K. (1985). Notes on blow fly (Diptera: Calliphoridae) succession on carrion in northern Mississippi. *Journal of the Entomological Society*, **20**, 312–17.

Gold, W. G. & Bliss, L. C. (1995). Water limitations and plant community development in a polar desert. *Ecology*, **76**, 1558–68.

Goldberg, D. E., Rajaniemi, T., Gurevitch, J. & Stewart-Oaten, A. (1999). Empirical approaches to quantifying interaction intensity: Competition and facilitation along productivity gradients. *Ecology*, **80**, 1118–31.

Goldthwait, R. P. (1966). Glacial history. In *Soil Development and Ecological Succession in a Deglaciated Area of Muir Inlet, Southeast Alaska*, ed. A. Mirsky, pp. 1–18. Institute of Polar Studies, Report Number 20. Ohio, U.S.A.: Ohio State University.

Golley, F. B. (ed.) (1977). *Ecological Succession*. Stroudsburg, Pennsylvania, U.S.A.: Dowden, Hutchison & Ross.

Goralczyk, K. (1998). Nematodes in a coastal dune succession: indicators of soil properties? *Applied Soil Ecology*, **9**, 465–9.

Grace, J. B. (1993). The effects of habitat productivity on competition intensity. *Trends in Ecology and Evolution*, **8**, 229–30.

Grandin, U. & Rydin, H. (1998). Attributes of the seed bank after a century of primary succession on islands in Lake Hjalmaren, Sweden. *Journal of Ecology*, **86**, 293–303.

Grant, C. D. & Loneragan, W. A. (1999). The effects of burning on the understorey composition of 11–13 year-old rehabilitated bauxite mines in Western Australia. *Plant Ecology*, **145**, 291–305.

Grassle, J. F. & Morse-Porteous, L. S. (1987). Macrofaunal colonization of disturbed deep-sea environments and the structure of deep-sea benthic communities. *Deep-Sea Research*, **34**, 1911–50.

Greacen, E. L. & Sands, R. (1980). Compaction of forest soils. A review. *Australian Journal of Soil Research*, **18**, 163–89.

Greenslade, P. (1999). Long distance migration of insects to a subantarctic island. *Journal of Biogeography*, **26**, 1161–7.

Greipsson, S. & Davy, A. J. (1997). Responses of *Leymus arenarius* to nutrients: Improvement of seed production and seedling establishment for land reclamation. *Journal of Applied Ecology*, **34**, 1165–76.

Greipsson, S. & El-Mayas, H. (1999). Large-scale reclamation of barren lands in Iceland by aerial seeding. *Land Degradation and Development*, **10**, 185–93.

Grice, A. C. & McIntyre, S. (1995). Speargrass (*Heteropogon controtus*) in Australia: dynamics of species and community. *Rangeland Journal*, **17**, 3–25.

Griffin, G., Smith, D., Morton, S., Allan, G. & Masters, K. (1989). Status and implications of the invasion of tamarisk (*Tamarix aphylla*) on the Finke River, Northern Territory, Australia. *Journal of Environmental Management*, **29**, 297–315.

Griggs, R. F. (1933). The colonization of the Katmai ash, a new and inorganic "soil." *American Journal of Botany*, **20**, 92–113.

Griggs, R. F. (1956). Competition and succession on a rocky mountain fellfield. *Ecology*, **37**, 8–20.

Grime, J. P. (1977). Evidence for the existence of three primary strategies in plants and its relevance to ecological and evolutionary theory. *The American Naturalist*, **111**, 1169–94.

Grime, J. P. (1979). *Plant Strategies and Vegetation Processes*. New York: Wiley.

Grime, J. P. (1998). Benefits of plant diversity to ecosystems: Immediate, filter and founder effects. *Journal of Ecology*, **86**, 902–10.

Grimm, N. B. & Petrone, K. C. (1997). Nitrogen fixation in a desert stream ecosystem. *Biogeochemistry*, **37**, 33–61.

Grishin, S. Yu., del Moral, R., Krestov, P. & Verkholat, V. P. (1996). Succession following the catastrophic eruption of Ksudach volcano (Kamchatka, 1907). *Vegetatio*, **127**, 129–53.

Grootjans, A. P., Ernst, W. H. O. & Stuyfzand, P. J. (1998). European dune slacks: Strong interactions of biology, pedogenesis and hydrology. *Trends in Ecology and Evolution*, **13**, 96–100.

Grosvernier, P., Matthey, Y. & Buttler, A. (1997). Growth potential of three *Sphagnum* species in relation to water table level and peat properties with implications for their restoration in cut-over bogs. *Journal of Applied Ecology*, **34**, 471–83.

Grubb, P. J. (1977). The maintenance of species-richness in plant communities: The importance of the regeneration niche. *Biological Reviews*, **52**, 107–52.

Grubb, P. J. (1986). The ecology of establishment. In *Ecology and Design in Landscape*, ed. A. D. Bradshaw, D. A. Goode & E. Thorp, pp. 83–98. Oxford: Blackwell.

Grubb, P. J. (1987). Some generalizing ideas about colonization and succession in green plants and fungi. In *Colonization, Succession and Stability*, ed. A. J. Gray, M. J. Crawley & P. J. Edwards, pp. 81–102. Oxford: Blackwell.

Grubb, P. J. (1988). The uncoupling of disturbance and recruitment, two kinds of seed banks, and persistence of plant populations at the regional and local scales. *Annales Zoologici Fennici*, **25**, 23–36.

Guariguata, M. R. (1990). Landslide disturbance and forest regeneration in the Upper Luquillo Mountains of Puerto Rico. *Journal of Ecology*, **78**, 814–32.

Guariguata, M. R. & Larsen, M. C. (1990). Preliminary map showing locations of landslides in El Yunque Quadrangle, Puerto Rico. *U.S. Geological Survey Open File Report, 89–257*. San Juan, Puerto Rico.

Guerrero-Campo, J. & Montserrat-Marti, G. (2000). Effects of soil erosion on the floristic composition of plant communities on marl in northeast Spain. *Journal of Vegetation Science*, **11**, 329–36.

Gunn, J. M. (1995). *Restoration and Recovery of an Industrial Region*. New York: Springer.

Gunn, J., Keller, W., Negusanti, J., Potvin, R., Beckett, P. & Winterhalder, K. (1995). Ecosystem recovery after emission reductions: Sudbury Canada. *Water, Air and Soil Pollution*, **85**, 1783–8.

Gutschick, V. P. (1981). Evolved strategies in nitrogen acquisition by plants. *The American Naturalist*, **118**, 607–37.

Guzmán-Grajales, S. M. & Walker, L. R. (1991). Differential seedling responses to litter after Hurricane Hugo in the Luquillo Experimental Forest. *Biotropica*, **23**, 407–13.

Hacker, S. D. & Gaines, S. D. (1997). Some implications of direct positive interactions for community species diversity. *Ecology*, **78**, 1990–2003.

Haeberli, W., Frauenfelder, R., Hoelzle, M. & Maisch, M. (1999). On rates and acceleration trends of global glacier mass changes. *Geografiska Annaler*, **81A**, 585–91.

Hagen, J. B. (1992). *An Entangled Bank: The Origins of Ecosystem Ecology*. New Brunswick, New Jersey: Rutgers University Press.

Haigh, M. J., Rawat, J. S., Bartarya, S. K. & Rawat, M. S. (1993). Environmental influences on landslide activity: Almora Bypass, Kumaun Lesser Himalaya. *Natural Hazards*, **8**, 153–70.

Haldorsen, S. (1981). Grain-size distribution of subglacial till and its relation to glacial crushing and abrasion. *Boreas*, **10**, 91–105.

Halpern, C. B. & Harmon, M. E. (1983). Early plant succession on the Muddy River mudflow, Mount St. Helens, Washington. *American Midland Naturalist*, **110**, 97–106.

Halvorson, J. J., Franz, E. H., Smith, J. L. & Black, R. A. (1992). Nitrogenase activity, nitrogen fixation, and nitrogen inputs by lupines at Mount St. Helens. *Ecology*, **73**, 87–98.

Halvorson, J. J., Smith, J. L. & Franz, E. H. (1991). Lupine influence on soil carbon, nitrogen and microbial activity in developing ecosystems at Mount St. Helens. *Oecologia*, **87**, 162–70.

Halwagy, R. (1963). Studies on the succession of vegetation on some islands and sand banks in the Nile near Khartoum, Sudan. *Vegetatio*, **11**, 217–34.

Hambler, D. J., Dixon, J. M. & Hale, W. H. G. (1995). Ten years in rehabilitation of spoil: Appearance, plant colonists, and the dominant herbivore. *Environmental Conservation*, **22**, 322–34.

Hamerlynck, E. P., McAuliffe, J. R. & Smith, S. D. (2000). Effects of surface and sub-surface soil horizons on the seasonal performance of *Larrea tridentata* (creosotebush). *Functional Ecology*, **14**, 596–606.

Hamilton, D. A. & Auble, G. T. (1993). *Wetland Modeling and Intermountain Needs at Stillwater National Wildlife Refuge*. Fallon, Nevada: U.S. Fish & Wildlife Service.

Hammon, D. A. (1997). *Creating Freshwater Wetlands*. Boca Raton, Florida: CRC Press.

Handel, S. N., Robinson, G. R., Parsons, W. F. J. & Mattei, J. H. (1997). Restoration of woody plants to capped landfills: Root dynamics in an engineered soil. *Restoration Ecology*, **5**, 178–86.

Hansen, K. & Jensen, J. (1972). The vegetation on a roadside in Denmark. *Dansk Botanisk Arkiv*, **28**, 7–61.

Harden, C. P. (1996). Interrelationships between land abandonment and land degradation: A case from the Ecuadorian Andes. *Mountain Research and Development*, **16**, 274–80.

Harper, J. L. (1977). *Population Biology of Plants*. New York: Academic Press.

Harper, K. A. & Kershaw, G. P. (1997). Soil characteristics of 48-year old borrow pits and vehicle tracks in a shrub tundra along the CANOL No. 1 pipeline corridor, Northwest Territories, Canada. *Arctic and Alpine Research*, **29**, 105–11.

Harris, E., Mack, R. N. & Ku, M. S. B. (1987). Death of steppe cryptogams under the ash from Mount St. Helens. *American Journal of Botany*, **74**, 1249–53.

Harris, L. G., Ebeling, A. W., Laur, D. R. & Rowley, R. J. (1984). Community recovery after storm damage: A case of facilitation in primary succession. *Science*, **224**, 1336–8.

Harrison, R. D., Banka, R., Thornton, I. W. B., Shanahan, M. & Yamuna, R. (2001). Colonization of an island volcano, Long Island, Papua New Guinea, and an emergent island, Motmot, in its caldera lake. II. The vascular flora. *Journal of Biogeography*, **28**, 1311–37.

Hastings, J. R. & Turner, R. M. (1980). *The Changing Mile: An Ecological Study of Vegetation Change With Time in the Lower Mile of an Arid and Semiarid Region*. Tucson, U.S.A.: University of Arizona Press.

Hatton, T. J. & West, N. E. (1987). Early seral trends in plant community diversity on a recontoured surface mine. *Vegetatio*, **73**, 21–9.

Heath, J. A. & Huebert, B. J. (1999). Cloudwater deposition as a source of fixed nitrogen in a Hawaiian montane forest. *Biogeochemistry*, **44**, 119–34.

Hegazy, A. K. (1997). Plant succession and its optimization on tar-polluted coasts in the Arabian Gulf region. *Environmental Conservation*, **24**, 149–58.

Hejkal, J. (1985). The development of a carabid fauna (Coleoptera, Carabidae) on spoil banks under conditions of primary succession. *Acta Entomologica Bohemoslovaca*, **82**, 321–46.

Helm, D. J., Allen, E. B. & Trappe, J. M. (1996). Mycorrhizal chronosequence near Exit Glacier, Alaska. *Canadian Journal of Botany*, **74**, 1496–506.

Helm, D. J. & Collins, W. B. (1997). Vegetation succession and disturbance on a boreal forest floodplain, Susitna River, Alaska. *Canadian Field-Naturalist*, **111**, 553–66.

Hendrix, L. B. (1981). Post-eruption succession on Isla Fernandina, Galápagos. *Madroño*, **28**, 242–54.

Henriksson, E., Henriksson, L. E. Norrman, J. O. & Nyman, P. O. (1987). Biological dinitrogen fixation (acetylene reduction) exhibited by blue-green algae (cyanobacteria) in association with mosses gathered on Surtsey, Iceland. *Arctic and Alpine Research*, **19**, 432–6.

Henriques, R. P. B. & Hay, J. D. (1992). Nutrient content and the structure of a plant community on a tropical beach-dune system in Brazil. *Acta Oecologica*, **13**, 101–17.

Henriques, R. P. B. & Hay, J. D. (1998). The plant communities of a foredune in southeastern Brazil. *Canadian Journal of Botany*, **76**, 1323–30.

Henry, C. P., Amoros, C. & Bornette, G. (1996). Species traits and recolonization processes after flood disturbance in riverine macrophytes. *Vegetatio*, **122**, 13–27.

Herben, T. (1996). Permanent plots as tools for plant community ecology. *Journal of Vegetation Science*, **7**, 195–202.

Hester, M. W., Wilsey, B. J. & Mendelssohn, I. A. (1994). Grazing of *Panicum amarum* in a Louisiana barrier island dune plant community – management implications for dune restoration projects. *Ocean and Coastal Management*, **23**, 213–24.

Heusser, C. J. (1956). Post-glacial environments in the Canadian Rocky Mountains. *Ecological Monographs*, **26**, 263–302.

Hewitt, K. (1965). Glacier surges in the Karakoram Himalaya (Central Asia). *Canadian Journal of Earth Sciences*, **6**, 1009–18.

Heyne, C. M. (2000). Soil and vegetation recovery on abandoned paved roads in a humid tropical rain forest, Puerto Rico. M.S. thesis, University of Nevada, Las Vegas.

Higgins, S. I., Coetzee, M. A. S., Marneweck, G. C. & Rogers, K. H. (1996). The Nyl River floodplain, South Africa, as a functional unit of the landscape: A review of current information. *African Journal of Ecology*, **34**, 131–45.

Hill, M. O. (1979). *DECORANA – A FORTRAN program for detrended correspondence analysis and reciprocal averaging*. Ecology and systematics. Ithaca, New York, U.S.A.: Cornell University.

Hilton, J. L. & Boyd, R. S. (1996). Microhabitat requirements and seed/microsite limitation of the rare granite outcrop endemic *Amphianthus pusillus* (Scrophulariaceae). *Bulletin of the Torrey Botanical Club*, **123**, 189–96.

Hirata, T. (1986). Succession of sessile organisms on experimental plates immersed in Habeta Bay, Izu Peninsula, Japan. I. Algal succession. *Marine Ecology Progress Series*, **34**, 51–61.

Hirata, T. (1992). Succession of sessile organisms on experimental plates immersed in Habeta Bay, Izu Peninsula, Japan. V. An integrated consideration on the definition and prediction of succession. *Ecological Research*, **7**, 31–42.

Hiroki, S. & Ichino, K. (1993). Differences of invasion behaviour between two climax species, *Castanopsis cuspidata* var. *sieboldii* and *Mechilus thunbergii*, on lava flows on Miyakejima, Japan. *Ecological Research*, **8**, 167–72.

Hirose, T. & Tateno, M. (1984). Soil nitrogen patterns induced by colonization of *Polygonum cuspidatum* on Mt. Fuji. *Oecologia*, **61**, 218–23.

Hobbie, E. A., Macko, S. A. & Shugart, H. H. (1999). Insights into nitrogen and carbon dynamics of ectomycorrhizal and saprotrophic fungi from isotopic evidence. *Oecologia*, **118**, 353–60.

Hobbie, S. E. (1992). Effects of plant species on nutrient cycling. *Trends in Ecology and Evolution*, **7**, 336–9.

Hobbs, R. J. (1999). Restoration of disturbed ecosystems. In *Ecosystems of Disturbed Ground, Ecosystems of the World 16*, ed. L. R. Walker, pp. 673–87. Amsterdam: Elsevier.

Hobbs, R. J. & Norton, D. A. (1996). Towards a conceptual framework for restoration ecology. *Restoration Ecology*, **4**, 93–110.

Hodgkin, S. E. (1984). Scrub encroachment and its effects on soil fertility on Newborough Warren, Anglesey, Wales. *Biological Conservation*, **29**, 99–119.

Hodkinson, I. D., Webb, N. R. & Coulson, S. J. (2002). Primary community assembly on land – the missing stages: why are the heterotrophic organisms always there first? *Journal of Ecology*, **90**, 569–77.

Hogg, P., Squires, P. & Fitter, A. H. (1995). Acidification, nitrogen deposition and rapid vegetational change in a small valley mire in Yorkshire. *Biological Conservation*, **71**, 143–53.

Holl, K. D. & Cairns, J. Jr. (1994). Vegetational community development on reclaimed coal surface mines in Virginia. *Bulletin of the Torrey Botanical Club*, **121**, 327–37.

Holling, C. S. (1994). An ecologist's view of the Malthusian conflict. In *Population, Economic Development and the Environment*, ed. K. Lindahl-Kiessling & H. Landberg, pp. 79–103. Oxford: Oxford University Press.

Hollinger, D. Y. (1986). Herbivory and the cycling of nitrogen and phosphorus in isolated California oak trees. *Oecologia*, **70**, 291–7.

Holmes, P. M. & Richardson, D. M. (1999). Protocols for restoration based on recruitment dynamics, community structure, and ecosystem functions: Perspectives from South African fynbos. *Restoration Ecology*, **7**, 215–30.

Holmgren, M., Scheffer, M., & Huston, M. A. (1997). The interplay of facilitation and competition in plant communities. *Ecology*, **78**, 1966–75.

Holter, V. (1979). Distribution of Leguminosae in early stages of a primary succession in relation to edaphic factors. *Botaniske Tidsskrift*, **74**, 79–87.

Holter, V. (1984). $N_2(C_2H_2)$-fixation in early stages of primary succession on a reclaimed salt marsh. *Holarctic Ecology*, **7**, 165–70.

Holzapfel, C. & Mahall, B. E. (1999). Bi-directional facilitation and interference between shrubs and annuals in the Mojave Desert. *Ecology*, **80**, 1747–61.

Honnay, O., Verhaeghe, W. & Hermy, M. (2001). Plant community assembly along dendritic networks of small forest streams. *Ecology*, **82**, 1691–702.

Hooper, D. U., Bignell, D. E., Brown, V. K., Brussaard, L., Dangerfield, J. M., Wall, D. H., Wardle, D. A., Coleman, D. C., Giller, K. E., Lavelle, P., Van der Putten, W. H., De Ruiter, P. C., Rusek, J., Silver, W. L., Tiedje, J. M. & Wolters, V. (2000). Interactions between aboveground and belowground biodiversity in terrestrial ecosystems: patterns, mechanisms, and feedbacks. *BioScience*, **50**, 1049–61.

Hooper, D. U. & Johnson, L. (1999). Nitrogen limitation in dryland ecosystems: Responses to geographical and temporal variation in precipitation. *Biogeochemistry*, **46**, 247–93.

Horn, H. S. (1976). Succession. In *Theoretical Ecology*, ed. R. M. May, pp. 253–71. Sunderland, Massachusetts: Sinauer.

Horton, B. K. (1999). Erosional control on the geometry and kinematics of thrust belt development in the central Andes. *Tectonics*, **18**, 1292–304.

Hosner, J. F. & Minckler, L. S. (1963). Bottomland hardwood forests of southern Illinois – regeneration and succession. *Ecology*, **44**, 29–41.

Houle, G. (1990). Species-area relationship during primary succession in granite outcrop plant communities. *American Journal of Botany*, **77**, 1433–9.

Houle, G. (1996). Environmental filters and seedling recruitment on a coastal dune in subarctic Quebec (Canada). *Canadian Journal of Botany*, **74**, 1507–13.

Houle, G. (1997). No evidence for interspecific interactions between plants in the first stage of succession on coastal dunes in subarctic Quebec, Canada. *Canadian Journal of Botany*, **75**, 902–15.

Houle, G. & Phillips, D. L. (1988). The soil seed bank of granite outcrop plant communities. *Oikos*, **52**, 87–93.

Houle, G. & Phillips, D. L. (1989). Seed availability and biotic interactions in granite outcrop plant communities. *Ecology*, **70**, 1307–16.

Howarth, F. (1979). Neogeoaeolian habitats on new lava flows on Hawaii island: an ecosystem supported by wind-borne debris. *Pacific Insects*, **20**, 133–44.

Howarth, F. G. (1987). Evoluution and ecology of aeolian subterranean habitats in Hawaii. *Trends in Ecology and Evolution*, **2**, 220–3.

Huebert, B., Vitousek, P., Sutton, J, Elias, T., Heath, J., Coeppicus, S., Howell, S. & Blomquist, B. (1999). Volcano fixes nitrogen into plant-available forms. *Biogeochemistry*, **47**, 111–18.

Hughes, F. M. R. (1997). Floodplain biogeomorphology. *Progress in Physical Geography*, **21**, 501–29.

Hughes, L. & Westoby, M. (1992a). Fate of seeds adapted for dispersal by ants in Australian sclerophyll vegetation. *Ecology*, **73**, 1285–99.

Hughes, L. & Westoby, M. (1992b). Effect of diaspore characteristics on removal of seeds adapted for dispersal by ants. *Ecology*, **73**, 1300–12.

Hulberg, L. W. & Oliver, J. S. (1980). Caging manipulations in marine soft-bottom communities: Importance of animal interactions to sedimentary habitat modifications. *Canadian Journal of Botany*, **37**, 1130–9.

Humphries, R. N. (1980). The development of wildlife interest in limestone quarries. *Reclamation Review*, **3**, 197–207.

Hupp, C. R. & Osterkamp, W. R. (1985). Bottomland vegetation distribution along Passage Creek, Virginia, in relation to fluvial landforms. *Ecology*, **66**, 670–81.

Huss-Danell, K., Sverrisson, H., Hahlin, A.-S. & Danell, K. (1999). Occurrence of *Alnus*-infective *Frankia* and *Trifolium*-infective *Rhizobium* in circumpolar soils. *Arctic, Antarctic and Alpine Research*, **31**, 400–6.

Huss-Danell, K., Uliassi, D. & Renberg, I. (1997). River and lake sediments as sources of infective *Frankia* (*Alnus*). *Plant and Soil*, **197**, 35–9.

Huston, M. & Smith, T. (1987). Plant succession: Life history and competition. *The American Naturalist*, **130**, 168–98.

Hüttl, R. H. and Bradshaw, A. (2001). Ecology of post-mining landscapes. *Restoration Ecology*, **9**, 339–40.

Imbert, É. & Houle, G. (2000). Persistence of colonizing plant species along an inferred successional sequence on a subarctic coastal dune (Québec, Canada). *Ecoscience*, **7**, 370–8.

Inderjit & del Moral, R. (1997). Is separating resource competition from allelopathy realistic? *Botanical Reviews*, **63**, 221–30.

Inouye, R. S. & Tilman, D. (1995). Convergence and divergence of old-field vegetation after 11 yr of nitrogen addition. *Ecology*, **76**, 1872–87.

Insam, H. & Haselwandter, K. (1989). Metabolic quotient of the soil microflora in relation to plant succession. *Oecologia*, **79**, 174–8.

Ishikawa, S. I., Furukawa, A. & Oikawa, T. (1995). Zonal plant distribution and edaphic and micrometeorolgical conditions on a coastal sand dune. *Ecological Research*, **10**, 259–66.

Ishikawa, S. K. & Kachi, N. (1998). Shoot population dynamics of *Carex homomugi* on a coastal sand dune in relation to its zonal distribution. *Australian Journal of Botany*, **46**, 111–21.

Itoh, S. & Barber, S. A. (1983). Phosphorus uptake by six plant species as related to root hairs. *Agronomy Journal*, **75**, 457–61.

Iverson, L. R. & Wali, M. K. (1982). The role of *Kochia scoparia* and other pioneers in the revegetation process of surface mined lands. *Reclamation and Revegetation Research*, **1**, 123–60.

Ives, A. R. (1991). Aggregation and coexistence in a carrion fly community. *Ecological Monographs*, **61**, 75–94.

Jackson, C. R., Churchill, P. F. & Roden, E. E. (2001). Successional changes in bacterial assemblage structure during epilithic biofilm development. *Ecology*, **82**, 555–66.

Jackson, R. B. & Caldwell, M. M. (1993). The scale of nutrient heterogeneity around individual plants and its quantification with geo-statistics. *Ecology*, **74**, 612–14.

Jackson, S. T., Futyma, R. P. & Wilcox, D. A. (1988). A paleoecological test of a classical hydrosere in the Lake Michigan dunes. *Ecology*, **69**, 928–36.

Jacobsen, G. L. Jr. & Birks, H. J. B. (1980). Soil development on recent end moraines of the Klutlan Glacier, Yukon Territory, Canada. *Quaternary Research*, **14**, 87–100.

Jacquemyn, H., Butaye, J., Dumortier, M., Hermy, M., & Lust, N. (2001). Effects of age and distance on the composition of mixed deciduous forest fragments in an agricultural landscape. *Journal of Vegetation Science*, **12**, 635–42.

Jacquez, G. M. & Patten, D. T. (1996). *Chesneya nubigena* on a Himalayan glacial moraine, a case of facilitation in primary succession? *Mountain Research and Development*, **16**, 265–73.

James, I. L. (1973). Mass movements in the upper Pohangina catchment, Ruahine Range. *Journal of Hydrology (New Zealand)*, **12**, 92–102.

Janos, D. P. (1980). Mycorrhizae influence tropical succession. *Biotropica*, **12**, 56–64.

Jansson, R., Nilsson, C. & Renofalt, B. (2000). Fragmentation of riparian floras in rivers with multiple dams. *Ecology*, **81**, 899–903.

Jefferson, R. G. & Usher, M. B. (1989). Seed rain dynamics in disused chalk quarries in the Yorkshire Wolds, England, with special reference to nature conservation. *Biological Conservation*, **47**, 123–36.

Jehne, W. & Thompson, C. H. (1981). Endomycorrhizae in plant colonization on coastal sand-dunes at Cooloola, Queensland. *Australian Journal of Ecology*, **6**, 221–30.

Jenny, H. (1941). *Factors of Soil Formation*. New York: McGraw-Hill.

Jenny, H. (1961). Derivation of state factor equations of soils and ecosystems. *Proceedings of the Soil Science Society of America*, **25**, 385–8.

Jenny, H. (1980). *The Soil Resource: Origin and Behavior.* New York: Springer.

Jensen, A. (1993). Dry coastal ecosystems of Denmark. In *Dry Coastal Ecosystems: Polar Regions and Europe, Ecosystems of the World 2A*, ed. E. van der Maarel, pp. 183–96. Amsterdam: Elsevier.

Jochimsen, M. E. A. (1996). Reclamation of colliery mine spoil founded on natural succession. *Water and Soil Pollution*, **91**, 99–108.

Johansson, M. E., Nilsson, C. & Nilsson, E. (1996). Do rivers function as corridors for plant dispersal? *Journal of Vegetation Science*, **7**, 593–8.

Johnson, F. L., Gibson, D. J. & Risser, P. J. (1982). Revegetation of unreclaimed coal strip-mines in Oklahoma. *Journal of Applied Ecology*, **19**, 453–63.

Johnson, N. C. (1998). Responses of *Salsola kali* and *Panicum virgatum* to mycorrhizal fungi, phosphorus and soil organic matter: Implications for reclamation. *Journal of Applied Ecology*, **35**, 86–94.

Johnson, W. B., Sasser, C. E. & Gosselink, J. G. (1985). Succession of vegetation in an evolving river delta, Atchafalaya Bay, Louisiana. *Journal of Ecology*, **73**, 973–86.

Jones, C. G., Lawton, J. H. & Shachak, M. (1994). Organisms as ecosystem engineers. *Oikos*, **69**, 373–86.

Jones, C. G., Lawton, J. H & Shachak, M. (1997). Positive and negative effects of organisms as physical ecosystem engineers. *Ecology*, **78**, 1946–57.

Jones, E. G., Collins, M. A., Bagley, P. M., Addison, S. & Priede, I. G. (1998). The fate of cetacean carcasses in the deep sea: Observations on consumption rates and succession of scavenging species in the abyssal north-east Atlantic Ocean. *Proceedings of the Royal Society of London*, B**265**, 1119–27.

Jónsson, V. K. & Matthíasson, M. (1993). Lava-cooling operations during the 1973: eruption of Eldfell Volcano, Heimaey, Vestmannaeyjar, Iceland. *U.S. Geological Survey Open-File Report*, 97–724 (translation from the original).

Jorgenson, S. E. (1997). *Integration of Ecosystem Theories: A Pattern*, 2nd edition. Dordrecht, The Netherlands: Kluwer.

Judd, K. W. & Mason, C. F. (1995). Colonization of a restored landfill site by invertebrates, with particular reference to the Coleoptera. *Pedobiologia*, **39**, 116–25.

Jumpponen, A., Mattson, K. G. & Trappe, J. M. (1998a). Mycorrhizal functioning of *Phialocephala fortinii* with *Pinus contorta* on glacier forefront soil: Interactions with soil nitrogen and organic matter. *Mycorrhiza*, **7**, 261–5.

Jumpponen, A., Mattson, K., Trappe, J. M. & Ohtonen, R. (1998b). Effects of established willows on primary succession on Lyman Glacier forefront, North Cascade Range, Washington, U.S.A.: Evidence for simultaneous canopy inhibition and soil facilitation. *Arctic and Alpine Research*, **30**, 31–9.

Jumpponen, A., Várem, H., Mattson, K. G., Ohtonen, R. & Trappe, J. M. (1999). Characterization of "safe sites" for pioneers in primary succession on recently deglaciated terrain. *Journal of Ecology*, **87**, 98–105.

Kadmon, R. & Pulliam, H. R. (1995). Effects of isolation, logging and dispersal on woody species richness of islands. *Vegetatio*, **116**, 63–8.

Kalliola, R., Salo, J., Pukakka, M. & Rajasilta, M. (1991). New site formation and colonizing vegetation in primary succession on the western Amazon floodplains. *Journal of Ecology*, **79**, 877–901.

Kaltenecker, J. H., Wicklow-Howard, M. & Pellant, M. (1999). Biological soil crusts: natural barriers to *Bromus tectorum* L. establishment in the northern Great Basin, USA. In *Proceedings of the VI International Rangeland Congress*, Aitkenvale, Queensland, Australia, ed. D. Eldridge & D. Freudenberger, pp. 109–11. Boise, Idaho, U.S.A.: Bureau of Land Management.

Kamijo, T. & Okutomi, K. (1995). Seedling establishment of *Castanopsis cuspidata* var. *sieboldii* and *Persia thunbergii* on lava and scoria of the 1962 eruption on Miyake-Jim island, the Izu Islands. *Ecological Research*, **10**, 235–42.

Kamnev, A. A. & van der Lelie, D. (2000). Chemical and biological parameters as tools to evaluate and improve heavy metal phytoremediation. *BioScience Reports*, **20**, 239–58.

Kaufmann, R. (2001). Invertebrate succession on an alpine glacier foreland. *Ecology*, **82**, 2261–78.

Kearey, P. & Vine, F. J. (1996). *Global Tectonics*. Oxford: Blackwell.

Keddy, P. (1992). Assembly and response rules: Two goals for predictive community ecology. *Journal of Vegetation Science*, **3**, 1957–64.

Keddy, P. (1999). Wetland restoration: The potential for assembly rules in the service of conservation. *Wetlands*, **19**, 716–32.

Keeland, B. D. & Conner, W. H. (1999). Natural regeneration and growth of *Taxodium distichum* (L.) Rich in Lake Chicot, Louisiana after 44 years of flooding. *Wetlands*, **19**, 149–55.

Keever, C. (1950). Causes of succession on old fields of the piedmont, North Carolina. *Ecological Monographs*, **20**, 230–50.

Keever, C. (1979). Mechanisms of plant succession on old fields of Lancaster County, Pennsylvania. *Bulletin of the Torrey Botanical Club*, **106**, 299–308.

Kershaw, G. P. & Kershaw, L. J. (1987). Successful plant colonizers on disturbances in tundra areas of northwestern Canada. *Arctic and Alpine Research*, **19**, 451–60.

Keylock, C. (1997). Snow avalanches. *Progress in Physical Geography*, **21**, 481–500.

Keys, D. (2000). *Catastrophe: An Investigation into the Origins of the Modern World*. New York: Ballantine Books.

Khalequzzaman, M. D. (1994). Recent floods in Bangladesh: Possible causes and solutions. *Natural Hazards*, **9**, 65–80.

Khan, A. G., Kuek, C., Chaudhry, T. M., Khoo, C. S. & Hayes, W. J. (2000). Role of plants, mycorrhizae and phytochelators in heavy metal contaminated land remediation. *Chemosphere*, **41**, 197–207.

Khurana, E. & Singh, J. S. (2001). Ecology of seed and seedling growth for conservation and restoration of tropical dry forest: a review. *Environmental Conservation*, **28**, 39–52.

Kielland, K. (1994). Amino acid absorption by Arctic plants: Implications for plant nutrition and nitrogen cycling. *Ecology*, **75**, 2373–83.

Kielland, K. & Bryant, J. P. (1998). Moose herbivory in taiga: Effects on biogeochemistry and vegetation dynamics in primary succession. *Oikos*, **82**, 377–83.

Kielland, K., Bryant, J. P. & Ruess, R. W. (1997). Moose herbivory and carbon turnover of early successional stands in interior Alaska. *Oikos*, **80**, 25–30.

Kiilsgaard, C. W., Greene, S. E. Stafford, S. G. & McKee, W. A. (1986). Recovery of riparian vegetation in the NE region of Mount St. Helens. In *Mount St. Helens: Five Years Later*, ed. S. A. C. Keller, pp. 222–30. Cheney, Washington, U.S.A.: Eastern Washington University Press.

Killham, K. (1994). *Soil Ecology*. Cambridge: Cambridge University Press.

Kim, J. H. & DeWreede, R. E. (1996). Effects of size and season of disturbance on algal patch recovery in a rocky intertidal community. *Marine Ecology Progress Series*, **133**, 217–28.

Kitayama, K. (1996a). Soil nitrogen dynamics along a gradient of long-term soil development in a Hawaiian wet montane rainforest. *Plant and Soil*, **183**, 253–62.

Kitayama, K. (1996b). Patterns of species diversity on an oceanic versus a continental island mountain: A hypothesis on species diversification. *Journal of Vegetation Science*, **7**, 879–88.

Kitayama, K. & Mueller-Dombois, D. (1995a). Vegetation changes along gradients of long-term soil development in the Hawaiian montane rainforest zone. *Vegetatio*, **120**, 1–20.

Kitayama, K. & Mueller-Dombois, D. (1995b). Biological invasion on an oceanic island mountain: Do alien plant species have wider ecological ranges than native species? *Journal of Vegetation Science*, **6**, 667–74.

Kitayama, K., Mueller-Dombois, D.& Vitousek, P. M. (1995). Primary succession of Hawaiian montane rain forest on a chronosequence of eight lava flows. *Journal of Vegetation Science*, **6**, 211–22.

Klemmedson, J. O. & Barth, R. C. (1975). Distribution and balance of biomass and nutrients in desert shrub ecosystems. *US/IBP Desert Biome Research Memo*, 75–8. Logan: Utah State University Press.

Klimes, L. (1987). Succession in road bank vegetation. *Folia Geobotanica et Phytotaxonomica*, **22**, 435–40.

Klinger, L. F. (1996). The myth of the classic hydrosere model of bog succession. *Arctic and Alpine Research*, **28**, 1–9.

Klironomos, J. N., Widden, P. & Deslandes, I. (1992). Feeding preferences of the Collembolan *Folsomia candida* in relation to microfungal successions on decaying litter. *Soil Biology and Biochemistry*, **24**, 685–92.

Koch, J. M., Ward, S. C., Grant, C. D. & Ainsworth, G. L. (1996). Effects of bauxite mine restoration operations on topsoil seed reserves in the jarrah forest of Western Australia. *Restoration Ecology*, **4**, 368–76.

Kochy, M. & Rydin, H. (1997). Biogeography of vascular plants on habitat islands, peninsulas and main lands in an east-central Swedish agricultural landscape. *Nordic Journal of Botany*, **17**, 215–23.

Kohls, S. J., van Kessel, C., Baker, D. D., Grigal, D. F. & Lawrence, D. B. (1994). Assessment of N_2 fixation and N cycling by *Dryas* along a chronosequence within the forelands of the Athabasca Glacier, Canada. *Soil Biology and Biochemistry*, **26**, 623–32.

Komárková, V. & Wielgolaski, F. E. (1999). Stress and disturbance in cold region ecosystems. In *Ecosystems of Disturbed Ground, Ecosystems of the World 16*, ed. L. R. Walker, pp. 39–122. Amsterdam: Elsevier.

Komulainen, V. M., Tuittila, E. S., Vasander H. & Laine, J. (1999). Restoration of drained peatlands in southern Finland, initial effects on vegetation change and CO_2 balance. *Journal of Applied Ecology*, **36**, 634–48.

Kondolf, G. (1995). Five elements for effective evaluation of stream restoration. *Restoration Ecology*, **3**, 133–6.

Koons, P. O., Craw D., Cox, S. C., Upton, P., Templeton, A. S. & Chamberlain, C. P. (1998). Fluid flow during active oblique convergence: A Southern Alps model from mechanical and geochemical observations. *Geology*, **26**, 159–62.

Korte, V. L. & Blinn, D. W. (1983). Diatom colonization on artificial substrata in pool and riffle zones studied by light and scanning electron microscopy. *Journal of Phycology*, **19**, 332–42.

Koske, R. E. & Gemma, J. N. (1990). VA Mycorrhizae in strand vegetation of Hawaii: Evidence for long-distance codispersal of plants and fungi. *American Journal of Botany*, **77**, 466–74.

Koske, R. E., Bonin, C., Kelly, J. & Martinez, C. (1996). Effects of sea water on spore germination of a sand-dune-inhabiting arbuscular mycorrhizal fungus. *Mycologia*, **88**, 947–50.

Krasny, M. E., Vogt, K. A. & Zasada, J. C. (1984). Root and shoot biomass and mycorrhizal development of white spruce seedlings naturally regenerating in interior Alaskan floodplain communities. *Canadian Journal of Forest Research*, **14**, 554–8.

Krasny, M. E., Vogt, K. A. & Zasada, J. C. (1988). Establishment of four Salicaceae species on river bars in interior Alaska. *Holarctic Ecology*, **11**, 210–19.

Krebs, C. J., Boonstra, R., Boutin, S. & Sinclair, A. R. E. (2001). What drives the 10-year cycle of snowshoe hares? *BioScience*, **51**, 25–35.

Kritzinger, J. J. & van Aarde, R. J. (1998). The bird communities of rehabilitating coastal dunes at Richards Bay, KwaZulu-Natal. *South African Journal of Science*, **94**, 71–8.

Kroh, G. C., White, J. D., Heath, S. K. & Pinder, J. E. (2000). Colonization of a volcanic mudflow by an upper montane coniferous forest at Lassen Volcanic National Park, California. *American Midland Naturalist*, **143**, 126–40.

Kurina, L. M. & Vitousek, P. M. (1999). Controls over the accumulation and decline of a nitrogen-fixing lichen, *Stereocaulon vulcani*, on young Hawaiian lava flows. *Journal of Ecology*, **87**, 784–99.

Kutiel, P., Peled, Y. & Geffen, E. (2000). The effect of removing shrub cover on annual plants and small mammals in a coastal sand dune ecosystem. *Biological Conservation*, **94**, 235–42.

Lammerts, E. J. & Grootjans, A. P. (1998). Key environmental variables determining the occurrence and life span of basiphilous dune slack vegetation. *Acta Botanica Neerlandica*, **47**, 369–92.

Lammerts, E. J., Pegtel, D. M., Grootjans, A. P. & van der Veen, A. (1999). Nutrient limitation and vegetation changes in a coastal dune slack. *Journal of Vegetation Science*, **10**, 111–22.

Landeweert, R., Hoffland, E., Finlay, R. D., Kuyper, T. W. & van Breemen, N. (2001). Linking plants to rocks: Ectomycorrhizal fungi mobilize nutrient from minerals. *Trends in Ecology and Evolution*, **16**, 248–54.

Langenheim, J. H. (1956). Plant succession on a sub-alpine earthflow in Colorado. *Ecology*, **37**, 301–17.

Larsen, M. C. & Simon, A. (1993). A rainfall intensity-duration threshold for landslides in a humid-tropical environment, Puerto Rico. *Geografiska Annaler*, **75A**, 13–23.

Larsen, M. C. & Torres-Sánchez, A. J. (1996). Geographic relations of landslide distributions and assessment of landslide hazards in the Blanco, Cibuco and Coamo Basins, Puerto Rico. *U.S. Geological Survey Water Resources Investigations Report* 95-4029. San Juan, Puerto Rico.

Larson, B. M. H. & Barrett, S. C. H. (2000). A comparative analysis of pollen limitation in flowering plants. *Biological Journal of the Linnean Society*, **69**, 503–20.

Larson, D. W., Matthes, U. & Kelly, P. E. (eds) (2000). *Cliff Ecology: Pattern and Process in Cliff Ecosystems*. Cambridge: Cambridge University Press.

Larson, M. M. (1984). Invasion of volunteer tree species on stripmine plantation in east-central Ohio. *Research Bulletin of the Ohio State University Agricultural Research and Development Center*, **1158**, 1–10.

Larson, M. M. & Vimmerstedt, J. P. (1983). Evaluation of 30-year-old plantations on stripmined land in East Central Ohio. *Research Bulletin of the Ohio State University Agricultural Research and Development Center*, **1149**, 1–20.

Law, R. & Morton, R. D. (1993). Alternative permanent states of ecological communities. *Ecology*, **74**, 1347–61.

Lawlor, T. E. (1986). Comparative biogeography of mammals on islands. *Biological Journal of the Linnaean Society*, **28**, 99–125.

Lawrence, D. B. (1958). Glaciers and vegetation in southeastern Alaska. *American Scientist*, **46**, 89–122.

Lawrence, D. B. (1979). Primary versus secondary succession at Glacier Bay National Monument, southeastern Alaska. *Proceedings 1st Conference on Scientific Research in the National Parks, Transactions and Proceedings Series 5*, ed. J. D. Wood Jr., pp. 213–24. U.S. Department of the Interior, National Park Service.

Lawrence, D. B., Schoenike, R. E., Quispel, A. & Bond, G. (1967). The role of *Dryas drummondii* in vegetation development following ice recession at Glacier Bay, Alaska, with special reference to its nitrogen fixation by root nodules. *Journal of Ecology*, **55**, 793–813.

Lawson, D. E. (1986). Response of permafrost terrain to disturbance: A synthesis of observations from northern Alaska, U.S.A. *Arctic and Alpine Research*, **18**, 1–17.

Leakey, R. E. (1979). *The Illustrated Origin of Species by Charles Darwin*. Abridged and introduced by R. E. Leakey. New York: Hill & Wang.

Lefèbvre, C. & Simon, E. (1979). Plant spacing in open communities from old zinc-lead mine wastes. *Oecologia Plantarum*, **14**, 461–73.

Le Houérou, H. N. (1986). The desert and arid zones of northern Africa. In *Hot Deserts and Arid Shrublands, Ecosystems of the World* 12B, ed. M. Evenari, I. Noy-Meir & D. W. Goodall, pp. 101–47. Amsterdam: Elsevier.

Le Houérou, H. N. (2000). Restoration and rehabilitation of arid and semiarid Mediterranean ecosystems in North Africa and west Asia: A review. *Arid Soil Research and Rehabilitation*, **14**, 3–14.

Le Houérou, H. N., Popov, G. F. & See, L. (1993). *Agro-bioclimatic Classification of Africa*. Agrometeorology Series Working Paper, No. 6, F.A.O. Rome, Italy. (Not seen but cited in Carretero (1999).)

Lei, S. A. & Walker, L. R. (1997). Classification and ordination of *Coleogyne* communities in southern Nevada. *Great Basin Naturalist*, **57**, 155–62.

Leisman, G. (1957). A vegetation and soil chronosequence on the Mesabi Iron Range spoil banks, Minnesota. *Ecological Monographs*, **27**, 221–44.

Le Maitre, D. C., van Wilgen, B. W., Chapman, R. A. & McKelly, D. H. (1996). Invasive plants and water resources in the Western Cape Province, South Africa: Modeling the consequences of a lack of management. *Journal of Applied Ecology*, **33**, 161–72.

Leopold, D. J. & Wali, M. K. (1992). The rehabilitation of forest ecosystems in the eastern United States and Canada. In *Ecosystem Rehabilitation*, vol. 2, *Ecosystem Analysis and Synthesis*, ed. M. K. Wali, pp. 187–232. The Hague: SPB Academic Publishing.

Lepš, J. S. & Rejmánek, M. (1991). Convergence or divergence: What should we expect from vegetation succession? *Oikos*, **62**, 261–4.

Lesica, P. & Cooper, S. V. (1999). Succession and disturbance in sand hills vegetation: Constructing models for managing biological diversity. *Conservation Biology*, **13**, 293–302.

Letnic, M. & Fox, B. J. (1997a). The impact of industrial fluoride fallout on faunal succession following sand-mining of dry sclerophyll forest at Tomago, NSW, I. Lizard recolonization. *Biological Conservation*, **80**, 63–81.

Letnic, M. & Fox, B. J. (1997b). The impact of industrial fluoride fallout on faunal succession following sand-mining of dry sclerophyll forest at Tomago, NSW, II. Myobatrachid frog recolonization. *Biological Conservation*, **82**, 137–46.

Levin, L. A., Talley, D. & Thayer, G. (1996). Succession of macrobenthos in a created salt marsh. *Marine Ecology Progress Series*, **141**, 67–82.

Levine, J. M. (1999). Indirect facilitation, evidence and predictions from a riparian community. *Ecology*, **80**, 1762–9.

Levine, J. M. (2000). Complex interactions in a streamside plant community. *Ecology*, **81**, 3431–44.

Levine, J. M., Brewer, J. S. & Bertness, M. D. (1998). Nutrients, competition and plant zonation in a New England salt marsh. *Journal of Ecology*, **86**, 285–92.

Levins, R. (1969). Some demographic and genetic consequences of environmental heterogeneity for biological control. *Bulletin of the Entomological Society of America*, **15**, 237–40.

Lewis, N. K. (1998). Landslide-driven distribution of aspen and steppe on Kathul Mountain, Alaska. *Journal of Arid Environments*, **38**, 421–35.

Lewis Smith, R. I. (1993). The role of bryophyte propagule banks in primary succession: Case study of an Antarctic fellfield soil. In *Primary Succession on Land*, ed. J. Miles & D. W. H. Walton, pp. 55–78. Oxford: Blackwell.

Ley, R. E. & D'Antonio, C. M. (1998). Exotic grass invasion alters potential rates of N fixation in Hawaiian woodlands. *Oecologia*, **113**, 179–87.

Lichter, J. (1998). Primary succession and forest development on coastal Lake Michigan sand dunes. *Ecological Monographs*, **68**, 487–510.

Lichter, J. (2000). Colonization constraints during primary succession on coastal Lake Michigan sand dunes. *Journal of Ecology*, **88**, 825–39.

Lindroth, C. H. (1965). Skaftafell, Iceland: A living glacial refugium. *Oikos*, **6**, 1–142.

Lockwood, J. L. (1997). An alternative to succession: Assembly rules offer guide to restoration efforts. *Restoration and Management Notes*, **15**, 45–50.

Lockwood, J. L. & Pimm, S. L. (1999). When does restoration succeed? In *Ecological Assembly Rules: Perspectives, Advances, Retreats*, ed. S. E. Weiher & P. Keddy, pp. 363–92. Cambridge: Cambridge University Press.

Londo, G. (1974). Successive mapping of dune slack vegetation. *Vegetatio*, **29**, 51–61.

Looney, P. B. & Gibson, D. J. (1995). The relationship between the soil seed bank and above-ground vegetation of a coastal barrier island. *Journal of Vegetation Science*, **6**, 825–36.

Lu, L. & Wu, R. S. S. (1998). Recolonization and succession of marine macroben-thos in organic-enriched sediment deposited from fish farms. *Environmental Pollution*, **101**, 241–51.

Lubchenco, J. (1983). *Littorina* and *Fucus*: Effects of herbivores, substratum heterogeneity and plant escapes during succession. *Ecology*, **64**, 1116–23.

Lugo, A. E. & Scatena, F. N. (1999). Background and catastrophic tree mortality in tropical moist, wet, and rain forests. *Biotropica*, **28**, 585–99.

Luken, J. O. 1990. *Directing Ecological Succession*. London: Chapman & Hall.

Luken, J. O. & Fonda, R. W. (1983). Nitrogen accumulation in a chronosequence of red alder communities along the Hoh River, Olympic National Park, Washington. *Canadian Journal of Forest Research*, **13**, 1228–37.

Luken, J. O. & Thieret, J. W. (1997). *Assessment and Management of Plant Invasions*. New York: Springer.

Lukešová, A. & Komárek, J. (1987). Succession of soil algae on dumps from strip coal-mining in the Most Region (Czechoslovakia). *Folia Geobotanica et Phytotaxonomica*, **22**, 355–62.

Lumini, E., Bosco, M., Puppi, G., Isopi, R., Frattegiani, M., Buresti, E. & Favilli, F. (1994). Field performance of *Alnus cordata* Loisel (Italian alder) inoculated with *Frankia* and VA-mycorrhizal strains in mine-spoil afforestation plots. *Soil Biology and Biochemistry*, **26**, 659–61.

Lundgren, L. (1978). Studies of soil and vegetation development on fresh landslide scars in the Mgeta Valley, Western Ulugura Mountains, Tanzania. *Geografiska Annaler*, **60A**, 91–127.

Lyell, C. (1850). *Principles of Geology*, 8th edition. London: John Murray.

MacArthur, R. & Wilson, E. O. (1967). *The Theory of Island Biogeography*. Princeton: Princeton University Press.

Macek, T., Mackova, M. & Kas, J. (2000). Exploitation of plants for the removal of organics in environmental remediation. *Biotechnology Advances*, **18**, 23–34.

MacDonald, I. A. W., Thebaud, C., Strahm, W. A. & Strasberg, D. (1991). Effects of alien plant invasions on native vegetation remnants on La Reunion (Mascarene Islands, Indian Ocean). *Environmental Conservation*, **18**, 51–61.

MacLeod, N. D., Brown, J. L. & Noble, J. C. (1993). Ecological and economic considerations for the management of shrub encroachment in Australian rangelands. In *Proceedings of the 10th Australian Weeds Conference and 14th Asian Pacific Weed Society Conference, 6–10. September 1993,* vol. 2, ed. J. T. Swarbrick, pp. 118–21. Brisbane, Australia: Weed Society of Queensland.

MacMahon, J. A. (1999). Disturbance in deserts. In *Ecosystems of Disturbed Ground, Ecosystems of the World 16*, ed. L. R. Walker, pp. 307–30. Amsterdam: Elsevier.

Mahall, B. E. & Callaway, R. M. (1992). Root communication mechanisms and intracommunity distributions of two Mojave Desert shrubs. *Ecology*, **73**, 2145–51.

Maheswaran, J. & Gunatilleke, I. A. U. N. (1988). Litter decomposition in a lowland rain forest and a deforested area in Sri Lanka. *Biotropica*, **20**, 90–9.

Majer, J. D. (1989a). Fauna studies and land reclamation technology – a review of the history and need for such studies. In *Animals in Primary Succession: The Role of Fauna in Reclaimed Lands*, ed. J. D. Majer, pp. 5–34. Cambridge: Cambridge University Press.

Majer, J. D. (ed.) (1989b). *Animals in Primary Succession: The Role of Fauna in Reclaimed Lands*. Cambridge: Cambridge University Press.

Majer, J. D. (1992). Ant recolonization of rehabilitated bauxite mines of Poços de Caldas, Brazil. *Journal of Tropical Ecology*, **8**, 97–108.

Majer, J. D. & Kock, A. E. de (1992). Ant recolonization of sand mines near Richards Bay, South Africa: An evaluation of progress with rehabilitation. *South African Journal of Science*, **88**, 31–6.

Malanson, G. P. (1993). *Riparian Landscapes*. Cambridge: Cambridge University Press.

Malanson, G. P. & Butler, D. R. (1991). Floristic variation among gravel bars in a subalpine river, Montana, USA. *Arctic and Alpine Research*, **23**, 273–8.

Malanson, G. P. & Cairns, D. M. (1997). Effects of dispersal, population delays, and forest fragmentation on tree migration rates. *Plant Ecology*, **131**, 67–79.

Mann, D. H., Fastie, C. L., Rowland, E. L. & Bigelow, N. H. (1995). Spruce succession, disturbance, and geomorphology on the Tanana River floodplain, Alaska. *Ecoscience*, **2**, 184–99.

Margalef, R. (1968a). *Perspectives in Ecological Theory*. Chicago: University of Chicago Press.

Margalef, R. (1968b). *Ecología*. S.A., Barcelona, Spain: Ediciones Omega.

Marks, A. F., Dickinson, K. J. M. & Fife, A. J. (1989). Forest succession on landslides in the Fiord Ecological Region, southwestern New Zealand. *New Zealand Journal of Botany*, **27**, 369–90.

Marks, P. L. & Bormann, F. H. (1972). Revegetation following forest cutting mechanisms for return to steady-state nutrient cycling. *Science*, **176**, 914–15.

Marquez, V. J. & Allen, E. B. (1996). Ineffectiveness of two annual legumes as nurse plants for establishment of *Artemisia californica* in coastal sage scrub. *Restoration Ecology*, **4**, 42–50.

Maron, J. L. & Jeffries, R. L. (1998). Restoring enriched grasslands: Effects of mowing on species richness, productivity, and nitrogen retention. *Ecological Applications*, **11**, 1088–100.

Marrs, R. H. & Bradshaw, A. D. (1993). Primary succession on man-made wastes: The importance of resource acquisition. In *Primary Succession on Land*, ed. J. Miles & D. H. Walton, pp. 221–48. Oxford: Blackwell.

Marrs, R. H., Roberts, R. D., Skeffington, R. A. & Bradshaw, A. D. (1983). Nitrogen and the development of ecosystems. In *Nitrogen as an Ecological Factor*, ed. J. A. Lee, S. McNeill & I. H. Rorison, pp. 113–36. *Symposium of the British Ecological Society*, vol. 22. Oxford: Blackwell.

Martínez, M. L., Vázquez, G. & Sánchez Colón, S. (2001). Spatial and temporal variability during primary succession on tropical coastal sand dunes. *Journal of Vegetation Science*, **12**, 361–72.

Masuzawa, T. (1985). Ecological studies on the timberline of Mount Fuji. I. Structure of plant community and soil development on the timberline. *Botanical Magazine of Tokyo*, **98**, 15–28.

Mathis, M. J. & Middleton, B. A. (1999). Simulated herbivory and vegetation dynammics in coal slurry ponds reclaimed as wetlands. *Restoration Ecology*, **7**, 392–8.

Matthews, J. A. (1992). *The Ecology of Recently-Deglaciated Terrain: A Geoecological Approach to Glacier Forelands and Primary Succession*. Cambridge: Cambridge University Press.

Matthews, J. A. (1996). Classics in physical geography revisited. *Progress in Physical Geography*, **20**, 193–203.

Matthews, J. A. (1999). Disturbance regimes and ecosystem recovery on recently-deglaciated substrates. In *Ecosystems of Disturbed Ground, Ecosystems of the World 16*, ed. L. R. Walker, pp. 17–37. Amsterdam: Elsevier.

Matthews, J. A. & Whittaker, R. J. (1987). Vegetation succession on the Storbreen glacier foreland, Jotunheimen, Norway: A review. *Arctic and Alpine Research*, **19**, 385–95.

Maun, M. A. (1994). Adaptations enhancing survival and establishment of seedlings on coastal dune systems. *Vegetatio*, **111**, 59–70.

May, V. J. (1997). Physiography of coastal cliffs. In *Dry Coastal Ecosystems*, ed. E. van der Maarel, pp. 29–41. Amsterdam: Elsevier.

McAuliffe, J. R. (1986). Herbivore-limited establishment of a Sonoran Desert tree, *Cercidium microphyllum*. *Ecology*, **67**, 276–80.

McAuliffe, J. R. (1988). Markovian dynamics in simple and complex desert plant communities. *The American Naturalist*, **131**, 459–90.

McBride, J. &. Stone, E. C. (1976). Plant succession on the sand dunes of the Monterey Peninsula, California. *American Midland Naturalist*, **96**, 118–32.

McCabe, O. M. & Otte, M. L. (2000). The wetland grass *Glyceria fluitans* for revegetation of metal mine tailings. *Wetlands*, **20**, 545–59.

McClanahan, T. R. & Wolfe, R. W. (1993). Accelerating forest succession in a fragmented landscape – the role of birds and perches. *Conservation Biology*, **7**, 279–88.

McCook, L.J. (1994). Understanding ecological community succession. *Vegetatio*, **110**, 115–47.

McCook, L. J. & Chapman, A. R. O. (1993). Community succession following massive ice-scour on a rocky intertidal shore: Recruitment, competition and predation during early, primary succession. *Marine Biology*, **115**, 565–75.

McCormick, J. (1968). Succession. *Via*, **1**, 22–35, 131–2.

McCormick, P. V., Smith, E. P. & Cairns, J. Jr. (1991). The relative importance of population versus community processes in microbial primary succession. *Hydrobiologia*, **213**, 83–98.

McCune, B. & Allen, T. F. H. (1985). Will similar forests develop on similar sites? *Canadian Journal of Botany*, **63**, 367–76.

McIntosh, R. P. (1967). The continuum concept of vegetation. *Botanical Review*, **33**, 130–87.

McIntosh, R. P. (1980). The relationship between succession and the recovery process in ecosystems. In *The Recovery Process in Damaged Ecosystems*, ed. J. Cairns Jr., pp. 11–62. Ann Arbor, Michigan: Ann Arbor Science Publishers.

McIntosh, R. P. (1985). *The Background of Ecology*. Cambridge: Cambridge University Press.

McIntosh, R. P. (1999). The succession of succession: a lexical chronology. *Bulletin of the Ecological Society of America*, **80**, 256–65.

McKee, K. L. & Faulkner, P. I. (2000). Restoration of biogeochemical function in mangrove forests. *Restoration Ecology*, **8**, 247–59.

McKendrick, J. D. (1987). Plant succession on disturbed sites, North Slope, Alaska, U.S.A. *Arctic and Alpine Research*, **19**, 554–65.

McKibben, B. (1989). *The End of Nature*. New York: Random House.

McKinstry, M. C. & Anderson, S. H. (1994). Evaluation of wetland creation and waterfowl use in conjunction with abandoned mine lands in northeast Wyoming. *Wetlands*, **14**, 284–92.

McLeod, K. W. (2000). Species selection trials and silvicultural techniques for the restoration of bottomland hardwood forests. *Ecological Engineering*, **15** (supplement), S35–S46.

McLachlan, A., Ascaray, C. & du Toit, P. (1987). Sand movement, vegetation succession and biomass spectrum in a coastal dune slack in Algoa Bay, South Africa. *Journal of Arid Environments*, **12**, 9–25.

McVaugh, R. (1943). The vegetation of the granitic flat-rocks of the southeastern United States. *Ecological Monographs*, **13**, 121–66.

Meirelles, S. T., Pivello, R. R. & Joly, C. A. (1999). The vegetation of granite rock outcrops in Rio de Janeiro, Brazil, and the need for its protection. *Environmental Conservation*, **26**, 10–20.

Melick, D. R. & Ashton, D. H. (1991). The effects of natural disturbances on warm temperate rainforests in South-eastern Australia. *Australian Journal of Botany*, **39**, 1–30.

Mellor, A. (1986). A micromorphological examination of two alpine soil chronosequences, southern Norway. *Geoderma*, **39**, 41–57.

Menges, E. S. & Waller, D. M. (1983). Plant strategies in relation to elevation and light in floodplain herbs. *The American Naturalist*, **122**, 454–73.

Messer, A. C. (1988). Regional variations in rates of pedogenesis and the influence of climatic factors on moraine chronosequences, southern Norway. *Arctic and Alpine Research*, **20**, 31–9.

Michelangeli, F. A. (2000). Species composition and species-area relationships in vegetation isolates on the summit of a sandstone mountain in southern Venezuela. *Journal of Tropical Ecology*, **16**, 69–82.

Middleton, B. A. (1995). Seed banks and species richness potential of coal slurry ponds reclaimed as wetlands. *Restoration Ecology*, **3**, 311–18.

Miles, J. (1979). *Vegetation Dynamics*. London: Chapman & Hall.

Miles, J. (1987). Vegetation succession: Past and present perceptions. In *Colonization, Succession and Stability*, ed. A. J. Gray, M. J. Crawley & P. J. Edwards, pp. 1–30. Oxford: Blackwell.

Miles, J. & Walton, D. W. H. (1993). Primary succession revisited. In *Primary Succession on Land*, ed. J. Miles & D. W. H. Walton, pp. 295–302. Oxford: Blackwell.

Miller, R. M. (1987). Mycorrhizae and succession. In *Restoration Ecology: A Synthetic Approach to Ecological Research*, ed. W. R. Jordan III, M. E. Gilpin & J. D. Aber, pp. 205–20. Cambridge: Cambridge University Press.

Mitchell, R. J., Marrs, R. H., Le Duc, M. G. & Auld, M. H. D. (1997). A study of succession on lowland heaths in Dorset, southern England: Changes in vegetation and soil chemical properties. *Journal of Applied Ecology*, **34**, 1426–44.

Mitchley, J., Buckley, G. Y. & Helliwell, D. R. (1996). Vegetation establishment on chalk marl spoil: The role of nurse grass species and fertilizer application. *Journal of Vegetation Science*, **7**, 543–8.

Mitsch, W. J. & Wilson, R. F. (1996). Improving the success of wetland creation and restoring the know-how, time, and self-design. *Ecological Applications*, **6**, 77–83.

Monastersky, R. (1996). Volcanoes under ice: recipe for a flood. *Science News*, **150**, 327.

Moreno-Casasola, P. (1986). Sand movement as a factor in the distribution of plant communities in a coastal dune system. *Vegetatio*, **65**, 67–76.

Moro, M. J., Pugnaire, F. I., Haase, P. & Puigdefábregas, J. (1997). Mechanisms of interaction between a leguminous shrub and its understory in a semi-arid environment. *Ecography*, **20**, 175–84.

Morrell, P. L., Porter, J. M. & Friar, E. A. (2000). Intercontinental dispersal: The origin of the widespread South American plant species *Gilea laciniata* (Polemoniaceae) from a rare California and Oregon coastal endemic. *Plant Systematics and Evolution*, **224**, 12–32.

Morris, M., Eveleigh, D. E., Riggs, S. C. & Tiffney, W. N. Jr. (1974). Nitrogen fixation in the bayberry (*Myrica pensylvanica*) and its role in coastal succession. *American Journal of Botany*, **61**, 867–70.

Morris, W. F. & Wood, D. M. (1989). The role of lupine in succession on Mount St. Helens: Facilitation or inhibition? *Ecology*, **70**, 697–703.

Moseley, M. E. (1999). Convergent catastrophe: Past patterns and future implications of collateral natural disasters in the Andes. In *The Angry Earth: Disaster in Anthropological Perspective*, ed. A. Oliver-Smith & S. M. Hoffman, pp. 59–72. New York: Routledge.

Motzkin, G., Wilson, P., Foster, D. R. & Allen, A. (1999). Vegetation patterns in heterogeneous landscapes, the importance of history and environment. *Journal of Vegetation Science*, **10**, 903–20.

Mrzljak, J. & Wiegleb, G. (2000). Spider colonization of former brown coal mining areas – time or structure dependent? *Landscape and Urban Planning*, **51**, 131–46.

Mueller-Dombois, D. (1985). 'Ōhi'a dieback in Hawaii: 1984 synthesis and evaluation. *Pacific Science*, **39**, 150–70.

Mueller-Dombois, D. (1986). Perspectives for an etiology of stand-level dieback. *Annual Review of Ecology and Systematics*, **17**, 221–43.

Mueller-Dombois, D. (1987a). Forest dynamics in Hawaii. *Trends in Ecology and Evolution*, **2**, 216–19.

Mueller-Dombois, D. (1987b). Natural dieback in forests. *BioScience*, **37**, 575–83.

Mueller-Dombois, D. (2000). Rain forest establishment and succession in the Hawaiian Islands. *Landscape and Urban Planning*, **51**, 147–57.

Mueller-Dombois, D. & Ellenberg, H. (1974). *Aims and Methods of Vegetation Ecology*. New York: Wiley & Sons.

Mueller-Dombois, D. & Whiteaker, L. D. (1990). Plants associated with *Myrica faya* and two other pioneer trees on a recent volcanic surface in Hawaii Volcanoes National Park. *Phytocoenologia*, **19**, 29–41.

Mullineaux, L. S., Fisher, C. R., Peterson, C. H. & Schaeffer, S. W. (2000). Tubeworm succession at hydrothermal vents: Use of biogenic cues to reduce habitat selection error? *Oecologia*, **123**, 275–84.

Munshower, F. F. (1993). *Practical Handbook of Disturbed Land Revegetation*. Boca Raton, Florida: Lewis Publishers.

Murdy, W. H., Johnson, T. M. & Wright, V. K. (1970). Competitive replacement of *Talinum mengesii* by *T. teretifolium* in granite outcrop communities of Georgia. *Botanical Gazette*, **131**, 186–92.

Muzika, R. M., Gladden, J. B. & Haddock, J. D. (1987). Structural and functional aspects of succession in southeastern floodplain forests following a major disturbance. *American Midland Naturalist*, **117**, 1–9.

Myster, R. W. (1997). Seed predation, disease and germination on landslides in Neotropical lower montane wet forest. *Journal of Vegetation Science*, **8**, 55–64.

Myster, R. W. & Fernández, D. S. (1995). Spatial gradients and patch structure on two Puerto Rican landslides. *Biotropica*, **27**, 149–59.

Nakamura, F., Yajima, T. & Kikuchi, S. (1997). Structure and composition of riparian forests with special reference to geomorphic site conditions along the Tokachi River, northern Japan. *Plant Ecology*, **133**, 209–19.

Nakamura, T. (1984). Seed dispersal on a landslide scar on the upper reaches of the Oi River, Central Japan. *Journal of the Japanese Forestry Society*, **66**, 375–9.

Nakashizuka, T., Iida, S., Suzuki, W. & Tanimoto, T. (1993). Seed dispersal and vegetation development on a debris avalanche on the Ontake volcano, Central Japan. *Journal of Vegetation Science*, **4**, 537–42.

Nakatsubo, T. (1995). Factors limiting the distribution of the annual legume *Kummerowia striata* in a river floodplain. *Ecological Research*, **10**, 179–87.

Nanson, G. C. & Beach, H. F. (1977). Forest succession and sedimentation on a meandering river floodplain, northeast British Columbia, Canada. *Journal of Biogeography*, **4**, 229–51.

Nathan, R., Safriel, U. N. & Noy-Meir, I. (2001). Field validation and sensitivity analysis of a mechanistic model for tree seed dispersal by wind. *Ecology*, **82**, 374–88.

Naveh, Z. & Lieberman, A. S. (1984). *Landscape Ecology: Theory and Application*. New York: Springer.

Nechaev, A. P. (1967). Seed regeneration of willow on the pebble-bed shoals of the Bureya River. *Lesobedenie*, **1**, 54–64. (Translated from Russian in 1971 by Environment Canada.)

Nelson, E. A., Dulohery, N. C., Kolka, R. K. & McKee, W. H. (2000). Operational restoration of the Pen Branch bottomland hardwood and swamp wetlands – the research setting. *Ecological Engineering*, Supplement **15**, S23–S33.

New, T. R. & Thornton, I. W. B. (1988). A pre-vegetation population of crickets subsisting on allochthonous aeolian debris on Anak-Krakatau. *Philosophical Transactions of the Royal Society of London*, B**322**, 481–5.

Newhall, C. G. & Punongbayan, R. S. (1997). *Fire and Mud: Eruptions and Lahars of Mount Pinatubo, Philippines*. Seattle, U.S.A.: University of Washington Press.

Newland, C. (1986). Do herbivores limit the effectiveness of artificial revegetation of sandstone soils? In *Environmental Science 1986*, Chapter 3. Australia: Biology Department, University of Wollongong. (Not seen but cited in Whelan (1989).)

Nicolson, M. & McIntosh, R. P. (2002). H. A. Gleason and the individualistic hypothesis revisited. *Bulletin of the Ecological Society of America*, **83**, 133–42.

Niederfringiger-Schlag, R. & Erschbamer, B. (2000). Germination and establishment of seedlings on a glacier foreland in the Central Alps, Austria. *Arctic, Antarctic and Alpine Research*, **32**, 270–7.

Niering, W. A., Dreyer, G. D., Egler, F. E. & Anderson, J. P. Jr. (1986). Stability of a *Viburnum lentago* shrub community after 30 years. *Bulletin of the Torrey Botanical Club*, **113**, 23–7.

Niering, W. A., Whittaker, R. H. & Lowe, C. H. (1963). The saguaro: A population in relation to environment. *Science*, **142**, 15–23.

Nilsson, C. & Wilson, S. D. (1991). Convergence in plant community structure along a disparate gradient: are lakeshores inverted mountainsides? *American Naturalist*, **137**, 774–90.

Ninkovich, D. (1976). Late Cenozoic clockwise rotation of Sumatra. *Earth and Planetary Science Letters*, **29**, 269–75.

Ninot, J. M., Herrero, P., Ferré, A. & Guàrdia, R. (2001). Effects of reclamation measures on plant colonization on lignite waste in the eastern Pyrenees, Spain. *Applied Vegetation Science*, **4**, 29–34.

Noble, I. R. (1981). Predicting successional change. In *Proceedings of the Conference on Fire Regimes and Ecosystem Properties* (Honolulu, Hawaii), pp. 278–300. United States Department of Agriculture, Forest Service, General Technical Report WO-26.

Noble, I. R. & Slatyer, R. O. (1980). The use of vital attributes to predict successional changes in plant-communities subject to recurrent disturbances. *Vegetatio*, **43**, 5–21.

Noble, M. G., Lawrence, D. B. & Streveler, G. P. (1984). *Sphagnum* invasion beneath an evergreen forest canopy in southeastern Alaska. *Bryologist*, **87**, 119–27.

Nordstrom, K. F., Lampe, R. & Vandemark, L. M. (2000). Reestablishing naturally functioning dunes on developed coasts. *Environmental Management*, **25**, 37–51.

Noyd, R. K, Pfleger, F. L. & Norland, M. R. (1996). Field responses to added organic matter, arbuscular mycorrhizal fungi, and fertilizer reclamation of taconite iron ore tailing. *Plant and Soil*, **179**, 89–97.

Odland, A. (1997). Development of vegetation in created wetlands in western Norway. *Aquatic Botany*, **59**, 45–62.

Odland, A. & del Moral, R. (2002). Thirteen years of wetland vegetation succession following a permanent drawdown, Myrkdalen Lake, Norway. *Plant Ecology*, **162**, 185–98.

Odum, E. P. (1953). *Fundamentals of Ecology*. Philadelphia: Saunders.

Odum, E. P. (1969). The strategy of ecosystem development. *Science*, **164**, 262–70.

Odum, E. P. (1971). *Fundamentals of Ecology*, 3rd edition. Philadelphia: Saunders.

Odum, E. P. (1992). Great ideas in ecology for the 1990's. *BioScience*, **42**, 542–5.

Ohkawara, K. & Higashi, S. (1994). Relative importance of ballistic and ant dispersal in two diplochorous *Viola* species (Violaceae). *Oecologia*, **100**, 135–40.

Ohsawa, M. & Yamane, M. (1988). Patterns and population dynamics in patchy communities on a maritime rock outcrop. In *Diversity and Pattern in Plant Communities*, ed. H. J. During, M. J. Werger, A. U. Wilems, & J. H. Wilems, p. 209. New York: Academic Press.

Ohtonen, R., Fritze, H., Pennanen, T., Jumpponen, A. & Trappe, J. (1999). Ecosystem properties and microbial community changes in primary succession on a glacier forefront. *Oecologia*, **119**, 239–46.

Olafsson, E. B., Peterson, C. H. & Ambrose, W. G. (1994). Does recruitment limitation structure populations and communities of macroinvertebrates in marine soft sediments? *Oceanography and Marine Biology*, **32**, 65–109.

Olff, H. & Bakker, J. P. (1991). Long-term dynamics of standing crop and species composition after cessation of fertilizer application to mown grassland. *Journal of Applied Ecology*, **28**, 1040–52.

Olff, H., De Leeuw, J., Bakker, J. P., Platerink, R. J., Van Wijnen, H. J. & De Munck, W. (1997). Vegetational succession and herbivory in a salt marsh: Changes induced by sea level rise and silt deposition along an elevation gradient. *Journal of Ecology*, **85**, 799–814.

Olff, H., Huisman, J. & van Tooren, B. F. (1993). Species dynamics and nutrient accumulation during early primary succession in coastal sand dunes. *Journal of Ecology*, **81**, 693–706.

Oliveira-Filho, A. T. de, Ratter, J. A. & Shepherd, G. J. (1990). Floristic composition and community structure of a central Brazilian gallery forest. *Flora*, **184**, 103–17.

Oliveira-Filho, A. T., Vilela, E. A., Gavilanes, M. L. & Carvalho, D. A. (1994). Effect of flooding regime and understorey bamboos on the physiognomy and tree species composition of a tropical semideciduous forest in Southeastern Brazil. *Vegetatio*, **113**, 99–124.

Oliver-Smith, A. & Hoffman, S. M. (eds) (1999). *The Angry Earth*. New York: Routledge.

Olson, J. S. (1958). Rates of succession and soil changes on southern Lake Michigan sand dunes. *Botanical Gazette*, **119**, 125–70.

Olson, J. S. (1997). Organic and physical dune building. In *Dry Coastal Ecosystems: General Aspects*, ed. E. van der Maarel, pp. 63–91. Amsterdam: Elsevier. (Written with the collaboration of E. van der Maarel.)

Onaindia, M., Albizu, I. & Amezaga, I. (2001). Effect of time on the natural regeneration of salt marsh. *Applied Vegetation Science*, **4**, 247–56.

O'Neill, R. V., DeAngelis, D. L., Waide, J. B. & Allen, T. F. H. (1986). *A Hierarchical Concept of Ecosystems*. Princeton: Princeton University Press.

Oosting, H. J. (1948). *The Study of Plant Communities*. San Francisco: Freeman.

Opperman, J. J. & Merenlender, A. M. (2000). Deer herbivory as an ecological constraint to restoration of degraded riparian corridors. *Restoration Ecology*, **8**, 41–7.

Oremus, P. A. I. & Otten, H. (1981). Factors affects growth and nodulation of *Hippophaë rhamnoides* L. ssp. *rhamnoides* in soils from two successional stages of dune formation. *Plant and Soil*, **63**, 317–31.

Øvreås, L. (2000). Population and community level approaches for analyzing microbial diversity in natural environments. *Ecology Letters*, **3**, 236–51.

Pacala, S. W., Canham, C. D. & Silander, J. A. Jr. (1993). Forest models defined by field measurements. I. The design of a northeastern forest simulator. *Canadian Journal of Forest Science*, **23**, 1980–8.

Packham, J. R. & Willis, A. J. (1997). *Ecology of Dunes, Salt Marsh and Shingle*. London: Chapman & Hall.

Paine, R. T. (1966). Food web complexity and species diversity. *The American Naturalist*, **100**, 65–75.

Paine, R. T. & Levin, S. A. (1981). Intertidal landscapes: Disturbance and dynamics of pattern. *Ecological Monographs*, **51**, 145–78.

Palaniappan, V. M., Marrs, R. H. & Bradshaw, A. D. (1979). The effect of *Lupinus arboreus* on the nitrogen status of china day wastes. *Journal of Applied Ecology*, **16**, 825–31.

Palmer, M. A., Ambrose, R. F. & Poff, N. L. (1997). Ecological theory and community restoration ecology. *Restoration Ecology*, **5**, 291–300.

Pandey, A. N. & Singh, J. S. (1985). Mechanism of ecosystem recovery: A case study from Kumaun Himalaya. *Recreation and Revegetation Research*, **3**, 271–92.

Parker, M. A. (2001). Mutualism as a constraint on invasion success for legumes and rhizobia. *Diversity and Distributions*, **7**, 125–36.

Parker, V. T., Simpson, R. L. & Leck, M. A. (1989). Pattern and process in the dynamics of seed banks. In *Ecology of Seed Banks*, ed. M. A. Leck, V. G. T. Parker & R. L. Simpson, pp. 367–84. New York: Academic Press.

Parikh, A. & Gale, N. (1998). Vegetation monitoring of created dune swale wetlands, Vandenberg Air Force Base, California. *Restoration Ecology*, **6**, 83–93.

Parmenter, R. R. & MacMahon, J. A. (1987). Early successional patterns of arthropod recolonization on reclaimed strip mines in southwestern Wyoming: The ground dwelling beetle fauna (Coleoptera). *Environmental Entomology*, **16**, 168–77.

Parmenter, R. R., MacMahon, J. A. & Gilbert, C. A. B. (1991). Early successional patterns of arthropod recolonization on reclaimed Wyoming strip mines – the grasshoppers (Orthoptera, Acrididae) and allied faunas (Orthoptera, Gryllacrididae, Tettigoniidae). *Environmental Entomology*, **20**, 135–42.

Parmenter, R. P., MacMahon, J. A., Waaland, M. E., Stuebe, M. M., Landres, P. & Crisafuli, C. (1985). Reclamation of surface coal mines in western Wyoming for wildlife habitat: A preliminary analysis. *Reclamation and Revegetation Research*, **4**, 98–115.

Parrotta, J. A. (1995). Influence of overstory composition on understory colonization by native species in plantations on a degraded tropical site. *Journal of Vegetation Science*, **6**, 627–36.

Parrotta, J. A. & Knowles, O. H. (1999). Restoration of tropical moist forests on bauxite-mined lands in the Brazilian Amazon. *Restoration Ecology*, **7**, 103–16.

Parrotta, J. A., Knowles, O. H. & Wunderle, J. M. (1997). Development of floristic diversity in 10-year old restoration forests on a bauxite mined site in Amazonia. *Forest Ecology and Management*, **99**, 21–42.

Pärtel, M., Kalamees, R., Zobel, M. & Rosen, E. (1998). Restoration of species-rich limestone grassland communities from overgrown land: The importance of propagule availability. *Ecological Engineering*, **10**, 275–86.

Partomihardjo, T., Mirmanto, E. & Whittaker, R. J. (1992). Anak Krakatau's vegetation and flora circa 1991, with observations on a decade of development and change. *GeoJournal*, **28**, 233–48.

Partridge, T. R. (1992). Successional interactions between bracken and broom on the Port Hills, Canterbury, New Zealand. *Journal of Applied Ecology*, **29**, 85–91.

Paschke, M. W., DeLeo, C. & Redente, E. F. (2000). Revegetation of roadcut slopes in Mesa Verde National Park, U.S.A. *Restoration Ecology*, **8**, 274–82.

Paton, T. R., Humphreys, G. S. & Mitchell, P. B. (1995). *Soils: A New Global View*. New Haven, Connecticut: Yale University Press.

Patten, B. C. & Jorgenson, S. E. (1995). *Complex Ecology: The Part-Whole Relation in Ecosystems*. Englewood Cliffs, New Jersey: Prentice Hall.

Pearce, A. J. & O'Loughlin, C. L. (1985). Landsliding during a M7.7 earthquake: Influence of geology and topography. *Geology*, **13**, 855–8.

Pearlstine, L., McKellar, H. & Kitchens, W. (1985). Modelling the impacts of a river diversion on bottomland forest communities in the Santee River floodplain, South Carolina. *Ecological Modelling*, **29**, 283–302.

Peet, R. K. (1992). Community structure and ecosystem function. In *Plant Succession: Theory and Prediction*, ed. S. D. C. Glenn-Lewin, R. K. Peet & T. T. Veblen, pp. 103–51. London: Chapman & Hall.

Peet, R. K. & Christensen, N. L. (1980). Succession: a population process. *Vegetatio*, **43**, 131–40.

Peloquin, R. L. & Hiebert, R. D. (1999). The effects of black locust (*Robinia pseudoacacia* L.) on species diversity and composition of black oak/savanna woodland communities. *Natural Areas Journal*, **19**, 121–31.

Pennings, S. C. & Richards, C. L. (1998). Effects of wrack burial in salt-stressed habitats: *Batis maritima* in a southwest Atlantic salt marsh. *Ecography*, **21**, 630–8.

Persson, Å. (1964). The vegetation at the margin of the receding Glacier Skaftafellsjökull, southeastern Iceland. *Botaniska Notiser*, **117**, 323–54.

Peterken, G. F. & Game, M. (1984). Historical factors affecting the number and distribution of vascular plant species in the woodlands of central Lincolnshire. *Journal of Ecology*, **72**, 155–72.

Petersen, J. (2000). *Die Dünentalvegetationen der Wattenmeer-Inseln in der südlichen Nordsee: Eine pflanzensoziologische und ökologische Vergleichsuntersuchung unter Berücksichtigung von Nutzung und Naturschutz*. Husum (Germany): Husum Druck-und Verlagsgesellschaft.

Peterson, C. G. & Stevenson, R. J. (1989). Substratum conditioning and diatom colonization in different current regimes. *Journal of Phycology*, **25**, 790–3.

Peterson, C. G. & Stevenson, R. J. (1992). Resistance and resilience of lotic algal communities: Importance of disturbance timing and current. *Ecology*, **73**, 1445–61.

Peterson, C. J. & Rebertus, A. J. (1997). Tornado damage and initial recovery in three adjacent, lowland temperate forests in Missouri. *Journal of Vegetation Science*, **8**, 559–64.

Petraitis, P. S. & Dudgeon, S. R. (1999). Experimental evidence for the origin of alternative communities on rocky intertidal shores. *Oikos*, **84**, 239–45.

Pfadenhauer, J. & Grootjans, A. (1999). Wetland restoration in Central Europe: Aims and methods. *Applied Vegetation Science*, **2**, 95–106.

Phillips, D. L. & MacMahon, J. A. (1981). Competition and spacing patterns in desert shrubs. *Journal of Ecology*, **69**, 97–115.

Phillips, J. F. V. (1934–35). Succession, development, the climax, and the complex organism. Parts I–III. *Journal of Ecology*, **22**, 554–71; **23**, 210–43, 488–508.

Pianka, E. R. (1970). On r and K selection. *The American Naturalist*, **104**, 592–7.

Pickart, A. J., Miller, L. M. & Duebendorfer, T. E. (1998). Yellow bush lupione invasion in northern California coastal dunes - I. Ecological impacts and manual restoration techniques. *Restoration Ecology*, **6**, 59–68.

Pickett, S. T. A. (1976). Succession: an evolutionary perspective. *The American Naturalist*, **110**, 107–19.

Pickett, S. T. A. (1989). Space-for-time substitutions as an alternative to long-term studies. In *Long-term Studies in Ecology*, ed. G. E. Likens, pp. 110–35. New York: Springer.

Pickett, S. T. A., Collins, S. L. & Armesto, J. J. (1987). A hierarchical consideration of causes and mechanisms of succession. *Vegetatio*, **69**, 109–14.

Pickett, S. T. A. & White, P. S. (eds) (1985). *The Ecology of Natural Disturbance and Patch Dynamics*. New York: Academic Press.

Pickett, S. T. A., Wu, J. & Cadenasso, M. L. (1999). Patch dynamics and the ecology of disturbed ground: A framework for synthesis. In *Ecosystems of Disturbed Ground*, ed. L. R. Walker, pp. 707–22. Amsterdam: Elsevier.

Piha, M. I., Vallaack, H. W., Reeler, B. M. & Michael, N. (1995a). A low-input approach to vegetation establishment on mine and coal ash wastes in semiarid regions. 1. Tin mine tailings in Zimbabwe. *Journal of Applied Ecology*, **32**, 372–81.

Piha, M. I., Vallaack, H. W., Michael, N. & Reeler, B. M. (1995b). A low-input approach to vegetation establishment on mine and coal ash wastes in semiarid regions. 2. Lagooned pulverized fuel ash in Zimbabwe. *Journal of Applied Ecology*, **32**, 382–90.

Pimentel, D. & Harvey, C. (1999). Ecological effects of erosion. In *Ecosystems of Disturbed Ground*, ed. L. R. Walker, pp. 123–35. Amsterdam: Elsevier.

Pimentel, D., Harvey, C., Resosudarmo, P., Sinclair, K., Kurz, D., McNair, D., Crist, S., Spritz, L., Fitton, L. Saffouri, L. & Blair, R. (1995). Environmental and economic costs of soil erosion and conservation benefits. *Science*, **267**, 1117–23.

Piotrowska, H. (1988). The dynamics of the dune vegetation on the Polish Baltic coast. *Vegetatio*, **77**, 169–75.

Platt, W. J. & Weis, I. M. (1977). Resource partitioning and competition within a guild of fugitive prairie plants. *The American Naturalist*, **111**, 479–513.

Poli, E. (1965). La vegetazione altomontana dell' Etna. *Memoria No. 5, Flora et Vegetatio Italica*, Rome.

Poli, E. & Grillo, M. (1975). *La colonizzazione vegetale della colata lavica etnea del 1391*. Inst. Botanico, Univ. Pavia, Series 6, vol. X, pp. 127–86.

Poli Marchese, E. & Grillo, M. (2000). Primary succession on lava flows on Mt. Etna. *Acta Phytogeographica Suecica*, **85**, 61–70.

Posada, J. M., Aide, T. M. & Cavelier, J. (2000). Cattle and weedy shrubs as restoration tools of tropical montane rainforest. *Restoration Ecology*, **8**, 370–9.

Poulin, M., Rochefort, L. & Desrochers, A. (1999). Conservation of bog plant species assemblages, assessing the role of natural remnants in mined sites. *Applied Vegetation Science*, **2**, 169–80.

Poulson, T. L. (1999). Autogenic, allogenic, and individualistic mechanisms of dune succession at Miller, Indiana. *Natural Areas Journal*, **19**, 172–6.

Poulson, T. L. & McClung, C. (1999). Anthropogenic effects on early dune succession at Miller, Indiana. *Natural Areas Journal*, **19**, 177–9.

Powell, E. A. (1992). Life history, reproductive biology, and conservation of the Mauna Kea silversword, *Argyroxiphium sandwicense* DC (Asteraceae), an endangered plant of Hawaii. Ph.D. dissertation, University of Hawaii at Manoa, Honolulu, Hawaii.

Prach, K. (1987). Succession of vegetation on dumps from strip coal mining, N. W. Bohemia, Czechoslovakia. *Folia Geobotanica et Phytotaxonomica*, **22**, 349–54.

Prach, K. (1994a). Succession of woody species in derelict sites in central Europe. *Ecological Engineering*, **3**, 49–56.

Prach, K. (1994b). Vegetation succession on river gravel bars across the northwestern Himalayas. *Arctic and Alpine Research*, **26**, 349–53.

Prach, K. & Pyšek, P. (1994). Spontaneous establishment of woody plants in Central Europe derelict sites and their potential for reclamation. *Restoration Ecology*, **2**, 190–7.

Prach, K. & Pyšek, P. (1999). How do species dominating in succession differ from others? *Journal of Vegetation Science*, **10**, 383–92.

Prach, K., Pyšek, P. & Smilauer, P. (1993). On the rate of succession. *Oikos*, **66**, 343–6.

Prach, K., Pyšek, P. & Smilauer, P. (1997). Changes in species traits during succession: A search for pattern. *Oikos*, **79**, 201–5.

Prach, K., Pyšek, P. & Smilauer, P. (1999). Prediction of vegetation succession in human-disturbance habitats using an expert system. *Restoration Ecology*, **7**, 15–23.

Prentice, C. (1992). Climate change and long-term vegetation dynamics. In *Plant Succession: Theory and Prediction*, ed. D. C. Glenn-Lewin, R. K. Peet & T. T. Veblen, pp. 293–39. London: Chapman & Hall.

Prentice, I. C. & Werger, M. J. A. (1985). Clump spacing in a dwarf desert shrub community. *Vegetatio*, **63**, 133–9.

Puerto, A. & Rico, M. (1994). Differences between oligotrophic communities resulting from old-field succession in relation to bedrock. *Vegetatio*, **113**, 83–92.

Pugnaire, F. L., Haase, P. & Puigdefábregas, J. (1996). Facilitation between higher plant species in a semiarid environment. *Ecology*, **77**, 1420–26.

Pugnaire, F. I. & Valladares, F. (eds) (1999). *Handbook of Functional Plant Ecology*. New York: M. Dekker Publishers.

Pyšek, P. (1992). Dominant species exchange during succession in reclaimed habitats – a case study from areas deforested by air pollution. *Forest Ecology and Management*, **54**, 27–44.

Qadir, M., Qureshi, R. H., Ahmad, N. & Ilyas, M. (1996). Salt-tolerant forage cultivation on a saline-sodic field for biomass production and soil reclamation. *Land Degradation and Development*, **7**, 11–18.

Rabinowitz, D. & Rapp, J. K. (1980). Seed rain in a North American tall grass prairie. *Journal of Applied Ecology*, **17**, 793–802.

Raich, J. W., Russell, A. E. & Vitousek, P. M. (1997). Primary productivity and ecosystem development along an elevational gradient on Mauna Loa, Hawai'i. *Ecology*, **78**, 707–21.

Rajan, S. R. (1999). Bhopal: Vulnerability, routinization, and the chronic disaster. In *The Angry Earth*, ed. A. Oliver-Smith & S. M. Hoffman, pp. 257–77. New York: Routledge.

Ramsey, D. S. L. & Wilson, J. C. (1997). The impact of grazing by macropods on coastal foredune vegetation in southwest Queensland. *Australian Journal of Ecology*, **22**, 288–97.

Ranwell, D. (1960). Newborough Warren, Anglesey. II. Plant associes and successional cycles of the sand dune and dune slack vegetation. *Journal of Ecology*, **48**, 117–41.

Rathcke, B. (1983). Competition and facilitation among plants for pollination. In *Pollination Biology*, ed. L. Real, pp. 305–29. New York: Academic Press.

Raunkiaer, C. (1937). *Plant Life Forms*. Oxford: Clarendon.

Raup, H. M. (1971). The vegetational relations of weathering, frost action and patterned ground processes. *Meddelelser om Grønland*, **194**, 1–92.

Raup, H. M. (1981). Physical disturbance in the life of plants. In *Biotic Crises in Ecological and Evolutionary Time*, ed. M. H. Nitecki, pp. 39–52. New York: Academic Press.

Rawlinson, H., Dickinson, N. & Putwain, P. D. (2000). The establishment of trees on closed landfill sites for community forest use. In *Spontaneous Succession in Ecosystem Restoration (Symposium abstracts)*, ed. K. Prach, J. Müllerová & P. Pyšek, p. 15. Czech Republic: České Budějovica.

Rebele, F. (1992). Colonization and early succession on anthropogenic soils. *Journal of Vegetation Science*, **3**, 201–8.

Rebele, F. (1994). Urban ecology and special features of urban ecosystems. *Global Ecology and Biogeography Letters*, **4**, 173–87.

Rebele, F., Surma, A., Kuznik, C., Bornkamm, R. & Brej, T. (1993). Heavy metal contamination of spontaneous vegetation and soil around the copper smelter Legnica. *Acta Societatis Botanicorum Poloniae*, **62**, 53–7.

Recher, H. F. (1989). Colonization of reclaimed land by animals: An ecologist's overview. In *Animals in Primary Succession*, ed. J. D. Majer, pp. 441–8. Cambridge: Cambridge University Press.

Reddell, P., Gordon, V. & Hopkins, M. S. (1999). Ectomycorrhizas in *Eucalyptus tetrodonta* and *E. miniata* forest communities in tropical northern Australia and their role in the rehabilitation of these forests following mining. *Australian Journal of Botany*, **47**, 881–907.

Rees, M. & Bergelson, J. (1997). Asymmetric light competition and founder control in plant communities. *Journal of Theoretical Biology*, **184**, 353–8.

Reid, H. F. & Taber, S. (1919). The Puerto Rican earthquakes of October–November 1918. *Bulletin of the Seismological Society of America*, **9**, 95–127.

Reinartz, J. A. & Warne, E. L. (1993). Development of vegetation in small created wetlands in southeastern Wisconsin. *Wetlands*, **13**, 153–64.

Reiners, W. A., Worley, I. A. & Lawrence, D. B. (1971). Plant diversity in a chronosequence at Glacier Bay, Alaska. *Ecology*, **52**, 55–69.

Reissek, S. (1856). Vortrag ueber die Bildungsgeschichte der Donauinseln im mittleren Laufe dieses Stromes. *Flora*, **39**, 609–24. (Not seen but cited in Clements (1928).)

Rey, P. J. & Alcántara, J. M. (2000). Recruitment dynamics of a fleshy-fruited plant (*Olea europaea*): Connecting patterns of seed dispersal to seedling establishment. *Journal of Ecology*, **88**, 622–33.

Rice, E. L. (1984). *Allelopathy*, 2nd edition. New York: Academic Press.

Rice, R. J. (1988). *Fundamentals of Geomorphology*, 2nd edition. Harlow: Longman.

Richards, E. N. & Goff, M. L. (1997). Arthropod succession on exposed carrion in three contrasting tropical habitats on Hawaii Island, Hawaii. *Journal of Medical Entomology*, **34**, 328–39.

Richards, P. W. (1952). *The Tropical Rain Forest*. Cambridge: Cambridge University Press.

Rietkerk, M., van den Bosch, F. & van de Koppel, J. (1997). Site-specific properties and irreversible vegetation changes in semi-arid grazing systems. *Oikos*, **80**, 241–52.

Riley, R. H. & Vitousek, P. M. (1995). Nutrient dynamics and nitrogen trace gas flux during ecosystem development in montane rain forest. *Ecology*, **76**, 292–304.

Ritz, K., Dighton, J. & Giller, K. E. (eds) (1994). *Beyond the Biomass: Compositional and Functional Analysis of Soil Microbial Communities*. Chichester, New York: Wiley.

Roberts, T. L. & Vankat, J. L. (1991). Floristics of a chronosequence corresponding to old field-deciduous forest succession in southwestern Ohio. II. Seed banks. *Bulletin of the Torrey Botanical Club*, **118**, 377–85.

Robertson, G. P. (1987). Geostatistics in ecology: Interpolating with known variance. *Ecology*, **68**, 744–8.

Robertson, G. P. & Vitousek, P. M. (1981). Nitrification potentials in primary and secondary succession. *Ecology*, **62**, 376–86.

Robertson, J. M. & Augspurger, C. K. (1999). Geomorphic processes and spatial patterns of primary forest succession on the Bogue Chitto River, USA. *Journal of Ecology*, **87**, 1052–63.

Rochefort, L. & Bastien, D. F. (1998). Reintroduction of *Sphagnum* into harvested peatlands: Evaluation of various methods for protection against desiccation. *Ecoscience*, **5**, 117–27.

Roozen, J. M. & Westhoff, V. (1985). A study of long-term salt-marsh succession using permanent plots. *Vegetatio*, **61**, 23–32.

Rosales, J., Cuenca, G., Ramirex, N. & DeAndrade, Z. (1997). Native colonizing species and degraded land restoration in La Gran Sabana, Venezuela. *Restoration Ecology*, **5**, 147–55.

Rose, R. J., Webb, N. R., Clarke, R. T. & Traynor, C. H. (2000). Changes on the heathlands in Dorset, England, between 1987 and 1996. *Biological Conservation*, **93**, 117–25.

Ross, C. (1999). Mine revegetation in Nevada: the state of the art in the arid zone. In *Closure, Remediation and Management of Precious Metal Heap Leach Facilities*, ed. D. Kosich & G. Miller, pp. 71–6. Reno, Nevada: Center for Environmental Sciences and Engineering, University of Nevada.

Rossow, L., Bryant, J. P. & Kielland, K. (1997). Effects of above-ground browsing by mammals on mycorrhizal colonization in an early successional taiga ecosystem. *Oecologia*, **110**, 94–8.

Roxburgh, S. H., Wilson, J. B. & Mark, A. F. (1988). Succession after disturbance of a New Zealand high-alpine cushionfield. *Arctic and Alpine Research*, **20**, 230–6.

Russell, A. E., Raich, J. W. & Vitousek, P. M. (1998). The ecology of the climbing fern *Dicranopteris linearis* on windward Mauna Loa, Hawaii. *Journal of Ecology*, **86**, 765–79.

Russell, A. E., Ranker, T. A., Gemmill, C. E. C. & Farrar, D. R. (1999). Patterns of clonal diversity in *Dicranopteris linearis* on Mauna Loa, Hawaii. *Biotropica*, **31**, 449–59.

Russell, A. E. & Vitousek, P. M. (1997). Decomposition and potential nitrogen fixation in *Dicranopteris linearis* litter on Mauna Loa, Hawai'i. *Journal of Tropical Ecology*, **13**, 579–94.

Russell, W. B. & Laroi, G. H. (1986). Natural vegetation and ecology of abandoned coal mined land, Rocky Mountain foothills, Alberta, Canada. *Canadian Journal of Botany*, **64**, 1286–98.

Rychert, R., Skujins, J. Sorensen, D. & Porcella, D. (1978). Nitrogen fixation by lichens and free-living microorganisms in deserts. In *Nitrogen in Desert Ecosystems*, ed. N. E. West & J. Skujins, pp. 20–33. Stroudsburg, Pennsylvania, USA: Dowden, Hutchinson & Ross.

Rydin, H. & Borgegård, S.-O. (1988a). Plant species richness on islands over a century of primary succession: Lake Hjälmaren. *Ecology*, **69**, 916–27.

Rydin, H. & Borgegård, S.-O. (1988b). Primary succession over sixty years on hundred-year old islets in Lake Hjälmaren, Sweden. *Vegetatio*, **77**, 159–68.

Rydin, H. & Borgegård, S.-O. (1991). Plant characteristics over a century of primary succession on islands, Lake Hjälmaren. *Ecology*, **72**, 1089–101.

Ryvarden, L. (1971). Studies in seed dispersal. I. Trapping of diaspores in the alpine zone at Finse, Norway. *Norwegian Journal of Botany*, **18**, 215–26.

Ryvarden, L. (1975). Studies in seed dispersal. II. Winter-dispersed species at Finse, Norway. *Norwegian Journal of Botany*, **22**, 21–4.

Sader, S. A. (1995). Spatial characteristics of forest clearing and vegetation regrowth as detected by landsat thematic mapper imagery. *Photogrammetric Engineering and Remote Sensing*, **18**, 215–26.

Salonen, V., Penttinen, A. & Sarkka, A. (1992). Plant colonization of a bare peat surface, population changes and spatial patterns. *Journal of Vegetation Science*, **3**, 113–18.

Sampat, P. (2000). *Deep Trouble: The Hidden Threat of Groundwater Pollution*. Paper No. 154, World Watch Institute, Washington, D. C.

Samuels, C. L. & Drake, J. A. (1997). Divergent perspectives on community convergence. *Trends in Ecology and Evolution*, **12**, 427–32.

Sande, E. & Young, D. R. (1992). Effect of sodium chloride on growth and nitrogenase activity in seedlings of *Myrica cerifera* L. *New Phytologist*, **120**, 345–50.

Santas, R., Lianou, C. & Danielidis, D. (1997). UVB radiation and depth interaction during primary succession of marine diatom assemblages of Greece. *Limnology and Oceanography*, **42**, 986–91.

Sarrazin, J., Robigou, V., Juniper, S. K. & Delaney, J. (1997). Biological and geological dynamics over four years on a high-temperature sulfide structure at the Juan de Fuca Ridge hydrothermal observatory. *Marine Ecology Progress Series*, **153**, 5–24.

Savage, M., Sawhill, B. & Askenazi, M. (2000). Community dynamics: what happens when we rerun the tape? *Journal of Theoretical Biology*, **205**, 515–26.

Scarth, A. (1999). *Vulcan's Fury: Man Against the Volcano*. New Haven, Connecticut: Yale University Press.

Scheidegger, A. E. (1997). Complexity theory of natural disasters; boundaries of self-structured domains. *Natural Hazards*, **16**, 103–12.

Schimel, J. P., Cates, R. G. & Ruess, R. (1998). The role of balsam poplar secondary chemicals in controlling soil nutrient dynamics through succession in the Alaskan taiga. *Biogeochemistry*, **42**, 221–34.

Schimel, J. P., Van Cleve, K., Cates, R. G., Clausen, T. P. & Reichardt, P. B. (1996). Effects of balsam poplar (*Populus balsamifera*) tannins and low molecular weight phenolics on microbial activity in taiga floodplain soil: Implications for changes in N cycling during succession. *Canadian Journal of Botany*, **74**, 84–90.

Schipper, L. A., Degens, B. P., Sparlling, G. P. & Duncan, L. (2001). Changes in microbial heterotrophic diversity along five plant successional sequences. *Soil Biology and Biochemistry*, **33**, 2093–103.

Schlesinger, W. H., Bruijnzeel, L. A., Bush, M. B., Klein, E. M., Mace, K. A., Raikes, J. A. & Whittaker, R. J. (1998). The biogeochemistry of phosphorus after the first century of soil development on Rakata Island, Krakatau, Indonesia. *Biogeochemistry*, **40**, 37–55.

Schlesinger, W. H., Raikes, J. A., Hartley, A. E. & Cross, A. E. (1996). On the spatial pattern of soil nutrients in desert ecosystems. *Ecology*, **77**, 364–74.

Schmitt, S. F. & Whittaker, R. J. (1998). Disturbance and succession on the Krakatau Islands, Indonesia. In *Dynamics of Tropical Communities*, ed. D. M. Newberry, H. H. T. Prins & N. D. Brown, pp. 515–48. Symposium of the British Ecological Society, vol. 37. Oxford: Blackwell Science.

Schoenike, R. E. (1958). Influence of mountain avens (*Dryas drummondii*) on growth of young cottonwoods (*Populus trichocarpa*) at Glacier Bay, Alaska. *Proceedings of the Minnesota Academy of Science*, **25–26**, 55–8.

Schoenly, K. & Reid, W. (1987). Dynamics of heterotrophic succession in carrion arthropod assemblages: Discrete seres or a continuum of change? *Oecologia*, **73**, 192–202.

Schramm, J. R. (1966). Plant colonization studies on black wastes from anthracite mining in Pennsylvania. *Transactions of the American Philosophical Society*, **56**, 1–194.

Schubiger-Bossard, C. M. (1988). Die Vegetation des Rhonegletscher-vorfeldes, ihre Sukzession und naturráumliche Gliederung. *Beitragen Geobotanisch Landesaufnahme Schweiz*, **64**, 1–228.

Schuller, D., Brunken-Winkler, H., Busch, P., Forster, M., Janiesch, P., von Lemm, R., Niedringhaus, R. & Strasser, H. (2000). Sustainable land use in an agriculturally misused landscape in northwest Germany through ecotechnical restoration by a "Patch Network Concept". *Ecological Engineering*, **16**, 99–117.

Schuman, G. E. Booth, D. T. & Cockrell, J. R. (1998). Cultural methods for establishing Wyoming big sagebrush on mined lands. *Journal of Range Management*, **51**, 223–30.

Schuster, W. S. & Hutnick, R. J. (1987). Community development on 35-year-old planted minespoil banks in Pennsylvania. *Reclamation and Revegetation Research*, **6**, 109–20.

Shaffer, G. P., Sasser, C. E., Gosselink, J. G. & Rejmánek, M. (1992). Vegetation dynamics in the emerging Atchafalaya Delta, Louisiana, USA. *Journal of Ecology*, **80**, 677–87.

Shao, G. F., Shugart, H. H. & Hayden, B. P. (1996). Functional classifications of coastal barrier island vegetation. *Journal of Vegetation Science*, **7**, 391–6.

Sharitz, R. R. & McCormick, J. F. (1973). Population dynamics of two competing annual plant species. *Ecology*, **54**, 723–40.

Sharma, D. P., Singh, K. & Rao, R. V. G. K. (2000). Subsurface drainage for rehabilitation of waterlogged saline lands: Example of a soil in semiarid climate. *Arid Soil Research and Rehabilitation*, **14**, 373–86.

Sharpe, C. F. S. (1960). *Landslides and Related Phenomena: A Study of Mass-movements of Soil and Rock*. New Jersey: Pageant Books.

Sheets, P. D. (1999). The effects of explosive volcanism on ancient egalitarian, ranked, and stratified societies in Middle America. In *The Angry Earth*, ed. A. Oliver-Smith & S. M. Hoffman, pp. 36–58. New York: Routledge.

Shi, Y. L., Allis, R. & Davey, F. (1996). Thermal modeling of the Southern Alps, New Zealand. *Pure and Applied Geophysics*, **146**, 469–501.

Shilton, L. A., Altringham, J. D., Compton, S. G. & Whittaker, R. J. (1999). Old world fruit bats can be long-distance seed dispersers through extended retention of viable seeds in the gut. *Proceedings of the Royal Society of London*, B**266**, 219–23.

Shroder, J. F. Jr., Scheppy, R. A. & Bishop, M. P. (1999). Denudation of small alpine basins, Nanga Parbat Himalaya, Pakistan. *Arctic, Antarctic and Alpine Research*, **31**, 121–7.

Shugart, H. H. & West, D. C. (1980). Forest succession models. *BioScience*, **30**, 308–13.

Shumway, S. W. (2000). Facilitative effects of a sand dune shrub on species growing beneath the shrub canopy. *Oecologia*, **124**, 138–48.

Shure, D. J. (1999). Granite outcrops of the southeastern United States. In *Savannas, Barrens, and Rock Outcrop Plant Communities of North America*, ed. R. C. Anderson, J. S. Fralish & J. M. Baskin, pp. 99–118. Cambridge: Cambridge University Press.

Shure, D. J., Gottschalk, M. R. & Parsons, K. A. (1986). Litter decomposition processes in a floodplain forest. *American Midland Naturalist*, **115**, 314–27.

Shure, D. J. & Ragsdale, H. L. (1977). Patterns of primary succession on granite outcrop surfaces. *Ecology*, **58**, 993–1006.

Sidle, R. C., Pearce, A. J. & O'Loughlin, C. L. (1985). *Hillslope Stability and Land Use*. Washington, D. C.: American Geophysical Union.

Silvertown, J. & Antonovics, J. (2001). *Integrating Ecology and Evolution in a Spatial Context*. Oxford: Blackwell.

Silvester, W. B. (1989). Molybdenum limitation of asymbiotic nitrogen fixation in forests of Pacific Northwest America (1989). *Soil Biology and Biochemistry*, **21**, 283–9.

Simmons, E. (1999). Restoration of landfill sites for ecological diversity. *Waste Management and Research*, **17**, 511–19.

Simonett, D. S. (1967). Landslide distribution and earthquakes in the Bewani and Torricelli Mountains, New Guinea. In *Landform Studies from Australia and New Guinea*, ed. J. N. Jennings & J. A. Mabbutt, pp. 64–84. Canberra: Australian National University Press.

Singh, K. P. Singh, P. K. & Tripathi, S. K. (1999). Litterfall, litter decomposition and nutrient release patterns in four native tree species raised on coal mine spoil at Singrauli, India. *Biology and Fertility of Soils*, **29**, 371–8.

Sistani, K. R., Mays, D. A. & Taylor, R. W. (1995). Biogeochemical characteristics of wetlands developed after strip mining for coal. *Communications in Soil Science and Plant Analysis*, **26**, 3221–9.

Sival, F. P. (1996). Mesotrophic basiphilous communities affected by changes in soil properties in two dune slack chronosequences. *Acta Botanica Neerlandica*, **45**, 95–106.

Sival, F. P. & Grootjans, A. P. (1996). Dynamics of seasonal bicarbonate supply in a dune slack: Effects on organic matter, nitrogen pool and vegetation succession. *Vegetatio*, **126**, 39–50.

Skaggs, J. M. (1995). *The Great Guano Rush: Entrepreneurs and American Overseas Expansion*. New York: St. Martin's Press.

Smale, M. C. (1990). Ecological role of buddleia (*Buddleja davidii*) in streambeds in Te Urewera National Park. *New Zealand Journal of Ecology*, **14**, 1–6.

Smale, M. C., Hall, G. M. J. & Gardner, R. O. (1995). Dynamics of kanuka (*Kunzea ericoides*) forest on South Kaipara spit, New Zealand, and the impact of fallow deer (*Dama dama*). *New Zealand Journal of Ecology*, **19**, 131–41.

Smale, M. C., McLeod, M. & Smale, P. N. (1997). Vegetation and soil recovery on shallow landslide scars in tertiary hill country, East Cape region, New Zealand. *New Zealand Journal of Ecology*, **21**, 31–41.

Smathers, G. A. & Gardner, D. E. (1979). Stand analysis of an invading firetree (*Myrica faya* Aiton) population, Hawaii. *Pacific Science*, **33**, 239–55.

Smathers, G. A. & Mueller-Dombois, D. (1974). Invasion and recovery of vegetation after a volcanic eruption in Hawaii. Ecosystems IRP/IBP Hawaii. University of Hawaii, Technical Report 10.

Smith, R. & Olff, H. (1998). Woody species colonization in relationship to habitat productivity. *Plant Ecology*, **139**, 203–9.

Smith, R. B., Commandeur, P. R. & Ryan, M. W. (1986). *Soils, vegetation and forest growth on landslides and surrounding logged and old-growth areas on the Queen Charlotte Islands*. Victoria, British Columbia: Canadian Forestry Service.

Smith, R. I. L. (1993). The role of bryophyte propagule banks in primary succession: Case study of an Antarctic fellfield soil. In *Primary Succession on Land*, ed. J. Miles & D. W. H. Walton, pp. 55–78. Oxford: Blackwell.

Smith, S. D., Huxman, T. E., Zitzer, S. F., Charlet, T. N., Housman, D. C., Coleman, J. S., Fenstermaker, L. K., Seemann, J. R. & Nowak, R. S. (2000). Elevated CO_2 increases productivity and invasive species success in an arid ecosystem. *Nature*, **408**, 79–82.

Smith, S. D., Monson, R. K. & Anderson, J. E. (1997). *Physiological Ecology of North American Desert Plants*. New York: Springer.

Smith, S. D., Smith, W. E. & Pattern, D. T. (1987). Effects of artificially imposed shade on a Sonoran Desert ecosystem: Arthropod and soil chemistry resonses. *Journal of Arid Environments*, **13**, 245–57.

Smyth, C. R. (1997). Early succession patterns with a native species seed mix on amended and unamended coal mine spoil in the Rocky Mountains of Southeastern British Columbia, Canada. *Arctic and Alpine Research*, **29**, 184–95.

Sojka, R. E. (1999). Physical aspects of soils of disturbed ground. In *Ecosystems of Disturbed Ground*, ed. L. R. Walker, pp. 503–19. Amsterdam: Elsevier

Sollins, P., Spycher, G. & Topik, C. (1983). Processes of soil organic-matter accretion at a mudflow chronosequence, Mt. Shasta, California. *Ecology*, **64**, 1273–82.

Sollin, P., Spycher, G. & Glassman, C. A. (1984). New nitrogen mineralization from light and heavy fraction of forest soil organic matter. *Soil Biology and Biochemistry*, **16**, 31–7.

Sommerville, P., Mark, A. F. & Wilson, J. B. (1982). Plant succession on moraines of the upper Dart Valley, southern South Island, New Zealand. *New Zealand Journal of Botany*, **20**, 227–44.

Sousa, W. P. (1979). Experimental investigation of disturbance and ecological succession in a rocky intertidal algal community. *Ecological Monographs*, **49**, 227–54.

Sousa, W. P. (1984). Intertidal mosaics: Patch size, propagule availability and spatially variable patterns of succession. *Ecology*, **65**, 1918–35.

Sousa, W. P. (1985). Disturbance and patch dynamics on rocky intertidal shores. In *The Ecology of Natural Disturbance and Patch Dynamics*, ed. S. T. A. Pickett & P. S. White, pp. 101–24. New York: Academic Press.

Southwick, C. H. (1996). *Global Ecology in Human Perspective*. New York: Oxford University Press.

Specht, R. L. (1997). Ecosystem dynamics in coastal dunes of eastern Australia. In *Dry Coastal Ecosystems: General Aspects*, ed. E. van der Maarel, pp. 483–95. Amsterdam: Elsevier.

Spellerberg, I. (1998). Ecological effects of roads and traffic: A literature review. *Global Ecology and Biogeography Letters*, **7**, 317–33.

Spiller, D. A., Losos, J. B. & Schoener, T. W. (1998). Impact of a catastrophic hurricane on island populations. *Science*, **281**, 695–7.

Sprent, J. I. (1987). *The Ecology of the Nitrogen Cycle*. Cambridge: Cambridge University Press.

Sprent, J. I. & Silvester, W. B. (1973). Nitrogen fixation by *Lupinus arboreus* growth in the open and under different aged stands of *Pinus radiata*. *New Phytologist*, **72**, 991–1003.

Sprent, J. I. & Sprent, P. (1990). *Nitrogen Fixing Organisms: Pure and Applied Aspects*. London: Chapman & Hall.

Stachowicz, J. J. (2001). Mutualism, facilitation, and the structure of ecological communities. *BioScience*, **51**, 235–46.

Standen, V. & Owen, M. J. (1999). An evaluation of the use of translocated blanket bog vegetation for heathland restoration. *Applied Vegetation Science*, **2**, 181–8.

Stevens, P. R. & Walker, T. W. (1970). The chronosequence concept and soil formation. *Quarterly Review of Biology*, **45**, 333–50.

Stevenson, R. J. (1983). Effects of current and conditions simulating autogenically changing microhabitats on benthic diatom immigration. *Ecology*, **64**, 1514–24.

Stocklin, J. & Baumler, E. (1996). Seed rain, seedling establishment and clonal growth strategies on a glacier foreland. *Journal of Vegetation Science*, **7**, 45–56.

Stone, C. P. (1984). Alien animals in Hawai'i's native ecosystems: Toward controlling the adverse effects of introduced vertebrates. In *Hawai'i's Terrestrial Ecosystems: Preservation and Management*, ed. D. B. Stone and J. M. Scott, Cooperative National Park Resources Studies Unit, pp. 251–97. Honolulu: University of Hawaii.

Stone, G. N., Willmer, P. & Rowe, J. A. (1998). Partitioning of pollinators during flowering in an African Acacia community. *Ecology*, **79**, 2808–27.

Strykstra, R. J., Bekker, R. M. & Bakker, J. P. (1998). Assessment of dispersal availability: Its practical use in restoration management. *Acta Botanica Neerlandica*, **47**, 57–70.

Sugg, P. M. & Edwards, J. S. (1998). Pioneer aeolian community development on pyroclastic flows after the eruption of Mount St. Helens, Washington, U.S.A. *Arctic and Alpine Research*, **30**, 400–7.

Sukopp, H., Hejny, S. & Kowarik, I. (eds) (1990). *Urban Ecology*. The Hague: SPB Academic.

Sukopp, H. & Starfinger, U. (1999). Disturbance in urban ecosystems. In *Ecosystems of Disturbed Ground*, ed. L. R. Walker, pp. 397–412. Amsterdam: Elsevier.

Svitil, K. A. (2000). Bugs in space. *Discover*, **21**, 17.

Swift, M. J., Heal, O. W. & Andersen, J. M. (1979). *Decomposition in Terrestrial Ecosystems*. Studies in Ecology, vol. 5. Oxford: Blackwell.

Sydnor, R. S. & Redente, E. F. (2000). Long-term plant community development on topsoil treatments overlying a phytotoxic growth medium. *Journal of Environmental Quality*, **29**, 1778–86.

Syers, J. K. & Walker, T. W. (1969). Phosphorus transformations in a chronosequence of soils developed on wind-blown sand in New Zealand. *Journal of Soil Science*, **20**, 57–64.

Sykes, M. T. & Wilson, J. B. (1990). An experimental investigation into the response of New Zealand sand dune species to different depths of burial by sand. *Acta Botanica Neerlandica*, **39**, 171–81.

Tagawa, H. (1964). A study of volcanic vegetation in Sakurajima, southwest Japan. I. Dynamics of vegetation. *Memoirs, Faculty of Science, Kyushu University, Series E (Biology)*, **3**, 165–228.

Tagawa, H. (1992). Primary succession and the effect of first arrivals on subsequent development of forest types. *GeoJournal*, **28**, 175–83.

Tagawa, H., Suzuki, E., Partomikhardio, T. & Suriadarma, A. (1985). Vegetation and succession on the Krakatau Islands, Indonesia. *Vegetatio*, **60**, 131–45.

Tang, S. M., Franklin, J. F. & Montgomery, D. R. (1997). Forest harvest patterns and landscape disturance processes. *Landscape Ecology*, **12**, 349–63.

Tansley, A. G. (1920). The classification of vegetation and the concept of development. *Journal of Ecology*, **8**, 118–48.

Tansley, A. G. (1935). The use and abuse of vegetational concepts and terms. *Ecology*, **16**, 284–307.

Taylor, B. R. & Parkinson, D. (1988). Does repeated wetting and drying accelerate decay of leaf litter? *Soil Biology and Biochemistry*, **20**, 647–56.

Taylor, D. R., Aarssen, L. W. & Loehle, C. (1990). On the relationship between r/K selection and environmental carrying capacity: A new habitat templet for plant life history strategies. *Oikos*, **58**, 239–50.

Taylor, N. (1912). Some modern trends in ecology. *Torreya*, **12**, 110–17.

Tekle, K. & Bekele, T. (2000). The role of soil seed banks in the rehabilitation of degraded hill slopes in southern Wello, Ethiopia. *Biotropica*, **32**, 23–32.

Tezuka, Y. (1961). Development of vegetation in relation to soil formation in the volcanic island of Oshima, Izu, Japan. *Japan Journal of Botany*, **17**, 371–402.

Thiebaut, G. & Muller, S. (1995). Plant communities sequences in relation to eutrophication in weakly mineralized streams in the Northern Vosges. *Acta Botanica Gallica*, **142**, 627–38.

Thompson, C. H. (1983). Development and weathering of large parabolic dune systems along the subtropical coast of eastern Australia. *Zeitschrift für Geomorphologie* N.F., **45**, 205–25.

Thompson, C. H. (1992). Genesis of podzols on coastal dunes in southern Queensland. I. Field relationships and profile morphology. *Australian Journal of Soil Research*, **30**, 593–613.

Thompson, J. N., Reichman, O. J., Morin, P. J., Polis, G. A., Power, M. E., Sterner, R. W., Couch, C. A., Gough, L., Holt, R., Hooper, D. U., Keesing, F., Lovell, C. R., Milne, B. T., Molles, M. C., Roberts, D. W. & Strauss, S. Y. (2001). Frontiers of ecology. *BioScience*, **51**, 15–24.

Thompson, K. (1978). The occurrence of buried viable seeds in relation to environmental gradients. *Journal of Biogeography*, **5**, 425–30.

Thompson, K. (1987). The resource ratio hypothesis and the meaning of competition. *Functional Ecology*, **1**, 297–303.

Thoreau, H. D. (1860). Succession of forest trees. Massachusetts Board of Agriculture Eighth Annual Report. Boston: Wm. White Printer. (Not seen but cited in McIntosh (1999).)

Thornton, I. (1996). *Krakatau: The Destruction and Reassembly of an Island Ecosystem.* Cambridge, Massachusetts: Harvard University Press.

Thornton, I. W. B., New, T. R., McLaren, D. A., Sudarman, H. K. & Vaughan, P. J. (1988). Air-borne arthropod fallout on Anak Krakatau and a possible pre-vegetation pioneer community. *Philosophical Transactions of the Royal Society of London*, B**322**, 471–9.

Thornton, I. W. B., Cook, S., Edwards, J. S., Harrison, R. D., Schipper, C., Shanahan, M., Singadan, R. & Yamuna, R. (2001). Colonization of an island volcano, Long Island, Papua New Guinea, and an emergent island, Motmot, in its caldera lake. VII. Overview and discussion. *Journal of Biogeography*, **28**, 1389–408.

Tielbörger, K. (1997). The vegetation of linear desert dunes in the northwestern Negev, Israel. *Flora*, **192**, 261–78.

Tielbörger, K. & Kadmon, R. (2000). Temporal environmental variation tips the balance between facilitation and interference in desert plants. *Ecology*, **81**, 1544–53.

Tiffney, W. N. Jr. & Barrera, J. F. (1979). Comparative growth of pitch and Japanese black pine in clumps of the N_2-fixing shrub, bayberry. *Botanical Gazette*, **140**, S108–9.

Tikka, P. M., Hogmander, H. & Koski, P. S. (2001). Road and railway verges serve as dispersal corridors for grassland plants. *Landscape Ecology*, **16**, 659–66.

Tilman, D. (1985). The resource-ratio hypothesis of plant succession. *The American Naturalist*, **125**, 827–52.

Tilman, D. (1987). The importance of the mechanisms of interspecific competition. *The American Naturalist*, **129**, 769–74.

Tilman, D. (1988). *Plant Strategies and the Dynamics and Structure of Plant Communities.* Princeton: Princeton University Press.

Tilman, D. (1993). Species richness of experimental productivity gradients: How important is colonization limitation? *Ecology*, **74**, 2179–91.

Tilman, D. (1994). Competition and biodiversity in spatially structured habitats. *Ecology*, **75**, 2–16.

Tilman, D. & Kareiva, P. (1997). *Spatial Ecology: The Role of Space in Population Dynamics and Interspecific Interactions*. Princeton: Princeton University Press.

Titus, J. H. (1991). Seed bank of a hardwood floodplain swamp in Florida. *Castanea*, **56**, 117–27.

Titus, J. H. & del Moral, R. (1998a). Seedling establishment in different microsites on Mount St. Helens, Washington, USA. *Plant Ecology*, **134**, 13–29.

Titus, J. H. & del Moral, R. (1998b). The role of mycorrhizae in primary succession on Mount St. Helens. *American Journal of Botany*, **85**, 370–5.

Titus, J. H., del Moral, R. & Gamiet, S. (1998). The distribution of vesicular-arbuscular mycorrhizae on Mount St. Helens, Washington. *Madroño*, **45**, 162–70.

Titus, J. H., Titus, P. & del Moral, R. (1999). Wetland development in primary and secondary successional substrates fourteen years after the eruption of Mount St. Helens, Washington, U.S.A. *Northwest Science*, **73**, 186–204.

Tolliver, K. S., Colley, D. M. & Young, D. R. (1995). Inhibitory effects of *Myrica cerifera* on *Pinus taeda*. *American Midland Naturalist*, **133**, 256–63.

Tongway, D. J. & Ludwig, J. A. (1996). Rehabilitation of semiarid landscapes in Australia. I. Restoring productive soil patches. *Restoration Ecology*, **4**, 388–97.

Tonkin, P. J. & Basher, L. R. (2001). Soil chronosequence in subalpine superhumid Cropp Basin, western Southern Alps, New Zealand. *New Zealand Journal of Geology and Geophysics*, **44**, 37–45.

Tordoff, G. M., Baker, A. J. M. & Willis, A. J. (2000). Current approaches to the revegetation and reclamation of metalliferous mine wastes. *Chemosphere*, **41**, 219–28.

Torres, J. A. (1994). Wood decomposition of *Cyrilla racemiflora* in a tropical montane forest. *Biotropica*, **26**, 124–40.

Treub, M. (1888). Notice sur la novelle flore de Krakatau. *Annales du Jardin Botanique de Buitenzorg*, **7**, 213–22.

Tsuyuzaki, S. & Titus, J. T. (1996). Vegetation development patterns in erosive areas on the Pumice Plains of Mount St. Helens. *American Midland Naturalist*, **135**, 172–7.

Tsuyuzaki, S., Titus, J. T. & del Moral, R. (1997). Seedling establishment patterns on the Pumice Plain, Mount St. Helens, Washington. *Journal of Vegetation Science*, **8**, 727–34.

Tu, M., Titus, J. H., del Moral, R. & Tsuyuzaki, S. (1998). Composition and dynamics of wetland seed banks on Mount St. Helens, Washington, USA. *Folia Geobotanica*, **33**, 3–16.

Tucker, C. J., Dregne, H. E. & Newcomb, W. W. (1991). Expansion and contraction of the Sahara Desert from 1980 to 1990. *Science*, **253**, 299–301.

Turner, D. R. & Vitousek, P. M. (1987). Nodule biomass in the nitrogen-fixing alien *Myrica faya* in Hawaii Volcanoes National Park. *Pacific Science*, **41**, 186–90.

Turner, M. G., Baker, W. L., Peterson, C. J. & Peet, R. K. (1998). Factors influencing succession: Lessons from large, infrequent natural disturbances. *Ecosystems*, **1**, 511–23.

Turner, M. G., Dale, V. H. & Everham, E. H. III (1997). Fires, hurricanes, and volcanoes: comparing large disturbances. *BioScience*, **47**, 758–68.

Turner, T. (1983). Facilitation as a successional mechanism in a rocky intertidal community. *The American Naturalist*, **121**, 729–38.

Tybirk, K. & Strandberg, B. (1999). Oak forest development as a result of historical land-use patterns and present nitrogen deposition. *Forest Ecology and Management*, **114**, 97–106.

Tyler, G. (1996). Cover distributions of vascular plants in relation to soil chemistry and soil depth in a granite rock ecosystem. *Vegetatio*, **127**, 215–23.

Tyler, A. C. & Zieman, J. C. (1999). Patterns of development in the creek bank region of a barrier island *Spartina alterniflora* marsh. *Marine Ecology Progress Series*, **180**, 161–77.

Uliassi, D. D. & Ruess, R. W. (2002). Limitations to symbiotic nitrogen fixation in primary succession on the Tanana River floodplain. *Ecology*, **83**, 88–103.

Ullman, I., Bannister, P. & Wilson, J. B. (1995). The vegetation of roadside verges with respect to environmental gradients in southern New Zealand. *Journal of Vegetation Science*, **6**, 131–42.

Ullman, I. & Heindl, B. (1989). Geographical and ecological differentiation of roadside vegetation in temperate Europe. *Botanica Acta*, **4**, 261–9.

Uno, G. E. & Collins, S. L. (1987). Primary succession on granite outcrops of southwestern Oklahoma. *Bulletin of the Torrey Botanical Club*, **114**, 387–92.

Urban, O. L., O'Neill, R. V. & Shugart, H. H. (1987). Landscape ecology: A hierarchical perspective can help scientists understand spatial patterns. *BioScience*, **37**, 119–27.

Urban, D. L. & Shugart, H. H. (1992). Individual-based models of forest succession. In *Plant Succession: Theory and Prediction*, ed. D. C. Glenn-Lewin, R. K. Peet & T. T. Veblen, pp. 249–92. London: Chapman & Hall.

Urbanek, R. P. (1989). The influence of fauna on plant productivity. In *Animals in Primary Succession: The Role of Fauna in Reclaimed Lands*, ed. J. D. Majer, pp. 71–106. Cambridge: Cambridge University Press.

Ursic, K. A., Kenkel, N. C. & Larson, D. W. (1997). Revegetation dynamics of cliff faces in abandoned limestone quarries. *Journal of Applied Ecology*, **34**, 289–303.

Usher, M. B. (1981). Modeling ecological succession, with particular reference to Markovian models. *Vegetatio*, **46**, 11–18.

Usher, M. B. (1992). Statistical models of succession. In *Plant Succession: Theory and Prediction*, ed. D. C. Glenn-Lewin, R. K. Peet & T. T. Veblen, pp. 215–48. London: Chapman & Hall.

Valiente-Banuet, A. & Ezcurra, E. (1991). Shade as a cause of the association between the cactus *Neobuxbaumia tetezo* and the nurse plant *Mimosa luisana* in the Tehuacán Valley, Mexico. *Journal of Ecology*, **79**, 961–71.

van Aarde, R. J., Ferreira, S. M., Kritzinger, J. J., van Dyk, P. J., Vogt, M. & Wassenaar, T. D. (1996). An evaluation of habitat rehabilitation on coastal dune forests in northern KwaZulu-Natal, South Africa. *Restoration Ecology*, **4**, 334–45.

Van Andel, J., Bakker, J. P. & Grootjans, A. P. (1993). Mechanisms of vegetation succession: A review of concepts and perspectives. *Acta Botanica Neerlandica*, **42**, 413–33.

van Breemen, N. (1995). How *Sphagnum* bogs down other plants. *Trends in Ecology and Systematics*, **10**, 270–5.

van Breemen, N. & Finzi, A. C. (1998). Plant-soil interactions: Ecological aspects and evolutionary implications. *Biogeochemistry*, **42**, 1–19.

Van Cleve, K., Chapin, F. S. III, Dyrness, C. T. & Viereck, L. A. (1991). Element cycling in taiga forests: State-factor control. *BioScience*, **41**, 78–88.

Van Cleve, K., Dyrness, C. T., Marion, G. M. & Erickson, R. (1993). Control of soil development on the Tanana River floodplain, interior Alaska. *Canadian Journal of Forest Research*, **23**, 941–55.

Van Cleve, K. & Viereck, L. A. (1972). Distribution of selected chemical elements in even-aged alder (*Alnus*) ecosystems near Fairbanks, Alaska. *Arctic and Alpine Research*, **4**, 239–55.

Van Cleve, K., Viereck, L. A. & Schlentner, R. L. (1971). Accumulation of nitrogen in alder (*Alnus*) ecosystems near Fairbanks, Alaska. *Arctic and Alpine Research*, **3**, 101–14.

van der Heijden, E. W. & Vosatka, M. (1999). Mycorrhizal associations of *Salix repens* L. communities in succession of dune ecosystems. II. Mycorrhizal dynamics and interactions of ectomycorrhizal and arbuscular mycorrhizal fungi. *Canadian Journal of Botany*, **77**, 1833–41.

van der Maarel, E. (1988). Vegetation dynamics: patterns in time and space. *Vegetatio*, **77**, 7–19.

van der Maarel, E. (1998). Coastal dunes: Pattern and process, zonation and succession. In *Dry Coastal Ecosystems: Ecosystems of the World* 2C, ed. E. van der Maarel, pp. 505–17. The Hague: Elsevier.

van der Maarel, E. & Sykes, M. T. (1993). Small scale plant species turnover in a limestone grassland – the carousel model and some comments on the niche concept. *Journal of Vegetation Science*, **4**, 179–88.

van der Maarel, E. & Sykes, M. T. (1997). Rates of small-scale species mobility in alvar limestone grassland. *Journal of Vegetation Science*, **8**, 199–208.

Vandermeer, J. (1980). Saguaros and nurse trees: A new hypothesis to account for population fluctuations. *Southwestern Naturalist*, **25**, 357–60.

van der Putten, W. H. (1997). Plant-soil feedback as a selective force. *Trends in Ecology and Systematics*, **12**, 169–70.

van der Putten, W. H., Van Dijk, C. & Peters, B. A. M. (1993). Plant-specific soil-borne diseases contribute to succession in foredune vegetation. *Nature*, **362**, 53–6.

van der Valk, A. G. (1974). Environmental factors controlling the distribution of forbs on coastal foredunes in Cape Hatteras National Seashores. *Canadian Journal of Botany*, **52**, 1057–73.

van der Valk, A. G., Pederson, R. L. & Davis, C. B. (1992). Restoration and creation of freshwater wetlands using seed banks. *Wetlands Ecology and Management*, **1**, 191–7.

Van Dover, C. L. (2000). *The Ecology of Deep-Sea Hydrothermal Vents*. Princeton: Princeton University Press.

van Hulst, R. (1992). From population dynamics to community dynamics: Modeling succession as a species replacement process. In *Plant Succession: Theory and Prediction*, ed. D. C. Glenn-Lewin, R. K. Peet and T. T. Veblen, pp. 188–214. London: Chapman & Hall.

Vanier, C. H. & Walker, L. R. (1999). Impact of a non-native plant on seed dispersal of a native. *Madroño*, **46**, 46–8.

van Mierlo, J. E. M., Wilms, Y. J. C. & Berendse, F. (2000). Effects of soil organic matter and nitrogen supply on competition between *Festuca ovina* and *Deschampsia flexuosa* during inland dune succession. *Plant Ecology*, **148**, 51–9.

van Noordwijk-Puijk, K., Beeftink, W. G. & Hogeweg, P. (1979). Vegetational development on salt marsh flats after disappearance of the tidal factor. *Vegetatio*, **39**, 1–13.

Vasek, F. C. & Lund, J. L. (1980). Soil characteristics associated with a primary plant succession on a Mojave Desert dry lake. *Ecology*, **61**, 1013–18.

Vazquez, G., Moreno-Casasola, P. & Barrera, O. (1998). Interaction between algae and seed germination in tropical dune slack species: A facilitation process. *Aquatic Botany*, **60**, 409–16.

Veblen, T. T. (1985). Stand dynamics in Chilean *Nothofagus* forests. In *The Ecology of Natural Disturbance and Patch Dynamics*, ed. S. T. A. Pickett & P. S. White, pp. 35–52. New York: Academic Press.

Veblen, T. T. & Ashton, D. H. (1978). Catastrophic influences on the vegetation of the Valdivian Andes, Chile. *Vegetatio*, **36**, 149–67.

Veblen, T. T., Ashton, D. H., Rubulis, S., Lorenz, D. C. & Cortes, M. (1989). *Nothofagus* stand development on in-transit moraines, Casa Pangue Glacier, Chile. *Arctic and Alpine Research*, **21**, 144–55.

Veblen, T. T., Ashton, D. H., Schlegel, F. M. & Veblen, A. T. (1977). Plant succession in a timberline depressed by vulcanism in south-central Chile. *Journal of Biogeography*, **4**, 275–94.

Versfeld, D. B. & van Wilgen, B. W. (1986). Impact of woody aliens on ecosystem properties. In *The Ecology and Management of Biological Invasions in Southern Africa*, ed. I. A. W. Macdonald, F. J. Kruger & A. A. Ferrar, pp. 239–46. Cape Town: Oxford University Press.

Vetaas, O. R. (1994). Primary succession of plant assemblages on a glacier foreland – Bødalsbreen, southern Norway. *Journal of Biogeography*, **21**, 297–308.

Viereck, L. A. (1966). Plant succession and soil development on gravel outwash of the Muldrow Glacier, Alaska. *Ecological Monographs*, **36**, 181–99.

Viereck, L. A. (1970). Forest succession and soil development adjacent to the Chena River in interior Alaska. *Arctic and Alpine Research*, **2**, 1–26.

Viereck, L. A., Dyrness, C. T. & Foote, M. J. (1993). An overview of the vegetation and soils of the floodplain ecosystems of the Tanana River, interior Alaska. *Canadian Journal of Forest Research*, **23**, 889–98.

Viereck, L. A., Dyrness, C. T., Van Cleve, K. & Foote, K. J. (1983). Vegetation, soils, and forest productivity in selected forest types in interior Alaska. *Canadian Journal of Forest Research*, **13**, 703–20.

Vimmerstedt, J. P. & Finney, J. H. (1973). Impact of earthworm introduction on litter burial and nutrient distribution on Ohio strip-mine spoil banks. *Soil Science Society of America Proceedings*, **37**, 388–91.

Virginia, R. A., Jarrell, W. M., Whitford, W. G. & Freckman, D. W. (1992). Soil biota and soil properties associated with the surface rooting zone of mesquite (*Prosopis glandulosa*) in historical and recently desertified habitats. *Biology and Fertility of Soils*, **14**, 90–8.

Vitousek, P. M. (1994). Potential nitrogen fixation during primary succession in Hawai'i Volcanoes National Park. *Biotropica*, **26**, 234–40.

Vitousek, P. M. (1999). Nutrient limitation to nitrogen fixation in young volcanic sites. *Ecosystems*, **2**, 505–10.

Vitousek, P. M. & Denslow, J. S. (1986). Nitrogen and phosphorus availability in treefall gaps of a lowland tropical rainforest. *Journal of Ecology*, **74**, 1167–78.

Vitousek, P. M., Ehrlich, P. R., Ehrlich, A. H. & Matson, P. A. (1986). Human appropriation of the products of photosynthesis. *BioScience*, **36**, 368–73.

Vitousek, P. M. & Farrington, H. (1997). Nutrient limitation and soil development: Experimental test of a biogeochemical theory. *Biogeochemistry*, **37**, 63–75.

Vitousek, P. M. & Field, C. B. (1999). Ecosystem constraints to symbiotic nitrogen fixers: A simple model and its implications. *Biogeochemistry*, **46**, 179–202.

Vitousek, P. M. & Hobbie, S. (2000). Heterotrophic nitrogen fixation in decomposing litter: Patterns and regulation. *Ecology*, **81**, 2366–76.

Vitousek, P. M. & Howarth, R. W. (1991). Nitrogen limitation on land and in the sea: How can it occur? *Biogeochemistry*, **13**, 87–115.

Vitousek, P. M. & Reiners, W. A. (1975). Ecosystem succession and nutrient retention: A hypothesis. *BioScience*, **25**, 376–81.

Vitousek, P. M., Van Cleve, K., Balakrishnan, K. & Mueller-Dombois, D. (1983). Soil development and nitrogen turnover in montane rainforest soils on Hawaii. *Biotropica*, **15**, 268–74.

Vitousek, P. M. & Walker, L. R. (1987). Colonization, succession and resource availability: Ecosystem-level interactions. In *Colonization, Succession and Stability*, ed. A. J. Gray, M. J. Crawley & P. J. Edwards, pp. 207–24. Symposium of the British Ecological Society, vol. 26. Oxford: Blackwell.

Vitousek, P. M. & Walker, L. R. (1989). Biological invasion by *Myrica faya* in Hawaii: Plant demography, nitrogen fixation, and ecosystem effects. *Ecological Monographs*, **59**, 247–65.

Vitousek, P. M., Walker, L. R., Whiteaker, L. D., Mueller-Dombois, D. & Matson, P. A. (1987). Biological invasion by *Myrica faya* alters primary succession in Hawaii. *Science*, **238**, 802–4.

Vittoz, P., Stewart, G. H. & Duncan, R. P. (2001). Earthquake impacts in old-growth *Nothofagus* forests in New Zealand. *Journal of Vegetation Science*, **12**, 417–26.

VivianSmith, G. & Handel, S. N. (1996). Freshwater wetland restoration of an abandoned sand mine: Seed bank recruitment dynamics and plant colonization. *Wetlands*, **16**, 185–96.

Wada, N. (1999). Factors affecting the seed-setting success of *Dryas octopetala* in front of Broggerbreen Glacier in the high Arctic, Ny-Ålesund, Svalbard. *Polar Research*, **18**, 261–8.

Wagner, R. & Walker, R. B. (1986). Mineral nutrient availability in some Mount St. Helens surface samples. In *Mount St. Helens: Five Years Later*, ed. S. A. C. Keller, pp. 153–62. Cheney, Washington, U.S.A.: Eastern Washington State University Press.

Waide, R. B., Willig, M. R., Steiner, C. F., Mittelbach, G., Gough, L., Dodson, S. I., Juday, G. P. & Parmenter, R. (1999). The relationship between productivity and species richness. *Annual Review of Ecology and Systematics*, **30**, 257–300.

Wali, M. K. (ed.) (1992). *Ecosystem Rehabilitation*. The Hague: SPB Academic Press.

Wali, M. K. (1999a). Ecological succession and the rehabilitation of disturbed terrestrial systems. *Plant and Soil*, **213**, 195–220.

Wali, M. K. (1999b). Ecology today: Beyond the bounds of science. *Nature and Resources*, **35**, 38–50.

Wali, M. K. & Kannowski, P. B. (1975). Prairie ant mound ecology: interrelationships of microclimate, soils and vegetation. In *Prairie: A Multiple View*, ed. M. K. Wali, pp. 155–69. Grand Forks: University of North Dakota Press.

Walker, D. A. & Everett, K. R. (1987). Road dust and its environmental impact on Alaskan taiga and tundra. *Arctic and Alpine Research*, **19**, 479–89.

Walker, G. P. L. (1994). Geology and volcanology of the Hawaiian Islands. In *A Natural History of the Hawaiian Islands*, ed. E. A. Kay, pp. 53–85. Honolulu: University of Hawaii Press.

Walker, J., Thompson, C. H., Fergus, I. F. & Tunstall, B. R. (1981). Plant succession and soil development in coastal sand dunes of subtropical eastern Australia. In *Forest Succession. Concepts and Application*, ed. D. C. West, H. H. Shugart & D. B. Botkin, pp. 107–31. New York: Springer.

Walker, J., Thompson, C. H., Reddell, P. & Olley, J. (2000). Retrogressive succession on an old landscape. In *Proceedings of the 41st Symposium of the IUVS*, ed. P. White, pp. 21–3. Uppsala: Opulus Press.

Walker, J., Thompson, C. H., Reddell, P. & Rapport, D. J. (2001). The importance of landscape age in influencing landscape health. *Ecosystem Health*, **7**, 7–14.

Walker, L. R. (1989). Soil nitrogen changes during primary succession on a floodplain in Alaska, USA. *Arctic and Alpine Research*, **21**, 341–9.

Walker, L. R. (1993). Nitrogen fixers and species replacements in primary succession. In *Primary Succession on Land*, ed. J. Miles & D. W. H. Walton, pp. 249–72. Oxford: Blackwell.

Walker, L. R. (1994). Effects of fern thickets on woodland development on landslides in Puerto Rico. *Journal of Vegetation Science*, **5**, 525–32.

Walker, L. R. (1995). How unique is primary plant succession at Glacier Bay? In *Proceedings of the Third Glacier Bay Science Symposium, 1993*, ed. D. R. Engstrom, pp. 137–46. Anchorage, Alaska: National Park Service.

Walker, L. R. (ed.) (1999a). *Ecosystems of Disturbed Ground, Ecosystems of the World 16*. Amsterdam: Elsevier.

Walker, L. R. (1999b). Patterns and processes in primary succession. In *Ecosystems of Disturbed Ground, Ecosystems of the World 16*, ed. L. R. Walker, pp. 585–610. Amsterdam: Elsevier.

Walker, L. R. (2000). Seedling and sapling dynamics of treefall pits in Puerto Rico. *Biotropica*, **32**, 262–75.

Walker, L. R. & Chapin, F. S. III (1986). Physiological controls over seedling growth in primary succession on an Alaskan floodplain. *Ecology*, **67**, 1508–23.

Walker, L. R. & Chapin, F. S. III (1987). Interactions among processes controlling successional change. *Oikos*, **50**, 131–5.

Walker, L. R., Clarkson, B. D., Silvester, W. & Clarkson, B. R. (2003). Facilitation outweights inhibition in post-volcanic primary succession in New Zealand. *Journal of Vegetation Science*. (in press.)

Walker, L. R. & Neris, L. E. (1993). Posthurricane seed rain dynamics in Puerto Rico. *Biotropica*, **25**, 408–18.

Walker, L. R. & Powell, E. A. (1999a). Regeneration of the Mauna Kea silversword *Argyroxiphium sandwicense* (Asteraceae) Hawaii. *Biological Conservation*, **89**, 61–70.

Walker, L. R. & Powell, E. A. (1999b). Effects of seeding on road revegetation in the Mojave Desert, southern Nevada. *Ecological Restoration*, **17**, 150–5.

Walker, L. R. & Powell, E. A. (2001). Soil water retention on gold mine surfaces in the Mojave Desert. *Restoration Ecology*, **9**, 95–103.

Walker, L. R. & Smith, S. D. (1997). Impacts of invasive plants on community and ecosystem properties. In *Assessment and Management of Plant Invasions*, ed. J. O. Luken & J. W Thieret, pp. 69–86. New York: Springer.

Walker, L. R., Thompson, D. B. & Landau, F. H. (2001). Experimental manipulations of fertile islands and nurse plant effects in the Mojave Desert, USA. *Western North American Naturalist*, **61**, 25–35.

Walker, L. R. & Vitousek, P. M. (1991). An invader alters germination and growth of a native dominant tree in Hawai'i. *Ecology*, **72**, 1449–55.

Walker, L. R. & Willig, M. R. (1999). An introduction to terrestrial disturbances. In *Ecosystems of Disturbed Ground, Ecosystems of the World 16*, ed. L. R. Walker, pp. 1–16. Amsterdam: Elsevier.

Walker, L. R., Zarin, D. J., Fetcher, N., Myster, R. W. & Johnson, A. H. (1996). Ecosystem development and plant succession on landslides in the Caribbean. *Biotropica*, **28**, 566–76.

Walker, L. R., Zasada, J. C. & Chapin, F. S. III. (1986). The role of life history processes in primary succession on an Alaskan floodplain. *Ecology*, **67**, 1243–53.

Walker, S. (1997). Models of vegetation dynamics in semi-arid vegetation: application to lowland central Otago, New Zealand. *New Zealand Journal of Ecology*, **21**, 129–40.

Walker, T. W. & Syers, J. K. (1976). The fate of phosphorus during pedogenesis. *Geoderma*, **15**, 1–19.

Wallen, B. (1980). Changes in structure and function of *Ammophila* during primary succession. *Oikos*, **34**, 227–38.

Walton, D. W. H. (1990). Colonization of terrestrial habitats – organisms, opportunities and occurrence. In *Antarctic Ecosystems: Ecological Change and Conservation*, ed. K. Kerry & G. Hempel, pp. 51–60. Heidelberg: Springer-Verlag.

Walton, D. W. H. (1993). The effects of cryptogams on mineral substrates. In *Primary Succession on Land*, ed. J. Miles & D. W. H. Walton, pp. 33–54. Oxford: Blackwell.

Ward, S. A. & Thornton, I. W. B. (2000). Chance and determinism in the development of isolated communities. *Global Ecology and Biogeography*, **9**, 7–18.

Wardenaar, E. C. P. & Sevink, J. (1992). A comparative study of soil formation in primary stands of Scots pine (planted) and poplar (natural) on calcareous dune sands in the Netherlands. *Plant and Soil*, **140**, 109–20.

Wardlaw, C. W. (1931). Observations on the dominance of pteridophytes on some St. Lucia soils. *Journal of Ecology*, **19**, 60–3.

Wardle, D. A. (1992). A comparative assessment of factors which influence microbial biomass, carbon and nitrogen levels in soil. *Biological Review*, **67**, 321–58.

Wardle, D. A. & Ghani, A. (1995). A critique of the microbial metabolic quotient (qCO_2) as a bioindicator of disturbance and ecosystem development. *Soil Biology and Biochemistry*, **27**, 1601–10.

Wardle, D. A. Yeates, G. W., Watson, R. N. & Nicholson, K. S. (1995). Development of the decomposer food-web, trophic relationships, and ecosystem properties during a 3-year primary succession in sawdust. *Oikos*, **73**, 155–66.

Wardle, D. A., Zackrisson, O., Hörnberg, G. & Gallet, C. (1997). Influence of island area on ecosystem properties. *Science*, **277**, 1296–9.

Wardle, P. (1980). Plant succession in Westland National Park and its vicinity. *New Zealand Journal of Botany*, **18**, 221–32.

Ware, S. (1991). Influence of interspecific competition, light and moisture levels on growth of rock outcrop *Talinum* (Portulaceae). *Bulletin of the Torrey Botanical Club*, **118**, 1–5.

Ware, S. & Pinion, G. (1990). Substrate adaptation in rock outcrop plants: Eastern United States *Talinum* (Portulaceae). *Bulletin of the Torrey Botanical Club*, **117**, 284–90.

Warming, E. (1895). *Plantesamfund: Grundträk af den Ökologiska Plantegeografi.* Copenhagen: Philipsen.

Warming, E. (1909). *Oecology of Plants: An Introduction to the Study of Plant Communities.* Oxford: Clarendon Press. (Modified English translation of Warming, 1895.)

Wasilewska, L. (1970). Nematodes of the sand dunes in the Kampinos Forest. I. Species structure. *Ekologia Polska*, **18**, 429–43.

Watt, A. S. (1947). Pattern and process in the plant community. *Journal of Ecology*, **35**, 1–22.

Watt, A. S. (1955). Bracken versus heather: A study in plant sociology. *Journal of Ecology*, **43**, 490–506.

Watt, K. E. F. (1968). *Ecology and Resource Management.* New York: McGraw-Hill.

Weaver, J. E. & Clements, F. E. (1938). *Plant Ecology.* New York: McGraw-Hill.

Webb, L. J. (1958). Cyclones as an ecological factor in tropical lowland rainforest, North Queensland. *Australian Journal of Botany*, **6**, 220–8.

Webb, R. H. (1983). Compaction of desert soils by off-road vehicles. In *Environmental Effects of Off Road Vehicles: Impacts and Management in Arid Regions*, ed. R. H. Webb & H. G. Wilshire, pp. 51–79. New York: Springer.

Webb, R. H. (1996). *Grand Canyon, a Century of Change.* Tucson, U.S.A.: The University of Arizona Press.

Webb, R. H. & Wilshire, H. G. (eds) (1983). *Environmental Effects of Off Road Vehicles: Impacts and Management in Arid Regions.* New York: Springer.

Webb, S L. (1999). Disturbance by wind in temperate-zone forests. In *Ecosystems of Disturbed Ground, Ecosystems of the World 16*, ed. L. R. Walker, pp. 187–222. Amsterdam: Elsevier.

Wegener, A. (1922). *Die Entstehung der Kontinente und Ozeane (The Origin of Continents and Oceans).* Braunschweig, Germany: F. Vieweg & Son.

Weiher, E. & Keddy, P. (1995). The assembly of experimental wetland plant communities. *Oikos*, **73**, 323–35.

Weiher, E. & Keddy, P. (eds) (1999). *Ecological Assembly Rules: Perspectives, Advances, Retreats.* Cambridge: Cambridge University Press.

Welden, C. W. and Slauson, W. L. (1986). The intensity of competition versus its importance: An overlooked distinction and some implications. *Quarterly Review of Biology*, **61**, 23–44.

Wells, A., Duncan, R. P. & Stewart, G. H. (2001). Forest dynamics in Westland, New Zealand: the importance of large infrequent earthquake-induced disturbance. *Journal of Ecology*, **89**, 1006–18.

Wenny, D. G. (2001). Advantages of seed dispersal: a re-evaluation of directed dispersal. *Evolutionary Ecology Research*, **3**, 51–74.

West, J. M. & Zedler, J. B. (2000). Marsh-creek connectivity: Fish use of a tidal salt marsh in southern California. *Estuaries*, **23**, 699–710.

West, N. E. & Young, J. A. (2000). Intermountain valleys and lower mountain slopes. In *North American Terrestrial Vegetation*, 2nd edition, ed. M. G. Barbour & W. D. Billings, pp. 256–84. Cambridge: Cambridge University Press.

Wheater, C. P. & Cullen, W. R. (1997). The flora and invertebrate fauna of abandoned limestone quarries in Derbyshire, United Kingdom. *Restoration Ecology*, **5**, 77–84.

Wheelwright, J. (1994). *Degrees of Disaster. Prince William Sound: How Nature Reels and Rebounds*. New York: Simon & Schuster.

Whelan, R. J. (1989). Influence of fauna on plant species composition. In *Animals in Primary Succession: The Role of Fauna in Reclaimed Lands*, ed. J. D. Majer, pp. 107–42. Cambridge: Cambridge University Press.

Whigham, D. F., Dickinson, M. B. & Brokaw, N. V. L. (1999). Background canopy gap and catastrophic wind disturbances in tropical forests. In *Ecosystems of Disturbed Ground, Ecosystems of the World 16*, ed. L. R. Walker, pp. 223–52. Amsterdam: Elsevier.

White, J. S. & Bayley, S. E. (1999). Restoration of a Canadian prairie wetland with agricultural and municipal wastewater. *Environmental Management*, **24**, 25–37.

White, P. S. (1979). Pattern, process, and natural disturance in vegetation. *Botanical Review*, **45**, 229–99.

White, P. S. & Jentsch, A. (2001). The search for generality in studies of disturbance and ecosystem dynamics. *Progress in Botany*, **62**, 399–449.

White, P. S. & Pickett, S. T. A. (1985). Natural disturbance and patch dynamics: An introduction. In *The Ecology of Natural Disturbance and Patch Dynamics*, ed. S. T. A. Pickett & P. S. White, pp. 3–16. New York: Academic Press.

Whitford, W. G., Meentemeyer, V., Seastedt, T. R., Cromack, K. Jr., Crossley, D. A. Jr., Santos, P., Todd, R. L. & Waide, J. B. (1981). Exceptions to the AET Model: Deserts and clear-cut forest. *Ecology*, **62**, 275–7.

Whitlock, C. (1992). Vegetational and climatic history of the Pacific-Northwest during the last 20,000 years – implications for understanding present-day biodiversity. *Northwest Environmental Journal*, **8**, 5–28.

Whitlock, C. & Bartlein, P. J. (1997). Vegetation and climate change in northwest America during the past 125 kyr. *Nature*, **388**, 57–61.

Whittaker, R. H. (1953). A consideration of climax theory: The climax as a population and pattern. *Ecological Monographs*, **23**, 41–78.

Whittaker, R. H. (ed.) (1973). *Ordination and Classification of Communities*. The Hague: Junk.

Whittaker, R. H. (1974). Climax concepts and recognition. In *Vegetation Dynamics*, ed. R. Knapp, pp. 137–54. *Handbook of Vegetation Science*, part 8. The Hague: Junk.

Whittaker, R. H. (1975). *Community and Ecosystems*, 2nd edition. New York: MacMillan.

Whittaker, R. H. & Levin, S. A. (1977). The role of mosaic phenomena in natural communities. *Theoretical Population Biology*, **12**, 117–39.

Whittaker, R. J. (1992). Stochasticism and determinism in island ecology. *Journal of Biogeography*, **19**, 587–91.

Whittaker, R. J., Bush, M. B. & Richards, K. (1989). Plant recolonization and vegetation succession on the Krakatau Islands, Indonesia. *Ecological Monographs*, **59**, 59–123.

Whittaker, R. J. & Jones, S. H. (1994a). The role of frugivorous bats and birds in the rebuilding of a tropical forest ecosystem, Krakatau, Indonesia. *Journal of Biogeography*, **21**, 689–702.

Whittaker, R. J. & Jones, S. H. (1994b). Structure in re-building insular ecosystems – an empirically derived model. *Oikos*, **69**, 524–30.

Whittaker, R. J., Jones, S. H. & Partomihardjo, T. (1997). The rebuilding of an isolated rain forest assemblage, how disharmonic is the flora of Krakatau? *Biodiversity and Conservation*, **6**, 1671–96.

Whittaker, R. J., Partomihardjo, T. & Jones S. H. (1999). Interesting times on Krakatau: Stand dynamics in the 1990s. *Philosophical Transactions of the Royal Society of London*, B**354**, 1857–67.

Wiegleb, G. & Felinks, B. (2001). Predictability of early stages of primary succession in post-mining landscapes of Lower Lusatia, Germany. *Journal of Applied Vegetation Science*, **4**, 5–18.

Wijdeven, S. M. J., & Kuzee, M. E. (2000). Seed availability as a limiting factor in forest recovery processes in Costa Rica. *Restoration Ecology*, **8**, 414–24.

Wilkinson, D. M. (1997). Plant colonization: Are wind dispersed seeds really dispersed by birds at larger spatial and temporal scales? *Journal of Biogeography*, **24**, 61–5.

Willems, J. H. (1985). Growth form and species diversity in permanent grassland plots with different management. In *Sukzession auf Grünlandbrachen*, ed. K. F. Schreiber, pp. 35–43. Schöningh: Padeborn.

Willig, M. R. & McGinley, M. A. (1999). The response of animals to disturbance and their roles in patch dynamics. In *Ecosystems of Disturbed Ground, Ecosystems of the World 16*, ed. L. R. Walker, pp. 633–57. Amsterdam: Elsevier.

Willig, M. R. & Walker, L. R. (1999). Disturbance in terrestrial ecosystems: Salient themes, synthesis and future directions. In *Ecosystems of Disturbed Ground, Ecosystems of the World 16*, ed. L. R. Walker, pp. 747–67. Amsterdam: Elsevier.

Willson, M. F., Rice, B. L. & Westoby, M. (1990). Seed dispersal spectra: A comparison of temperate plant communities. *Journal of Vegetation Science*, **1**, 547–62.

Wilmshurst, J. M. (1997). The impact of human settlement on vegetation and soil stability in Hawke's Bay, New Zealand. *New Zealand Journal of Botany*, **35**, 97–111.

Wilmshurst, J. M. & McGlone, M. S. (1996). Forest disturbance in the central North Island, New Zealand, following the 1850 BP Taupo eruption. *The Holocene*, **6**, 399–411.

Wilson, J. B. (1994). Who makes the assembly rules? *Journal of Vegetation Science*, **2**, 289–90.

Wilson, J. B. (1999). Assembly rules in plant communities. In *Ecological Assembly Rules: Perspectives, Advances, Retreats*, ed. E. Weiher & P. Keddy, pp. 251–71. Cambridge: Cambridge University Press.

Wilson, J. B. & Agnew, A. D. Q. (1992). Positive-feedback switches in plant communities. *Advances in Ecological Research*, **23**, 263–336.

Wilson, J. B., Allen, R. B. & Lee, W. G. (1995). An assembly rule in the ground and herbaceous strata of a New Zealand rain forest. *Functional Ecology*, **9**, 61–4.

Wilson, J. B. & Lee, W. G. (2000). C-S-R triangle theory: Community-level predictions, tests, evaluations of criticisms, and relation to other theories. *Oikos*, **91**, 77–96.

Wilson, J. B., Ullman, I. & Bannister, P. (1996). Do species assemblages ever recur? *Journal of Ecology*, **84**, 471–4.

Wilson, J. B. & Whittaker, R. J. (1995). Assembly rules demonstrated in a salt marsh community. *Journal of Ecology*, **83**, 801–7.

Wilson, R. E. (1970). Succession in stands of *Populus deltoides* along the Missouri River in southeastern South Dakota. *American Midland Naturalist*, **83**, 330–42.

Wilson, S. D. (1999). Plant interactions during secondary succession. In *Ecosystems of Disturbed Ground, Ecosystems of the World 16*, ed. L. R. Walker, pp. 611–32. Amsterdam: Elsevier.

Winterringer, W. S. & Vestal, A. G. (1956). Rock-ledge vegetation in southern Illinois. *Ecological Monographs*, **26**, 105–30.

Wiser, S. K., Peet, R. K. & White, P. S. (1996). The high-elevation rock outcrop vegetation of the Southern Appalachian Mountains. *Journal of Vegetation Science*, **7**, 703–22.

Wiser, S. K. & White, P. S. (1999). High-elevation outcrops and barrens of the Southern Appalachian Mountains. In *Savannas, Barrens and Rock Outcrop Communities of North America*, ed. R. C. Anderson, J. S. Fralish & J. M. Baskin, pp. 119–32. Cambridge: Cambridge University Press.

Witkowski, E. T. F. (1991). Effects of invasive alien acacias on nutrient cycling in the coastal lowlands of the cape fynbos. *Journal of Applied Ecology*, **28**, 1–15.

Wolff, J. O. & Zasada, J. C. (1979). Moose habitat and forest succession on the Tanana River floodplain and Yukon-Tanana upland. In *Proceedings of the North American Moose Conference and Workshop*, vol. 15, ed. H. G. Cummings, pp. 213–45. Soldotna-Kenai, Alaska. Thunder Bay, Ontario, Canada: School of Forestry, Lakehead University.

Wood, D. M. & del Moral, R. (1987). Mechanisms of early primary succession in subalpine habitats on Mount St. Helens. *Ecology*, **68**, 780–90.

Wood, D. M. & del Moral, R. (1988). Colonizing plants on the Pumice Plains, Mount St. Helens, Washington. *American Journal of Botany*, **75**, 1228–37.

Wood, D. M. & del Moral, R. (2000). Seed rain during early primary succession on Mount St. Helens, Washington. *Madroño*, **47**, 1–9.

Wood, D. M. & Morris, W. F. (1990). Ecological constraints to seedling establishment on the Pumice Plains, Mount St. Helens, Washington. *American Journal of Botany*, **77**, 1411–18.

Wood, M. (1995). *Environmental Soil Biology*. London: Chapman & Hall.

Woolhouse, M. E. J., Harmsen, R. & Fahrig, L. (1985). On succession in a saxicolous lichen community. *Lichenologist*, **17**, 167–72.

World Resources Institute (1994). *World Resources 1994–95*. Washington, D. C.: World Resources Institute.

Worley, I. A. (1973). The "black crust" phenomenon in upper Glacier Bay, Alaska. *Northwest Science*, **47**, 20–9.

Worster, D. E. (1979). *Dustbowl: The Southern Plains in the 1930s.* Oxford: Oxford University Press.

Wright, R. A. & Mueller-Dombois, D. (1988). Relationships among shrub population structure, species associations, seedling root form and early volcanic succession, Hawaii. In *Plant Form and Vegetation Structure*, ed. M. J. A. Werger, P. J. M van der Aart, H. J. During & J. T. A. Verhoeven, pp. 87–104. The Hague: SPB Academic.

Wright, R. G. & Bunting, S. C. (1994). *The Landscapes of Craters of the Moon National Monument: An Evaluation of Environmental Change.* Moscow, Idaho: University of Idaho Press.

Wurmli, M. (1974). Biocoenoses and their successions on the lava and ash of Mount Etna. *Image Roche*, **59**, 32–40.

Wynn-Williams, D. D. (1993). Microbial processes and initial stabilization of fellfield soil. In *Primary Succession on Land*, ed. J. Miles & D. W. H. Walton, pp. 17–32. Oxford: Blackwell.

Yarranton, G. & Morrison, R. (1974). Spatial dynamics of a primary succession: Nucleation. *Journal of Ecology*, **62**, 417–28.

Ye, Z. H., Wong, J. W. C., Wong, M. H., Baker, A. J. M., Shu, W. S. & Lan, C. Y. (2000). Revegetation of Pb/Zn mine tailings, Guangdong Province, China. *Restoration Ecology*, **8**, 87–92.

Yeaton, I. R. (1978). A cyclical relationship between *Larrea tridentata* and *Opuntia leptocaulis* in the northern Chihuahuan desert. *Journal of Ecology*, **66**, 651–6.

Yeaton, R. I. & Esler, K. J. (1990). The dynamics of a succulent karoo vegetation. A study of species association and recruitment. *Vegetatio*, **88**, 103–13.

Young, D. R. (1992). Photosynthetic characteristics and potential moisture stress for the actinorhizal shrub, *Myrica cerifera* (Myricaceae), on a Virginia barrier island. *American Journal of Botany*, **79**, 2–7.

Young, D. R., Sande, E. & Peters, G. A. (1992). Spatial relationships of *Frankia* and *Myrica cerifera* on a Virginia, USA barrier island. *Symbiosis*, **12**, 209–20.

Young, D. R., Shao, G. & Porter, J. H. (1995). Spatial and temporal growth dynamics of barrier island shrub thickets. *American Journal of Botany*, **82**, 638–45.

Young, T. P., Chase, J. M. & Huddleston, R. T. (2001). Community succession and assembly. *Ecological Restoration*, **19**, 5–18.

Zaalishvili, G. V., Khatisashvili, G. A., Ugrkhelidze, D. S., Gordeziani, M. S. & Kvesitadze, G. I. (2000). Plant potential for detoxification. *Applied Biochemistry and Microbiology*, **36**, 443–51.

Zak, J. C. & Freckman, D. W. (1991). Soil communities in deserts: Microarthropods and nematodes. In *The Ecology of Desert Communities*, ed. G. A. Polis, pp. 55–88. Tucson, Arizona, U.S.A.: University of Arizona Press.

Zaman, M. Q. (1999). Vulnerability, disaster, and survival in Bangladesh: Three case studies. In *The Angry Earth*, ed. A. Oliver-Smith & S. M. Hoffman, pp. 192–212. New York: Routledge.

Zarin, D. J. & Johnson, A. H. (1995a). Nutrient accumulation during primary succession in a montane tropical forest, Puerto Rico. *Soil Science Society of America Journal*, **59**, 1444–52.

Zarin, D. J. & Johnson, A. H. (1995b). Base saturation, nutrient cation, and organic matter increases during early pedogenesis on landslide scars in the Luquillo Experimental Forest, Puerto Rico. *Geoderma*, **65**, 317–30.

Zavaleta, E. S., Hobbs, R. J. & Mooney, H. A. (2001). Viewing invasive species removal in a whole-ecosystem context. *Trends in Ecology and Evolution*, **16**, 454–9.

Zedler, J. B. (2000). Progress in wetland restoration ecology. *Trends in Ecology and Evolution*, **15**, 402–7.

Zedler, J. B. (2001). *Handbook for Restoring Tidal Wetlands*. Boca Raton, Florida, U.S.A.: CRC Press.

Zedler, J. B. & Callaway, J. C. (1999). Tracking wetland restoration: Do mitigation sites follow desired trajectories? *Restoration Ecology*, **7**, 69–73.

Zeff, M. L. (1999). Salt marsh tidal channel morphometry: Application for wetland creation and restoration. *Restoration Ecology*, **7**, 205–11.

Zhang, J. & Maun, M. A. (1994). Potential for seed bank formation in seven Great Lakes sand dune species. *American Journal of Botany*, **81**, 387–94.

Zink, T. A. & Allen, M. F. (1998). The effects of organic amendments on the restoration of a disturbed coastal sage scrub habitat. *Restoration Ecology*, **6**, 52–8.

Zobel, D. B. & Antos, J. A. (1991). 1980 tephra from Mount St. Helens: Spatial and temporal variation beneath forest canopies. *Biology and Fertility of Soils*, **12**, 60–6.

Zobel, D. B. & Antos, J. A. (1992). Survival of plants buried for eight growing seasons by volcanic tephra. *Ecology*, **73**, 698–701.

Zobel, D. B. & Antos, J. A. (1997). A decade of recovery of understory vegetation buried by volcanic tephra from Mount St. Helens. *Ecological Monographs*, **67**, 317–44.

Zobel, M., Suurkask, M., Rosen, E. & Partel, M. (1996). The dynamics of species richness in an experimentally restored calcareous grassland. *Journal of Vegetation Science*, **7**, 203–10.

Zoladeski, C. (1991). Vegetation zonation in dune slacks on the Leba Bar, Polish Baltic Sea coast. *Journal of Vegetation Science*, **2**, 255–8.

Index

DATE DUE